COOL STARS WITH EXCESSES OF HEAVY ELEMENTS

ASTROPHYSICS AND SPACE SCIENCE LIBRARY

A SERIES OF BOOKS ON THE RECENT DEVELOPMENTS OF SPACE SCIENCE AND OF GENERAL GEOPHYSICS AND ASTROPHYSICS PUBLISHED IN CONNECTION WITH THE JOURNAL SPACE SCIENCE REVIEWS

VOLUME 114

PROCEEDINGS

COOL STARS WITH EXCESSES OF HEAVY ELEMENTS

PROCEEDINGS OF THE
STRASBOURG OBSERVATORY COLLOQUIUM
HELD AT STRASBOURG, FRANCE, JULY 3–6, 1984

Edited by

MERCEDES JASCHEK
Strasbourg Observatory, Strasbourg, France

and

PHILIP C. KEENAN
Perkins Observatory, Delaware, Ohio, U.S.A.

D. REIDEL PUBLISHING COMPANY
A MEMBER OF THE KLUWER ACADEMIC PUBLISHERS GROUP
DORDRECHT / BOSTON / LANCASTER

Library of Congress Cataloging in Publication Data

CIP

Strasbourg Observatory Colloquium (1984: Strasbourg, France)
Cool stars with excesses of heavy elements.

(Astrophysics and space science library; v. 114)
Includes index.
1. Cool stars–Congresses. 2. Heavy elements–Congresses.
I. Jaschek, Mercedes, 1926– . II. Keenan, Philip C. (Philip Childs), 1908– III. Title. IV. Series.
QB843.C6S77 1984 523.8′7 85-1787
ISBN-13: 978-94-010-8851-0 e-ISBN-13: 978-94-009-5325-3
DOI: 10.1007/978-94-009-5325-3

Published by D. Reidel Publishing Company,
P.O. Box 17, 3300 AA Dordrecht, Holland.

Sold and distributed in the U.S.A. and Canada
by Kluwer Academic Publishers,
190 Old Derby Street, Hingham, MA 02043, U.S.A.

In all other countries, sold and distributed
by Kluwer Academic Publishers Group,
P.O. Box 322, 3300 AH Dordrecht, Holland.

TABLE OF CONTENTS

VI. - BINARY SYSTEMS. RADIAL VELOCITY CURVES.

VII. - EVOLUTIONARY TRACKS OF THE PECULIAR STARS. NUCLEAR REACTIONS. DREDGE-UP MECHANICS.

VIII. - CONCLUSIONS. DISCUSSIONS OF POSSIBLE COOPERATIVE PROGRAMS.

PREFACE

The idea of this colloquium was first put forward by Prof. P. Keenan when he visited the Centre de Données Stellaires, in September 1982. Dr. A. Florsch, Director of the Strasbourg Observatory, kindly accepted to held it at Strasbourg.

The scientific organization of the colloquium was in charge of a Committee chaired by Prof. P.C. Keenan and composed by M. Feast, A. Florsch, B. Gustafsson, D. Harmer, M. Jaschek (secretary), R. McClure, A. Maeder, W. Straizys and T. Tsuji.

The members of the Committee helped to translate the idea into a very successful meeting. It should be added that this colloquium was the first meeting devoted to late-type peculiar stars.

The program was organized by sessions devoted to the following topics : taxonomy, photometric properties, distribution and space motions, chemical composition, model atmosphere, binary systems and evolutionary tracks. Each session started with a review paper, followed by contributed papers presented as posters. The different sessions were chaired by W. Bidelman, R. Cayrel, Ch. Cowley, A. Heck, H.R. Johnson, J.P. Kaufmann and T. Tsuji, who did an excellent job.

The President of the "Louis Pasteur University of Strasbourg", Prof. H. Duranton, welcomed the participants to the University where the sessions were held ; Dr. A. Florsch welcomed the participation on behalf of the French astronomical community and Prof. P. Keenan spoke on behalf on the Scientific Organizing Committee.

The local organization was handed very efficiently by the Local Organizing Committee, chaired by Dr. A. Florsch and composed by P. Dubois, D. Egret, J.L. Halbwachs, M. Hamm and F. Ochsenbein. The work of the Committee was greatly helped by the collaboration of the members of the Observatory and the University.

The presence of participants from many countries was supported by financial aid from the "Institut National d'Astronomie et de Géophysique", from the

"Département des Relations Internationales du Ministère de l'Education Nationale", and from the "Conseil Général du Bas-Rhin".

The meeting took place at the "Institut Le Bel" of the University - Strasbourg I. We thank all these institutions for their support.

November 1984

M. JASCHEK

P. KEENAN

SCIENTIFIC ORGANIZING COMMITTEE

M. Feast, A. Florsch, B. Gustafsson, D. Harmer, M. Jaschek (secretary), P.C. Keenan (chairman), R. McClure, A. Maeder, W. Straizys, T. Tsuji.

LOCAL ORGANIZING COMMITTEE

P. Dubois, D. Egret, A. Florsch (chairman), J.L. Halbwachs, M. Hamm, F. Ochsenbein.

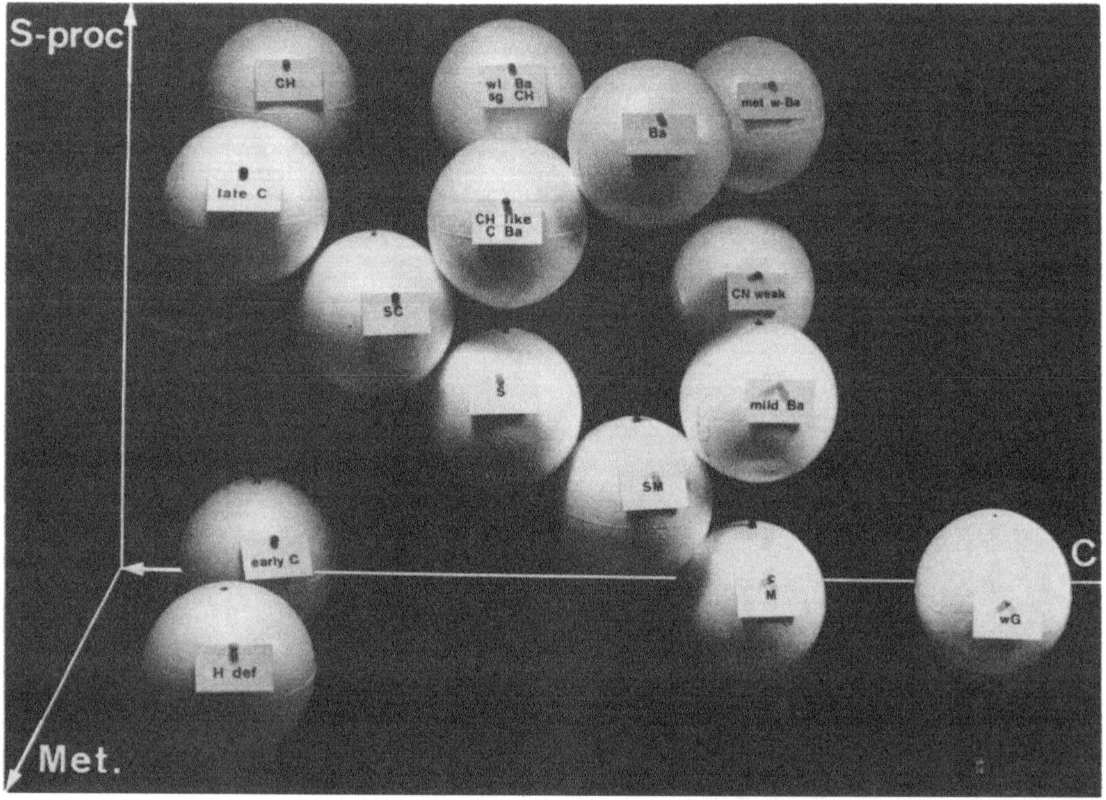

S-proc
CH
wl Ba
sg CH
met w-Ba
Ba
late C
CH like
C Ba
SC
CN weak
S
mild Ba
SM
early C
C
M
wG
H def
Met.

LIST OF PARTICIPANTS

ACKER A. - Observatoire de Strasbourg, France
BARBUY B. - Observatoire de Meudon, France
BEER H. - Kernforschungszentrum Karlsruhe, Germany F.R.
BENNETT P. - University of British Columbia, Canada
BERGEAT J. - Observatoire de Lyon, France
BIDELMAN W.P. - Case Western Reserve University, Cleveland, Ohio, U.S.A.
BISCHOFF M. - Observatoire de Strasbourg, France
CANNON R. - Royal Observatory Edinburgh, U.K.
CATCHPOLE R. - South African Astronomical Observatory, Cape, South Africa
CAYREL R. - Observatoire de Meudon, France
CHMIELEWSKI Y. - Observatoire de Genève, Switzerland
CORNIDE M. - Astrofisica Fac. Fisicas, Univ. Complutense, Madrid, Spain
COWLEY C. - University of Michigan, Ann Arbor, U.S.A.
DUBOIS P. - Observatoire de Strasbourg, France
DUSCHL W. - M P I für Astrophysik, Garching bei München, Germany F.R.
EGRET D. - Observatoire de Strasbourg, France
FERNANDEZ-FIGUEROA M.J. - Astrofisica. Fac. Fisicas. Univ. Complutense, Madrid, Spain
FLORSCH A. - Observatoire de Strasbourg, France
FREIRE R. - Observatoire de Strasbourg, France
FUJITA Y. - University of Tokyo, Japan
GAIL H.P. - Technische Universität Berlin, Germany F.R.
GRENON M. - Observatoire de Genève, Switzerland
GRIFFIN R.F. - The Observatories, Cambridge, U.K.
GRIFFIN - R.E.M. - The Observatories, Cambridge, U.K.
GRUDZINSKA S. - Institut of Astronomy, Torun, Poland
HALBWACHS J.L. - Observatoire de Strasbourg, France
HECK A. - Observatoire de Strasbourg, France
HINKLE K.H. - Kitt Peak National Observatory, Tucson, U.S.A.
HOLTZER M. - Observatoire de Strasbourg, France
JASCHEK C. - Observatoire de Strasbourg, France
JASCHEK M. - Observatoire de Strasbourg, France
JOHNSON H.R. - Indiana University, Astronomy Department, Bloomington, U.S.A.
JØRGENSEN U.G. - Astronomiska Observatoriet, Uppsala, Sweden

KAPPELER F. - Kernforschungszentrum Karlsruhe, Germany F.R.
KAUFMANN J.P. - Technische Universität Berlin, Germany F.R.
KEENAN P.C. - Perkins Observatory, Delaware, U.S.A.
KIEHLING R. - Astronomisches Institut, Bochum, Germany F.R.
KOVACS N. - Institut für Theoretische Phys. und Sternwarte Kiel, Germany F.R.
LABHARDT L. - Astronomical Institut, University of Basel, Switzerland
LAMBERT D.L. - Dept. of Astronomy, University of Texas, Austin, U.S.A.
LLOYD EVANS T.L. - S.A. Astronomical Observatory, South Africa
LOCANTHI DAVIS D. - Altadena, U.S.A.
LU P.R. - Western Conn. State University, Danbury, U.S.A.
MALANEY R.A. - University Observatory, St Andrews, U.K.
MANFROID J. - Inst. d'Astrophysique, University of Liège, Belgium
McCARTHY M. (S.J.) - Specola Vaticana, Vatican
McCLURE R.D. - Dominion Astrophysic. Observatory, Victoria, Canada
MULLER E.A. - Basel, Switzerland
OCHSENBEIN F. - Observatoire de Strasbourg, France
PEERY B.F. - Howard University, Washington, U.S.A.
QUERCI F. - Observatoire du Pic du Midi et de Toulouse, France
QUERCI M. - Observatoire du Pic du Midi et de Toulouse, France
REGO M. - Astrofisica Fac. Fisicas. Univ. Complutense, Madrid, Spain
RICHER H.B. - Dept. Geophys. - Astron., Univ. of B C, Vancouver, Canada
SCHATZ G. - Kernforschungszentrum Karlsruhe, Germany F.R.
SEDLMAYR E. - Technische Universität Berlin, Germany F.R.
STEENBOCK W. - Institut für Theoretische Phys. und Sternwarte, Kiel, Germany F.R.
TRAN MINH F. - Observatoire de Meudon, France
TSUJI T. - Toky Astronomical Observatory, Japan
TUTUKOV A. - Astronomical Council, Moscow, U.S.S.R.
VIEIRA T. - Astronomiska Observatoriet, Uppsala, Sweden
WALTER G. - Kernforschungszentrum Karlsruhe, Germany F.R.
WEHRSE R. - Intitut für Theoretische Astrophysik, Heidelberg, Germany F.R.
WEISS A. - Max Planck Institut für Astrophysik, München, Germany F.R.
WING R.F. - Astron. Dept., Ohio State University, Columbus, U.S.A.
WOOD P.R. - Mt Stromlo and Siding Spring Observatories, Australia
YORKA S.B. - Dept. of Physics and Astronomy, Denison University, Granville, U.S.A.

WELCOMING ADDRESS

P.C. Keenan
Perkins Observatory
Box 449
Delaware OH 43015 - U.S.A.

I just want to welcome you on behalf of the Scientific Organizing Committee, and to say a word about what we are trying to do here. The disentanglement of possible relations between the several groups of cool stars with apparent abundance anomalies is almost certainly not just one problem, but a whole series of problems. We hoped that by bringing you together here, and then locking the doors, you might answers to arrive at least a few of these questions. Perhaps that is too much to expect, but we form a group small enough so that it is not unreasonable to hope that we can make this colloquium an effective working conference -one that will result in some cooperative programs that will lead to solutions.

Now, in order to do that we have tried to leave ample time for discussion - of both the review papers and the contributed ones. The contributed papers are important, but it is always difficult to decide how they should be presented. You well know how often the papers near the end of a session are crowded together and given to a tired audience. That is not fair to the last speakers.

The best way to avoid this is to have them presented as poster papers, and for this meeting we finally decided to ask that they all be poster papers, especially since more papers were contributed than we origninally expected.

The question with poster papers is how to allow everyone to take part in the discussion of them. As at several recent symposia, we plan to follow the viewing of each group of posters with a session at which each author has 5 minutes to speak, after which the chairman will encourage discussion of that paper. The experiment that we are trying is to ask each author not to try to present his results in 5 minutes (the poster have done that), but to just make a few comments on what he or she thinks has been accomplished by the paper. Then, at the end of that whole session, the chairman will have a chance to make whatever comments he wants - perhaps to point out connections between some of the paper.

We do not know how this experiment will work but hope that, with the cooperation of all of you it will give all the paper a fair hearing.

Now, to get down to work, it seemed sensible to agree first on what the groups are that we are going to discuss. Since Carlos and Mercedes Jaschek are completing a book describing the groups of stellar spectra, we asked them to open the meeting by defining these groups of cool stars, and Carlos has agreed to do that in the opening paper.

I. – SPECTRAL CHARACTERISTICS THAT DEFINE THE GROUPS OF PECULIAR COOL STARS

TAXONOMY OF LATE-TYPE GIANTS

C. Jaschek
Centre de Données Stellaires
Observatoire de Strasbourg
11, Rue de l'Université
67000 Strasbourg
France

The giants of spectral types G, K, and M constitute the taxonomically most complex region of the HR diagram. In this rather small region is found a bewildering variety of phenomena, displayed by a small number of stars. This is one important point to be stressed from the start - in the solar neighboorhood, giants constitute about one percent of the stellar population, and all those peculiar objects which are the subject of this meeting make up at most 5% of this 1%-. The spectra of these stars are so complex that many astronomers have shied away from studying them. This can be clearly seen in the IAU sympcsia: of the 111 held to date, none was dedicated to late-type giants, whereas at least four were dedicated to early-type stars.

This situation was certainly due in part to the faintness in the blue spectral region, of many objects of these groups, and to the lack of sensitivity of receivers in the wavelength region where the objects radiate most strongly. The situation has now changed, thanks to improvements in the technology of infrared detectors. Late-type giants are very bright there (Wing and Yorka, 1977) and can be studied relatively easily. The next few years will thus present us with an explosive increase in knowledge about these objects; thus our meeting seems to be very timely.

The object of my talk is to summarize the taxonomy of late-type giants. Since this is a review and not a history of the subject, I shall provide only short descriptions, which mention just a few of the papers that have been written on the subject. A more detailed picture will be provided in a forthcoming book on stellar classification by Mercedes and me.

Let me start by recapitulating some well-known features cf normal giants, for comparison purposes. Stars later than G0 have a classical region ($\lambda\lambda$ 3800-4800) full of lines. The Balmer lines become progressively fainter toward the later types, and no helium line is seen. Of the lighter elements ($Z < 20$) only a few -magnesium, aluminium, and silicon- are represented by strong lines. Most of the visible lines come from iron peak elements ($23 < Z < 28$), which practically blot out all heavier elements, although no line is particularly strong. In μ Her (a G5 object), there are only seven lines between $\lambda\lambda$4000 and 4400 with

M. Jaschek and P. C. Keenan (eds.), Cool Stars with Excesses of Heavy Elements, 3–14.

an equivalent width larger than 1 A, the strongest being Hγ with a width of 3 A. The general blocking at λ4000 for G0 stars, on the other hand, exceeds $0^{m}.7$, implying that we see less than half the light of the star. In addition to atomic lines, there are also bands corresponding to di- and triatomic molecules in this spectral region. The most clearly visible molecules are listed in Table 1; these are all visible in the spectra illustrated in the Keenan-McNeil Atlas (1976).

TABLE I. Molecules observed in late-type stars

Molecule	Band heads	Notes
CH	4295, 4315	
CN	4216, 3883	luminosity-sensitive
TiO	4954, 4352, 4422, 4584, 4626	
C_2	4395, 4697, 4737	
SiC_2	4868, 4979	strong in C-type stars
C_3	4053	
ZrO	4641, 4620	strong in S-type stars
LaO	7900, 7404	
CO	1.56μ-2.35μ	

Spectral types between G0 and K7 are usually determined on the basis of the ratios λ 4143 Fe/λ 4101 H and λ 4045 Fe/λ 4101 H. However, if anomalies of composition are suspected, one uses λ 4254 Cr/λ 4250 Fe and λ 4274 Cr/λ 4271 Fe, which are practically independent of compositional anomalies, because all iron peak elements tend to behave similarly. From K5 to M8, spectral types are determined by the strength of the TiO bands until these become saturated at around M3. Then λ 4226 CaI is used; this increases rapidly in late M-type stars.

The luminosity criteria in G-, K-, and early M-type stars are the line-intensity ratios λ 4077 SrII / λ 4063 FeI and λ 4216 SrII/λ 4226 CaI plus some additional criteria for very luminous stars. Beyond M5, luminosity effects become less prominent. These luminosity criteria cannot be used if are suspects anomalies in the "s-type" elements (that is, elements producible by nucleosynthesis, as mentioned later), because Sr is such an element. Let us turn now to the stars with peculiar spectra.

Carbon stars were discovered long ago by Secchi (1868), who singled them out as a special type. They were introduced into the Henry Draper catalog as R and N stars (2,3) and the characteristic bands were attributed to carbon compounds. Rufus (1916) took the essential step in the interpretation of carbon stars, when he attributed the different appearances of the carbon stars and the normal stars to varying amounts of oxygen and carbon atoms in their atmospheres. The R and N types of star were decisively unified into a single group by Keenan and Morgan (1941), who proposed a single sequence of carbon stars running parallel to normal stars, as given in Table II.

TABLE II. Carbon-star sequence

Carbon stars		Normal stars
Former designation	KM designation	
	C0	G5
R3	C2	K0
	C3	K2
R5	C4	K4
N1	C5	M0
R9	C6	M2
N2	C7	M4

The number following the "C" is a temperature index, obtained from the atomic lines of metals (Cr and Fe). To this first index the authors added a second figure indicating the strength of the C_2 bands, with numbers running from 1 (weakest) to 5 (strongest). C2,3 is thus a K0 giant with medium-strong C_2 bands.

A carbon star can be defined as a late-type giant whose spectrum contains strong bands of carbon compounds such as C_2, CH, CN, and C_3, and no metallic oxide bands. At higher dispersions, the metallic lines can be seen to behave as in giants, but heavier elements of the fifth and sixth lines of the periodic table ($Z > 30$) are stronger in some carbon stars than in normal giants. These elements include Y, Zr, Mo, Ba, La, Ce, Nd, and Sm, which for the sake of conciseness we denote "s-type elements", that is, elements which can be produced by nucleosynthesis, adding slow electrons to matter having a solar-type composition. Dominy (1984) has shown that s-type elements are not enhanced in C0-C3 stars, so any such enhancement exists only in late-carbon stars.

In summary, carbon stars are characterized by at least three independent parameters, namely the oxygen-to-carbon ratio, the metallic line strength, and the enhancement of s-type elements, which for the sake of brevity, I shall call the x, y, and z parameters. As we have seen, carbon stars have different x and z from normal stars.

A second family of anomalous stars, the "barium stars", was defined by Bidelman and Keenan (1951). Barium stars are giants (G2-K4) with a strong λ 4554 Ba II line. Bidelman and Keenan also pointed out that some carbon bands such as C_2 and CN are stronger than in normal stars, although not as strong as in carbon stars. More detailed studies showed that in Ba stars "s-type elements" are also enhanced (see the references provided by McClure, 1984), specifically those elements with

$38 < Z < 56$. Carbon is the only lighter element that is definitely stronger. Since Ba stars are defined spectroscopically on the basis of the strength of the Ba II line(s), Warner (1965) and other authors have added to the spectral type an estimate of the strength of the barium line (as compared with the line in normal giants) for example, K0-Ba3. Since various scales are used in various publications, one should always go back to the original reference, to be sure what scale was used. When line-intensity estimates of a single feature are used, some scatter must result when the stars are analyzed in detail. This is exactly what happened to analyses of stars discovered on objective-prism plates. Much discussion has centered on this scatter, although in fairness it should be said that if two different atmospherists make abundance estimates for the same object, the disagreements are often of the same order as the anomalies found. Therefore, in the present state of the art, more analysis should be performed for characteristic objects of the group, rather than for marginal objects which may not even belong to the group.

From all these analyses, however, one point became clear, namely that the group of barium stars has no sharp boundaries. There is a gradual transition between normal and barium stars, and this fact argues against any hypothesis linking anomalies in the abundance of barium to something clear-cut like binarity.

Stars intermediate between Ba and normal are called "semi-barium stars" when the anomalies are definite because they have been detected on high-dispersion material. Those in which the phenomenon is not obvious are "marginal barium stars", and in practice these have been the most controversial. The word "marginal", however, is fuzzy, so that it is sometimes difficult to know exactly what the word means.

The third group of peculiar spectra is the one of S stars. An S-type star is a late-type (K5-M) giant whose spectrum contains distinct ZrO bands. Keenan defined the group (1954), although Merrill described them first (1922). Usually (but not always) the ZrO bands are accompanied by LaO bands. S-type stars present the same problem as C stars -the need for a temperature index (α) and an abundance index (β)-. Several ways of achieving these have been proposed, but I will take the latest one proposed by Keenan and Boeshaar (1980).

The temperature index (α) is defined, as in normal M-type stars, with the help of lines of singly ionized and neutral elements, and the two temperature scales are equivalent: S0 $\equiv$ M0, S5 $\equiv$ M5, and so on. The abundance index (β) indicates the strength of the ZrO bands relative to the TiO and C_2 bands. When ZrO is just visible, $\beta = 0$; ZrO = TiO at $\beta = 3$, and TiO disappears relative to ZrO at $\beta = 6$. $\beta = 8$ marks the point where $C \sim O$ and thus no C_2 or ZrO is visible. The star is then an SC star. When C_2 becomes stronger ($\beta = 10$) the star is a carbon star and thus $S\alpha/10 = C\alpha/2$.

Thus S stars constitute a natural bridge between M and C stars, with all kinds of intermediates. Detailed analysis has shown that S-type stars contain metals with about normal strength and definitely enhanced amounts of heavy elements ($Z > 38$), including Zr, Sr, Y, Ba, and La. Moreover some of these stars contain the unstable element Tc. There have been many published articles on the nucleosynthetic aspects of element production.

The temperature index of the S-type stars is somewhat more complex than that of the C-type stars. For convenience the spectral types (temperature scales) of the various groups of stars are shown in Table III. The index for C stars starts at G5, whereas that for S and MS stars starts at M0. So the transition from S to C through SC stars presents difficulties. Since SC0 $\simeq$ M0 on the C-side, the equivalent C type should be several subdivisions later than the earlier types. However for later types the differences is less, so that SC7 $\cong$ C7.

TABLE III. Spectral types (temperature scales) of late peculiar stars

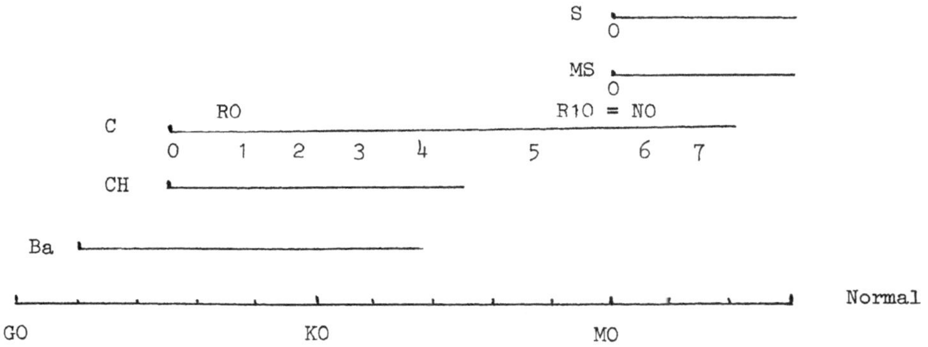

The fourth important group is the CH stars. The group was first described by Keenan (1942). A CH star is a G-type (G5–K5) giant in which the molecular CH bands are very strong, so strong that they blot out most of the features in the region below λ 4300. Most of the metals are weakened, but the spectra contain strong λ 4077 SrII and λ 4554 BaII lines. Later detailed analyses have shown that s-type elements are enhanced, but not the iron peak elements. Since the stars have weak metal lines and high radial velocity they can be considered to be both population-II objects and early carbon stars.

Since the other groups (C, S, and Ba) did not exhibit large radial velocities (and are thus not straightforward cases of population II stars), one can ask if there exist (other) CH-type stars which do not belong to population II. The answer is affirmative, there are the so-called CH-like stars. These are defined as low-velocity C0-C3 stars with an enhancement of λ 4554 Ba II (Yamashita, 1972). The stars are denoted C3,0 ch (ch = CH-like). These stars are like the Ba-C stars found by McConnell et al. (1972).

A group which also should be fitted in somewhere is the subgiant CH stars described by Bond (1974). They are defined as a group having

G-type spectra with weak metallic lines, very strong CH features, and strong Sr II lines (λλ 4077 and 4215). He adds that several, but not all of these stars also showed a strong λ 4554 Ba II line and were classified as "weak lined Ba stars" by McConnell et al. (1972).

Just to complete the different groups I shall add two more which are sometimes mentioned in the literature; these are the J-type stars and the Li stars.

A J star is a carbon star characterized by strong isotopic bands of carbon. The designation was introduced by Bouigue (1954) for stars in which λ 6260 of $C^{13}N^{14}$ is very conspicuous. Gordon (1971) defines a J star as having a λ 6168 $C^{12}C^{13}$ band at least half as strong as the λ 6122 band of $C^{12}C^{12}$. J stars are also called sometimes "λ 6168" stars. Yamashita writes the "J" after the carbon classification.

"Li stars" are carbon stars with an extreme enhancement of the resonance line of λ 6708 Li I. The name seems to have been introduced by Bidelman (1956). Boesgaard (1970) calls them "super lithium rich".

The position of the groups can be easily visualized in a three-dimensional representation, as pictured on page xii of this volume. For discussion however, it seems best to present a pair of two-dimensional plots with axes "strength of metals" (y) and "oxygen-carbon" (x) assuming that the third axis, the "strength of s-type elements", is perpendicular to the paper. In Fig. 1 we have cut the three dimensional representation at a high value of the third coordinate. This is thus the plane of the stars that are the subject of the present conference. The middle of each scale is marked "n", the "normal" value. Please notice that we are representing observations, so that the scales are line intensities, and not abundance. This is specially true for the "oxygen-carbon" axis, where stars on the left side show oxide bands and those on the right show increasingly stronger carbon-molecule bands.

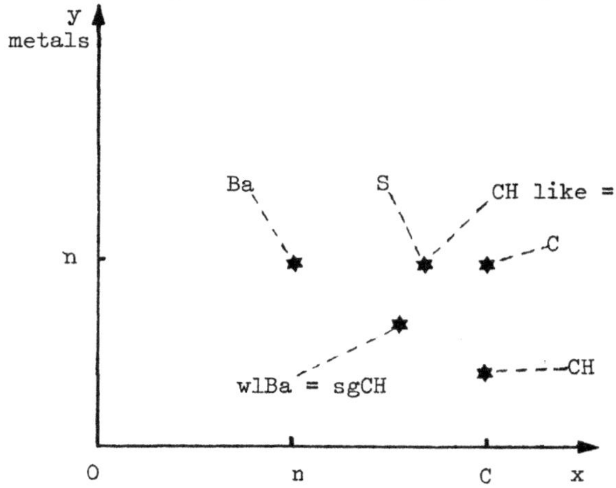

Figure 1
S stars lie in a plane below CH like stars.

The various groups are located according to the descriptions I have given. Since in another cut of the third axis, "normal" giants would occupy the place here occupied by Ba stars, S stars fall on a line joining M and C stars -with M5 and SC stars on the same string.

To complete the taxonomy of late-type giants, I shall next mention five additional groups which really fall outside the scope of the present conference. They have in common the normalcy of "s-type elements". They are :

a - The normal giants, upon which I shall not comment further.

b - The early carbon stars. I mentioned earlier that according to Dominy (1984) only the oxygen-carbon ratio is abnormal here.

c - The weak-G-band stars. These will be commented on below.

d - The H-deficient carbon stars.

e - The giants of population II with normal CH and weak CN.

"Weak-G-band" stars are late G- or early K-type giants in which the G band of CH is very weak or absent. The first star of this group was discussed by Cannon (1912) and afterwards by Bidelman (1950). Bidelman and McConnell (1973) then identified a large number of them, which furnished the basis for our present knowledge. In these objects, CN is generally but not always weak; and there is no correlation between CH faintness and metallic line strength.

This group is distinct from group e, metal-weak giants (i.e. old giants), in that high-velocity stars have weak CN but normal or even strong CH.

Finally stars of group d - hydrogen-deficient carbon stars - are defined as high-luminosity G or K giants whose spectra contain weak or absent hydrogen lines, no trace of CH, and strong C_2 features. The first star of this group was described by Ludendorff (1906), and the group as such was introduced by Bidelman (1953). Many of the stars are variables of the R Cr B types, but a few are non-variable; these are called Hd (hydrogen-deficient) - thus C2,2 Hd. Very few objects of this type are known.

For the location of these groups, see Fig. 2.

For the moment this seems to cover most of what is known about the taxonomy of the late-giant group. More advances are to be expected from the analysis of giants in clusters, such as the analyses by Lloyd Evans (1983, 1984) of the late giants in ω Cen and in the Large Magellanic Cloud globular clusters. These clusters contain stars with further differences in the CN and CH behavior, so that probably a future picture of the late-type giants would need an additional dimension, for instance the N/O ratio. But for the moment we are not sure that what we see in one cluster can be generalized to stars in the field: it could well be

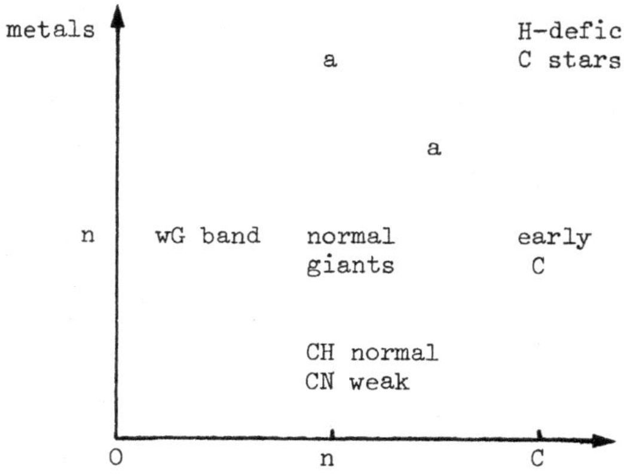

Figure 2

that there are variations from cluster to cluster which do not show up very much outside the particular cluster (or cluster system).

I shall now spend a few moments more on the three dimensional representation to ask why certain places are empty. Are there no stars with those characteristics, or have we overlooked them ?

Let's start with the diagram at "s elements" = n, that is at the "lower floor". Missing here are objects with strong metals and either normal or weak carbon. This is probably the place to locate the "CN-strong" stars (or λ 4150 stars) described by Roman (1952) and by Schmitt (1971). In these stars CN is strong, metals are about normal, and hydrogen is slightly weaker than in normal stars. These characteristics suggest that the stars are slightly (or moderately) metal-strong and perhaps slightly carbon-rich. They could thus occupy either of the places marked "a" in Fig. 2. I think that in view of the major debate in the literature on the "super-metal-rich stars" it is best not to be too definite about this group at present (see for instance Taylor, 1982).

Other empty place would correspond to the "carbon" stars with weak metals and to metal-rich carbon-poor stars. If the assumption made by theoreticians is true that in the atmosphere of a late giant there are products of the 3α cycle mixed with the outer atmosphere, one would expect that such stars do exist, where are they ?

Shifting now to the "upper floor" that is, the stars with "strong s-type elements", we find a lack of metal-strong stars with varying degrees of carbon enhancement, implying perhaps that no stars with enhancement metals are old enough to show an enhancement of heavy elements. Perhaps population-I S-type stars, if they exist,would show such enhancement (Yorka and Wing, 1979).

Also missing are weak-metal-line stars with weak carbon features and s-type elements. This lack probably implies that no stars of this type can exist. However, stars of this type may be represented by the few Ba stars with weak metals and normal CH which McConnell et al. (1972) described; in two of them Catchpole et al. (1977) found "very weak metals".

Further questions concern the homogeneity of the groups I have been discussing. Do they all belong to the same population ? We have clear evidence of membership in population II only for the CH stars (plus the metal-weak giants). All other groups seem to belong to an old population-group I or to an intermediate population, although probably all the types I have discussed represent a mixture of ages. No clear evidence seems to exist that any of these objects belongs to a young population I.

A similar question can be asked concerning the absolute magnitudes. Are all of the stars in question giants ? The answer would be yes, if one forgets about spectroscopic criteria which can be influenced by element anomalies. We thus have to rely only upon the absolute magnitude determinations, and according to this criterion all these stars are giants except two groups, namely the hydrogen-poor carbon stars, which are supergiants (and here the evidence seems convincing), and the CH stars, which are subgiants according to indirect evidence. However we should not be too confident because our methods of determining absolute magnitudes generally do not establish the value of the dispersion around the mean. There do exist, for instance, stars which are considerably brighter than the average -such as the well studied S-type star π^1Gru.

And to finish my talk I just want to ask: are you sure that there are no dwarfs with spectral anomalies like those of the groups under consideration ? I have grouped in Table IV all the troubling cases which I found in the literature.

TABLE IV. Non-giant carbon-enhanced stars

G 77-61	Dahn et al. (1977)	strong C_2, strong CH
2C24-SA51	Chiu and Kron (1982)	metal-weak, strong CH, moderately strong C_2 and CN, strong Sr II and Ba II - Ba star ?
GD 439	Liebert et al. (1979)	strong CH, no Swan bands
GH 7-21	Liebert (1976)	strong CH, C_2 bands

At present it is not clear how these stars should be classified. If these peculiarities develop in binary stars, sooner or later we are bound to find dwarf peculiars because they would have received the nuclear-processed material from their (now extinct) companions. On the other hand, Dahn et al. (1977) searched for nearby carbon stars, by looking for stars with a large proper motion. Since they found none, they concluded that such objects must be several orders of magnitude rarer than ordinary K and M dwarfs.

So let me conclude this survey by suggesting that we have not yet finished with the taxonomy of this group and that more classes are just around the corner.

Note : After giving my lecture I was told that Straizys has made a similar three-dimensional representation.

References

Bidelman, W.P. : 1950, Astrophys. J. **111**, 333
Bidelman, W.P. and Keenan, P.C. : 1951, Astrophys. J. **114**, 473
Bidelman, W.P. : 1953, Astrophys. J. **117**, 25
Bidelman, W.P. : 1956, Vistas in Astron. **2**, 1428
Bidelman, W.P. and McConnell, D.J. : 1973, Astron. J. **78**, 687
Boesgaard, A.M. : 1970, Astrophys. J. **161**, 1003
Bond, H.E. : 1974, Astrophys. J. **194**, 95
Bouigue, R. : 1954, Ann. d'Astrophys. **17**, 104
Cannon, A.J. : 1912, Harvard Ann. **56**, 65
Catchpole, R.M., Robertson, B.S.C. and Warren, P.R. : 1977, Monthly Not. Roy. Astron. Soc. **181**, 391
Chiu, L.T.G. and Kron, R. : 1982, Nature **299**, 702
Dahn, C.C., Liebert, J., Kron, R.G., Spinrad, H. and Hintzen, P.M. : 1977, Astrophys. J. **216**, 757
Dominy, J.E. : 1984, Astrophys. J. Suppl. **55**, 27
Gordon, C.P. : 1971, Publ. Astron. Soc. Pacific **83**, 667
Keenan, P.C. : 1942, Astrophys. J. **96**, 101
Keenan, P.C. : 1954, Astrophys. J. **120**, 484
Keenan, P.C. and Boeshaar, P.C. : 1980, Astrophys. J. Suppl. **43**, 379
Keenan, P.C. and McNeil, R. : 1976, An Atlas of spectra of the cooler stars, Ohio State University Press.
Keenan, P.C. and Morgan, W.W. : 1941, Astrophys. J. **94**, 501
Liebert, J. : 1976, Astrophys. J. L **204**, L 93
Liebert, J., Dahn, C.C., Gresham, M. and Strittmatter, P.A. : 1979, Astrophys. J. **233**, 226
Lloyd-Evans, Th. : 1983, Monthly Not. Roy. Astron. Soc. **204**, 985
Lloyd-Evans, Th. : 1984, Monthly Not. Roy. Astron. Soc. **208**, 447
Ludendorff, H. : 1906, Astron. Nachr. **173**, 1
McClure, R.D. : 1984, Publ. Astron. Soc. Pacific, **96**, 117
McConnell, D.J., Frye, R.L., Upgren, A. : 1972, Astron. J. **77**, 384
Merrill, P.W. : 1922, Astrophys. J. **56**, 457
Roman, N. : 1952, Astrophys. J. **116**, 122
Rufus, W.C. : 1916, Publ. Obs. Michigan **2**, 103
Schmitt, J.L. : 1971, Astrophys. J. **163**, 75
Secchi, A. : 1868, Comptes Rendus Acad. Sc. Paris **66**, 124
Taylor, B.J. : 1982, Vistas in Astron.**26**, 253
Warner, B. : 1965, Monthly Not. Roy. Astron. Soc. **129**, 263
Wing, R.F. and Yorka, S. : 1977, Monthly Not. Roy. Astron. Soc. **178**, 383
Yamashita, Y. : 1972, Ann. Tokyo **13**, 169
Yorka, S.B. and Wing, R.F. : 1979, Astron. J. **84**, 1010

DISCUSSION

KEENAN - Two comments. The scale of barium intensities that we use is essentially that introduced by Warner. The second comment is that the supergiant α Ori shows enhanced ZrO. And now a question: since Dr. W. Straizys is unfortunately not present, does your model differ from the three-dimensional diagram published by Straizys in the Vilnius Bulletin ?

JASCHEK - In answer to your first comment: yes, indeed, I forgot to mention this. With regard to your question: I was unaware of the Straizys's diagram so I cannot comment upon differences. I hope that both agree !

BIDELMAN - I believe that the field no-G-band stars that have been analyzed all have weakened or absent CN, as well as near absent CH. Also I think that one must consider the abundance of nitrogen as well as that of carbon in discussing these problems. A final comment: it is interesting to ask why the relationship of the C and S stars was not discussed long ago. Merrill observed the S stars mainly in the blue and Sanford observed the C stars mainly in the red; but they did not look at each other's plates ! If they had, the relationship might have been realized much earlier.

TSUJI - What is the basis of the correspondence between the temperature scale of K and M giants and that of C stars ? How reliable is it ?

JASCHEK - I think Prof. Keenan could answer this.

KEENAN - For early C stars (R-type), having temperatures of late-G and early-K stars, the C_2 bands are weak enough to allow the usual ratios of atomic lines to be used. These give fairly good temperature types. For late M-type C stars the temperature classification of the C system is very poor.

RICHER - Your suggestion that a range in the strength of the barium in Ba stars is an argument against the binary nature of these objects, is, I think, not entirely justified. After all binary separation which is a continuous parameter and presumably related to mass transfer rate would provide for a range in barium strength in these stars.

JASCHEK - Yes, that could be.

WING - The fact that the space defined by your three-composition parameters is nearly full indicates that what a star does on the inside is largely independent of the metallicity that it starts with, and that it takes two-composition parameters to describe the results of the internal processing. But you also briefly mentioned a third composition parameter, the hydrogen to helium ratio. Apart from the observational difficulty of measuring H/He in the colour stars, do you think we really should add this as a fourth dimension to your model, or is the degree of hydrogen deficiency perhaps determined by the values of O/C and s/Fe ?

JASCHEK - I agree with your first point. A fourth composition parameter is probably needed, but it is not clear to me which one would be the most convenient.

BIDELMAN - My feeling is that there are really not any M-type hydrogen-poor stars. I cannot believe that they would not look sufficiently different from normal stars to be recognized, and none have been.

CATCHPOLE - I would like to comment about population types of the stars. We have two examples; NT Tel a high velocity star of probably extreme population II and stars in η Car. These are indistinguishable at 80 A/mm, which is not to say that they would be so at high dispersion, but they nicely may explain an empty space in your diagram.

BIDELMAN - Stephenson has recently completed a new visual region survey of the entire northern sky, finding a substantial number of new S stars and is preparing a new general catalogue of S stars which will be ready in a few months.
McConnell's southern survey in which he found a large number of new S stars was done with the 10-degree prism, which enables stars with fairly weak ZrO λ 7474 bands to be detected. But these objects still deserve to be called S rather than MS in general.

McCLURE - Are the CH-like stars metal-poor like the CH stars ? Or, if a little carbon where added to a Ba star, would it become a CH-like star ?

JASCHEK - This is difficult to answer. The group is perhaps not homogeneous.

SPECTRAL CLASSIFICATION AND THE RELATIONS BETWEEN PECULIAR GIANTS

Sandra B. Yorka (Denison University)
Philip C. Keenan (Perkins Observatory)

Spectral classification may make a modest contribution to our understanding of the stars with anomalous atmospheric abundances. For one thing, the more spectra we examine the more evident it becomes that the groups that we are talking about here are practically never discrete and clearly bounded. There always seem to be marginal stars that blur the distinctions between the so-called groups.

Let us look first at four stars of types near G3, shown in Figure 1 as linear intensity traces made from typical classification spectrograms. In addition to their similar temperatures the feature that these stars have in common is the strength of the G-band of CH, which is greater than in ordinary giants. The two upper spectra represent hot barium stars. ζ Cap is the well-known supergiant, and perhaps should be termed a barium-carbon star in view of the strong C_2 absorption. Sneden et al. (1981) measured its C/O ratio as 1.02.

The bottom spectrum is of HD 96765, which was noted by Bidelman as having a strong G-band, and is similar to the group called subgiant CH stars by Bond (1974). Above it is shown the spectrum of BS 8626 which, like BS 7606, is similar to the subgiant CH stars except that the general appearance of the spectrum suggests somewhat higher luminosity. These two stars have low space velocities and relatively low galactic latitudes, in contrast to some of the other CH stars which are clearly Population II objects. The obvious conclusion is that objects which are called G-type CH stars may belong to quite different stellar populations, ranging from young disk stars to halo stars.

The major difference between the barium stars and the G-type CH stars is, of course, the greater strength of the lines of Sr II and Ba II in the spectra of the former. This is due in part to differences in luminosity, but must also reflect the known greater abundance of the s-process elements in the barium stars.

In the second set of spectra the top three are strong barium stars arranged in order of decreasing temperature. These all show traces of the C_2 λ4737 Swan band, which increases in strength as the type advances. This enhancement as the temperature decreases is shown by the spectrum of every variable star that we have observed to show

M. Jaschek and P. C. Keenan (eds.), Cool Stars with Excesses of Heavy Elements, 15–18.

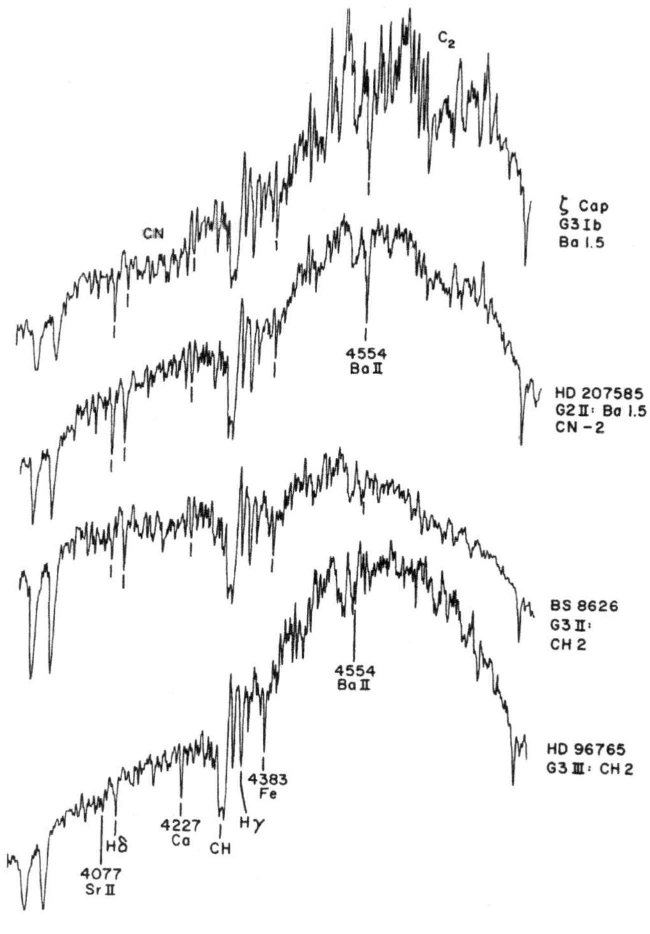

Figure 1 -- Comparison of G-type CH stars with barium stars of similar temperature. Original spectrograms taken at scale of 76 A/mm with Perkins 1.8 meter reflector at Lowell Observatory. The scales are matched for the continua near the λ4300 G-band of CH.

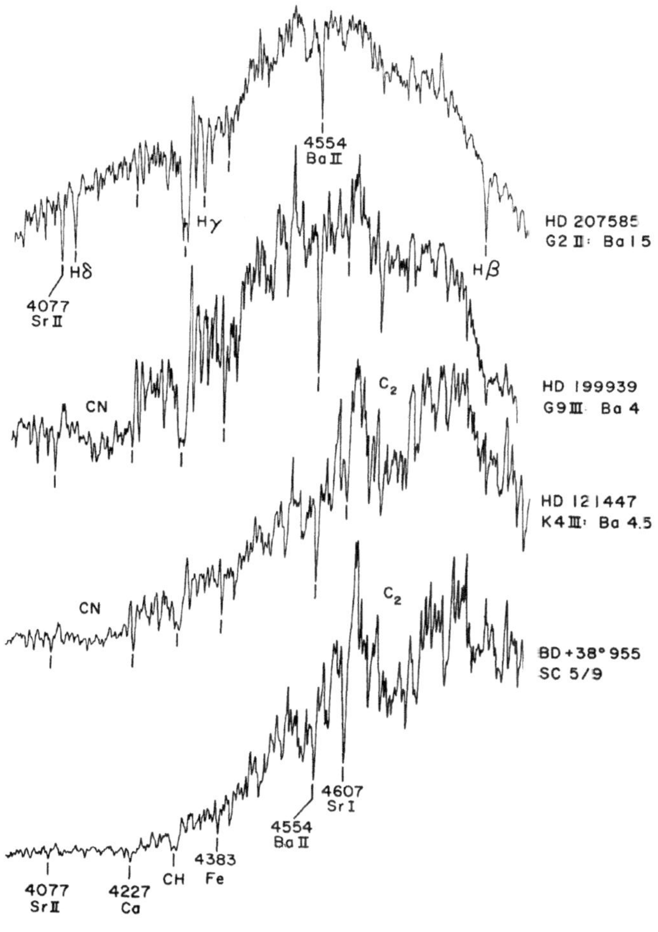

Figure 2 -- A temperature sequence of three giant barium stars and the SC star BD+38°955. As in Fig. 1, the resolution is 2 to 3 A. The scales are matched near the λ4737 Swan band of C_2.

the carbon bands at any phase. It therefore seems reasonable to look for chemical analogues of the strong barium stars among cooler stars with slightly stronger C_2 absorption. The SC 5/9 star BD +38°955 is shown as a promising example. The type of SC5 (Keenan and Boeshaar 1980) suggests a temperature perhaps comparable to that of an M5 star, while the abundance index 9 shows that the star is near the carbon end of the SC sequence and could be called a weak carbon star.

It can be seen in Figure 2 that although Ba II λ4554 is somewhat weaker in BD +38°955 than in the barium stars, the great strength of Sr I λ4607 indicates a comparable abundance of the s-process elements. This is borne out by Catchpole's (1982) analysis of the typical SC star UY Cen. The connection between barium and SC stars was earlier suggested by Straizys (1981) and McClure (1984).

If we are correct in identifying SC stars with carbon bands as the low-temperature analogues of the strong barium stars (practically all of which show C_2 bands), then it follows that if a pure S star had its temperature raised enough to cause the bands of ZrO, etc. to disappear, its spectrum would differ from that of a strong barium star in not showing C_2 bands. This seems to be confirmed by observations of the hot Se variables S UMa and V Cnc near maximum light.

References:

Bond, H. E., 1974, Astrophys. J., 194, 95.
Catchpole, R. M., 1982, M.N.R.A.S., 199, 1.
Keenan, P. C., and Boeshaar, P. C., 1980, Astrophys. J. Suppl., 43, 379.
McClure, R. D., 1984, Pub. Astr. Soc. Pacific, 96, 117.
Sneden, C., Lambert, D. L., and Pilachowski, C. A., 1981, Astrophys. J., 247, 1052.
Straizys, V., 1981, Vilnius Obs. Bull., No. 56, 38.

DISCUSSION

JASCHEK: Do you distinguish between CS and SC stars ?

YORKA: No

THE INFRARED SPECTRA OF RED VARIABLES: II THE SC AND CS STARS

R.M. Catchpole and P.A. Whitelock
South African Astronomical Observatory
P.O. Box 9
Observatory 7935
South Africa

ABSTRACT. Representative 1-4 μm spectra of cool stars accross the transition 1 > C/O > 1 are shown. The 3.1 μm absorption feature is seen to be a characteristic of CS as well as C stars, though not of SC or S stars and is therefore a sensitive indicator of C/O variation when C/O ≃ 1.0. The depths of the various absorption features within the specta of CS stars differ considerably from one star to the next while the SC stars all show very similar spectra. This is consistent with the pattern which has emerged from infrared photometry of these stars.

1. INTRODUCTION

The flux distribution of many cool stars reaches its maximum in the near infrared, a region which has, to date been rather poorly studied spectroscopically. We have started a programme of infrared spectroscopy in an attempt to relate the position of these stars in the two colour diagram, described by Feast et al. (1982), to their molecular band strengths. In particular to explain why the CS stars spread so widely in the two-colour diagram while the SC stars formed a very tight group. In the visible these two classes of star show rather small spectral differences e.g., Catchpole & Feast (1971). Although the literature contains a few examples of the 1-4 μm spectra of M, S and C stars (e.g., Hinkle 1978, Goebel et al. 1981) very little has been published on SC and CS stars in this wavelength range. Merrill & Stein (1976) show a spectrum of the SC star GP Ori.

2. OBSERVATIONS AND REDUCTIONS

The infrared spectra were obtained with the SAAO filter-wheel spectrometer, at a resolution of $\Delta\lambda/\lambda \sim 0.01$, using the 1.9m and 0.75m reflectors at Sutherland. No observations were made accross the very strong telluric water-vapour bands centered at 1.4, 1.9 and 2.7 μm. The reduction technique has been described by Whitelock and Catchpole (1984, paper I). In order to facilitate comparison and measurement the spectra were normalised so that the fluxes at 1.3 and 3.75 μm were

M. Jaschek and P. C. Keenan (eds.), Cool Stars with Excesses of Heavy Elements, 19–23.

equal; this is achieved by dividing by a suitable black-body; although no true continuum is visible between 1.2 and 4.0 μm the regions around 1.3 and 3.75 μm are free from the very deep absorptions that blanket the spectra of S and C stars.

3. INFRARED SPECTRA

Observations have been obtained of a number of S, SC, CS and C stars from which a sequence of spectra of stars with various C/O ratios can be defined and which form a useful scale on which to compare other stars. Figure 1 shows such a sequence on which the positions of the major molecular absorbers are marked.

The most striking change through the sequence of spectra in Figure 1 is the presence of the 3.1 μm absorption feature in C and CS stars and its absence in SC and S stars. This feature is generally attributed to the combined absorption of C_2H_2 and HCN and is considered a diagnostic of cool carbon stars (Ridgway, Carbon & Hall 1978) but it is not detected in the R-type carbon star HD19557 (Goebel et al. 1983). The 3.1 μm feature is also very weakly present in Merill & Stein's (1976) spectrum of the SC star GP Ori where it is seen with an absorption feature at 3.0 μm having a similar strength. BH Cru (Hen 120) is an interesting possible exception to the rule, in that although it has been classified as an SC star (Catchpole & Feast 1971, Keenan 1971) it clearly shows the 3.1 μm absorption feature in Figure 1. However recent work on BH Cru indicates that it has, at least occasionally, shown the characteristics of a CS star (Lloyd Evans this volume). Hence it seems possible that at the time of our observation (April 1981) BH Cru showed a CS star spectrum and therefore does follow the trend described above.

The absorption feature at 1.55 μm is attributed to a blend of CO (2nd overtone), C_2H_2 and HCN. It is this feature which is, at least partially, responsible for the considerable spread in the intrinsic infrared colours of CS stars. For instance Figure 1 illustrates a strong 1.55 μm feature in TV Vel and a much weaker one in TT Cen. Only one spectrum of an SC star, that of AM Cen, is shown as the others are very similar, consistent with their limited spread in the two colour diagram.

The variation of the 2.3 μm CO first overtone band strength amongst the stars of Figure 1 is also interesting. This band is almost certainly saturated in all these stars and the changes in apparent strength must be largely a result of variations of the local continuous opacity. It is well known that CO is weak in carbon stars, as a result of CN opacity, and that it is particularly strong in S stars. It can be seen here and in Figure 3 of paper I, that the SC stars show a very similar CO strength to that of the S stars while the CS stars cover the entire range of CO strengths. The bands of C_2 and CN are clearly stronger in the C stars than in the others.

The spectrum of the S star W Cet illustrates the appearance of H_2O absorption which occurs in some S Miras, though not with the same intensity as is common in M-type Miras.

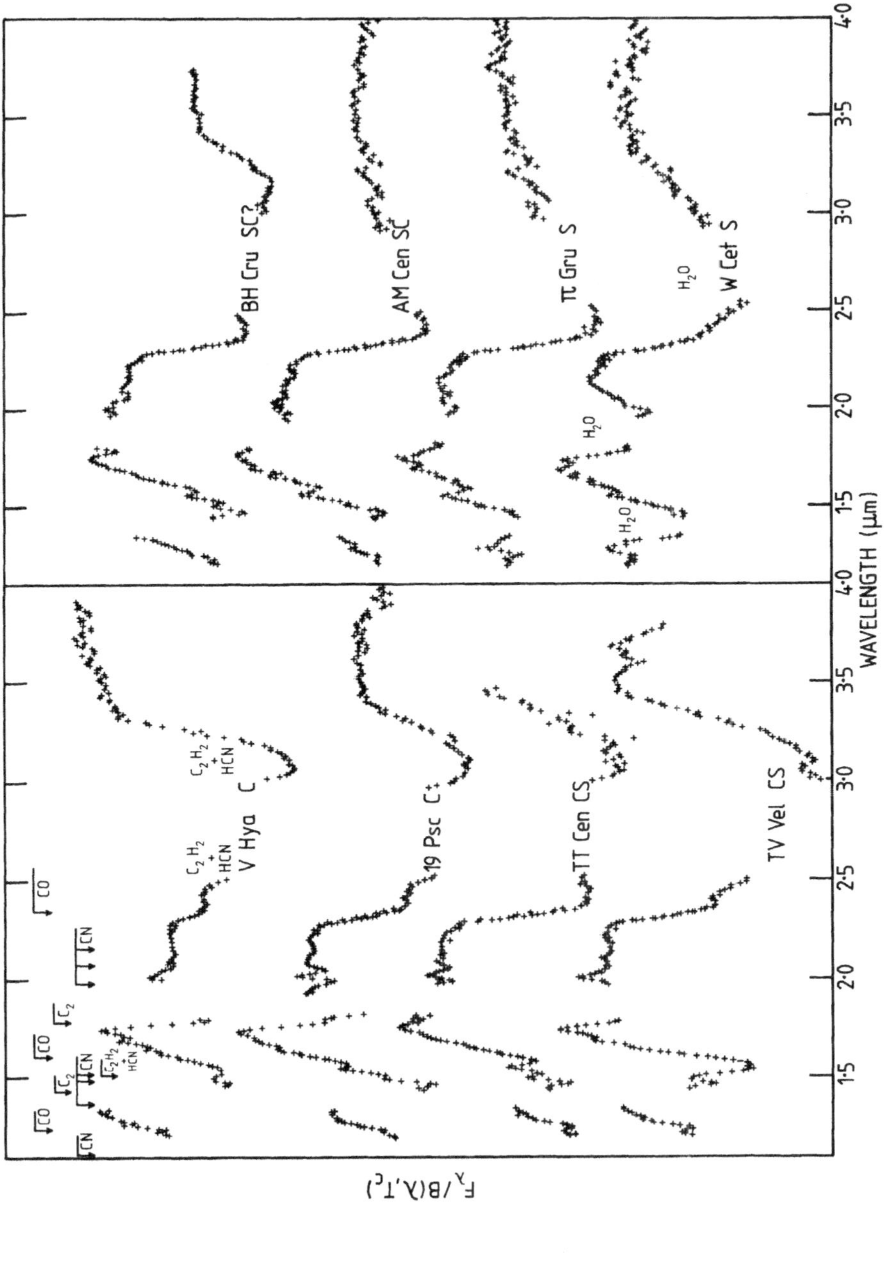

Figure 1. Infrared spectra of cool stars accross the transition 1 > C/O > 1.

4. CONCLUSION

The 1.4 μm spectra of S, SC, CS and C stars shown here illustrate qualitatively the changes in molecular band strength with C/O ratio and the great sensitivity of the 3.1 μm feature to C/O. Synthetic spectra derived from cool star model atmospheres would be invaluable in making full use of this sort of data to better understand the characteristics of these interesting stars.

REFERENCES

Catchpole, R.M. & Feast, M.W., 1971. Mon. Not. R. astr. Soc. **154**, 197.
Feast, M.W., Robertson, B.S.C., Catchpole, R.M. Lloyd Evans, T., Glass, I.S. & Carter, B.S., 1982. Mon. Not. R. astr. Soc. **201**, 439.
Goebel, J.H., Bregman, J.D., Witteborn, F.C., Taylor, B.J., Willner, S.P., 1981. Astrophys. J. **246**, 455.
Goebel, J.H., Bregman, J.D., Cooper, D.M., Goorvitch, D., Langhoff, S.R. & Witteborn, F.C., 1983. Astrophys. J. **270**, 190.
Hinkle, K.H., 1978. Astrophys. J. **220**, 210.
Keenan, P.C., 1971. Mon. Not. R. astr. Soc. **153**, Pl.
Merrill, K.M., Stein, W.A., 1976. Publ. astr. Soc. Pacif., **88**, 285.
Ridgway, S.T., Carbon, D.F. & Hall, D.N.B., 1978. Astrophys. J. **225**, 138.

DISCUSSION

KEENAN - Where is BH Cru in your CO - log T_e diagram ?

CATCHPOLE - I am sorry I do not remember exactly but I believe it lies with the other SC stars.

WING - The behavior you described in BH Cru -showing C_2 at some phases, and ZrO at others- seems similar to the behavior of TT Cen reported by Stephenson (Ap. J. 1973).

The role of opacity effects should be emphasized. The SC stars with O = C such as VX Aql (Greene and Wing, Ap. J. 1975) have the lowest general opacity of all late type stars since neither CN nor H_2O is very important ; as a result we see down to deep, high pressure-layers, and we find enhancements of easily ionized elements (Li, Na, Ca) and their hydrides.

CATCHPOLE - I agree in general but I don't think this explains the range of Li strength amongst the SC - CS stars.

YORKA - Have you looked for the 3.1 absorption band in a variety of stars ?

CATCHPOLE - Yes, in a number of C, S, SC or CS stars. We hope to publish our spectra eventually.

JASCHEK - a) Were all your S stars discovered through ZrO bands ? Or also through LaO bands ?

b) In view of the large number of variable stars, how accurate is the average apparente magnitude of each star ? How does that translate into $\overline{M}_V$?

CATCHPOLE - a) Yes, they are discovered using ZrO.

b) I use K (2,2μ) magnitudes and even for Mira variables the range is only ∿ 1 mag. For the S stars with known distances I mentioned, I think the uncertainty in M_{bol} due to variability will be very small, say $\pm$ 0.2, less than uncertainties in distances probably.

LLOYD EVANS - Does BH Cru show changes between SC and CS ?

CATCHPOLE - In 1965 - 1973, BH Cru was SC with weak ZrO and no C_2. In 1980 - 1984, the spectrum is CS with strong C_2 but no ZrO. This appearance seems independent of the phase of variability. This could be a real evolutionary change, but one cannot rule out change due to rotation of a spotted star or some effect due to circumstellar shell.

SPECTROPHOTOMETRIC CHARACTERISTICS OF BARIUM STARS

Phillip K. Lu
Western Connecticut State University

Arthur R. Upgren
Van Vleck Observatory

Image-tube spectra with dispersions of 43A/mm at H-gamma and intermediate-band photometry on the DDO system were obtained for about 300 barium stars. These stars with enhanced Ba II at 4554A are otherwise generally similar to late-type giant stars. The majority of them were identified in an objective-prism survey by MacConnell et al. (1972). They listed 150 "certain" and 90 "marginal" Ba stars but made no further estimates of barium intensity. A more complete spectrophotometric study was carried out by Lu et al. (1983, hereafter referred to as Paper I). For 221 stars whose Ba intensities are one or more on Warner's (1965) intensity scheme, the study has shown that the intensity of the line at 4554A correlates with both spectral type and the DDO color index C(41-42). In addition to the cyanogen band, this index measures the intensity of a broad depression feature appearing in one barium star according to Bond and Neff (1969) and found to exist in other barium stars as well by Lu and Sawyer (1979).

Statistical parallax solutions using Yale proper motions have shown that barium stars are of normal giant luminosity. However, the color-color diagram in the C(42-45) - C(45-48) plane shows that the majority of barium stars appear to be of luminosity class II (Lu and Upgren 1979). This present study intends to examine the extent to which the luminosity anomaly found for the certain or strong-barium stars occurs as well in the marginal or weak-barium stars. These latter stars include the semi-barium stars identified by Keenan and Pitts (1980) in their list of revised MK standard stars, for which they assign intensities of 0.1 to 0.9

Among the semi-barium stars in the list, 42 have DDO photometry. These barium stars with strengths less than one were not listed in Paper I for that reason, although some had DDO photometry at that time. In an earlier paper (Lu and Upgren 1979) the authors showed that among the Ba stars with intensities greater than one, there is little or no variation with intensity in the luminosity anomaly as revealed in the C(42-45) - C(45-48) diagram. On the other hand, the C(38-41) - C(42-45) diagram given in Paper I for these stars reveals a large range in loci between

M. Jaschek and P. C. Keenan (eds.), Cool Stars with Excesses of Heavy Elements, 25–29.

stars with intensities of one and those with 5, with the former coinciding with the normal giants.

These two diagrams are shown in figures 1 and 9 of the 1979 paper and Paper I, respectively. The 42 barium stars for which Keenen and Pitts have assigned an intensity of less than one are distributed in the same manner as are all barium stars in the C(42-45) - C(45-48) diagram, suggesting that they too appear to be of luminosity class II on average. In the C(38-41) - C(42-45) plane, they inhabit the same loci as do the normal giants and the stars of Ba intensities 1 and 2; that is, the weaker barium stars. They reveal that these semi-barium stars show the same overluminous anomaly as do stars with stronger barium. In Paper I, a mean color correction was computed from the differences between the Ba stars and the normal giants. When this correction is applied to the semi-barium stars, they are also reduced to the loci of the normal giants.

In addition to the DDO photometry, quantitative indices have been determined for four spectral lines in the spectra mentioned above. The lines include the 4554A and 4077A lines of ionized barium and strontium, respectively, and H-gamma and H- delta. Comparisons between the ratios of PDS counts were made between their centers and those of nearby iron lines in the spectra without reference to the continua or to the equivalent widths following a procedure used by Rose (1984). The comparisons can be used to form indices which are sensitive to both temperature and metal abundance (Keenan and McNeil 1976). This analysis provides more precise information on the intensities of both barium and strontium than do visual estimates made from the spectra.

The quantitative indices for the barium line intensities are given in Table I and Figure 1. In Figure 2, they are plotted against the C(41-42) index. The marginal barium stars of MacConnell et al., are indicated as those with Ba < 1. They appear to overlap the weaker portion of the range of stars with Ba > 1. If these marginal stars can be assumed to be equivalent to the semi-barium stars of Keenan and Pitts, it would suggest that the scale of intensities which they give may be systematically too low, and that they may in fact overlap those of intensities 1 and 2.

The relationship between the ratios Ba/Fe and Sr/Fe is shown in Figure 3. Itis clear that there is little or no correlation between these indices, although the close relationship between the Ba/Fe ratio and the estimated barium intensity class is evident. We conclude that the degrees of enhancement of the intensities of these ionized lines of barium and strontiumare not closely associated within the group of barium stars, although some enhancement of both lines may be present in the group as a whole when compared to the normal giants. In recent years it has been recognized that the Ba line at 4554A is not a reliable criterion for the luminosity classification of late-type stars as had once been believed. We cannot make the same statement about the Sr line

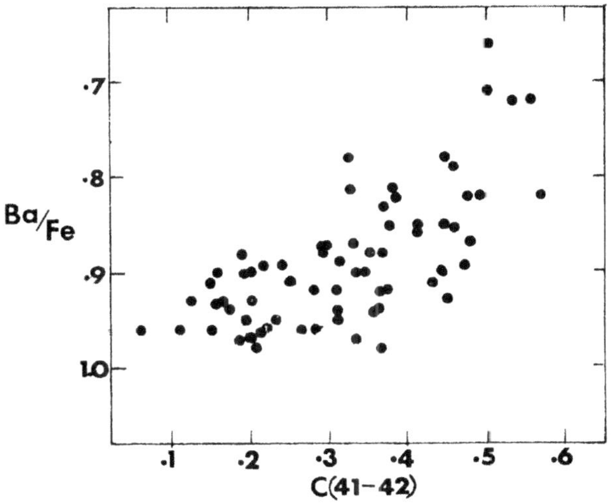

Fig. 2.

Comparison between Ba/Fe ratio and C(41-42) color index.

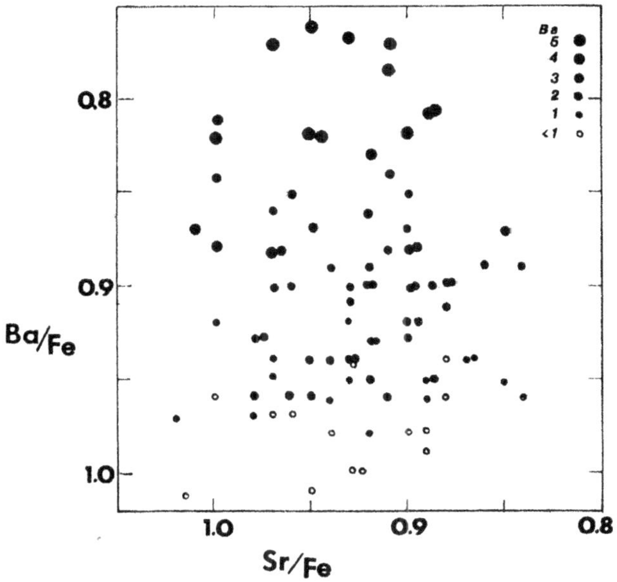

Fig. 3.

Relation between Ba/Fe vs Sr/Fe ratios.

Table I - Mean Spectral Indices of Barium Stars

Ba Class	Ba/Fe		n
5	0.777	± 0.047	13
4	0.842	0.035	10
3	0.857	0.039	12
2	0.907	0.030	41
1	0.941	0.022	23
< 1	0.985	0.029	15

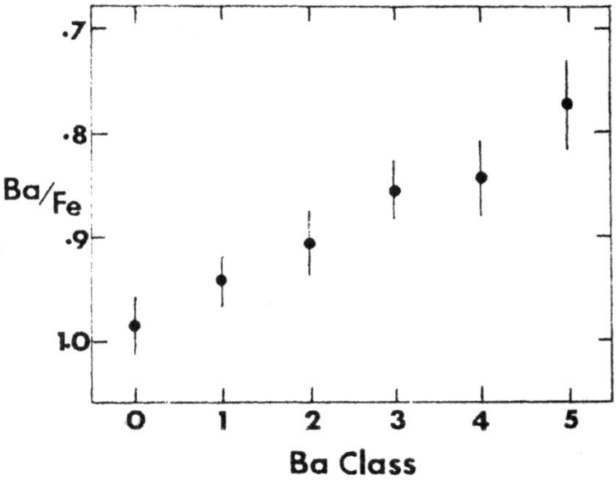

Fig. 1.

Ba intensity class vs mean Ba/Fe ratio. Errors are indicated.

at 4077A. Rose (1984) has found confirming evidence that the strontium line is closely correlated with luminosity for the normal giants. At this time, we cannot yet determine whether and to what degree the Sr line is also a reliable luminosity criterion in the group of barium stars.

REFERENCES

Bond, H. E. and Neff, J. S. 1969, Astrophys. J. 158, 1325.

Keenan, P. C. and McNeil, R. C. 1976, An Atlas of Spectra of the Cooler Stars, Ohio State Univ. Press, Columbus.

Keenen, P. C. and Pitts, R. E. 1980, Astrophys. J. Suppl. 42, 541.

Lu, P. K. and Sawyer, D. 1979, Astrophys. J. 231, 144.

Lu, P. K. and Upgren, A. R. 1979, Astron. J. 84, 101.

Lu, P. K., Dawson, D. W., Upgren, A. R. and Weis, E. W. 1983, Astrophys. J. Suppl. 52, 169.

MacConnell, D. J., Frye, R. L. eand Upgren, A. R. 1972, Astron. J. 77, 384.

Rose, J. A. 1984, Astron. J. 89, in press.

Warner, B. 1965, Mon. Not. R. Astron. Soc. 129, 263.

DISCUSSION

KEENAN - Is it not reasonable to say that your "semi-barium" stars have weaker Ba lines than the "marginal" barium stars just because we had spectra of better resolution than the people classifying from objective prism plates had ?

LU - I did not suggest that "semi-barium" stars are different from "marginal Ba", but only said that I do not have Ba/Fe values for semi-Ba stars, thus I cannot compare them. However, 0.8 and 0.9 class of semi-Ba stars may be the same as Ba class 1 to 1.5.

CAYREL - I am afraid I do not understand the ordinate in this diagram. Does it mean that the Ba/Fe ratio decreases from the Ba 1 class to the Ba 5 class ?

LU - Yes. Small ratio of Ba/Fe means strong Ba-intensity and large Ba/Fe ratio is for weak Ba - intensity. For normal giants the ratio $\lesssim$ 1 means that Ba $\lesssim$ Fe.

CATCHPOLE - I am unhappy about using the convertional DDO luminosity criteria when dealing with peculiar stars such as Ba II stars.

LU - I am not claiming that I do determine luminosity class since it has been done by Mc Clure and Janes. I am simply correcting the possible anomaly and suggest that by doing this, strong Ba stars will become normal luminosity and weak Ba stars will remain anomalous in color ; thus I suggest those weak Ba stars might be luminosity II.

IDENTIFICATION OF PHOTOGRAPHIC INFRARED SPECTRAL REGION IN CARBON STARS

Yoshio Fujita
Department of Astronomy, University of Tokyo
Bunkyo-ku, Tokyo 113, Japan

ABSTRACT. Microphotometric tracings of the spectrograms of carbon stars with the dispersion of 8A/mm from the spectral region λ 7800 to λ 8800 are discussed in connection with the identification of ^{12}CN, ^{13}CN, HCN, [CI] λ8727, and telluric H_2O lines. The microphotometric tracing of the same region of an M-type star μ Cep seems very helpful for the identification of telluric H_2O lines.

Though it is evident that the main contributors to the photographic infrared spectral region from λ7800 to λ 8800 in carbon stars are CN and C_2 molecules, there still remain many absorption lines which should be identified admitting some of them already have been assigned to telluric and atomic origins. In connection with the identification problems in this spectral region, we intend to discuss few items in some detail.

i) Identification of isotopic ^{12}CN and ^{13}CN lines and the abundance ratio $^{12}C/^{13}C$. The determination of $^{12}C/^{13}C$ abundance ratio in carbon stars has been one of interesting problems of these stars, because the role of ^{13}C in nucleosynthesis of cool stars is theoretically important. From the detailed investigation of the microphotometric tracing of the spectral region considered, two portions around λ7921 and λ7934 were found interesting(Fujita 1978,1980,1981). The wave-lengths and their identification around these two regions are listed in Table I. It revealed that in stas of low $^{12}C/^{13}C$ ratio(<10), the presence of the two ^{13}CN lines λ 7921.13-.32 and λ 7935.67 is clearly recognized. However, in stars of high $^{12}C/^{13}C$ ratio(>20) we cannot recognize the profiles of both isotopic lines. It seems that the behaviour of the profiles of these two lines acts as an indicator of $^{12}C/^{13}C$ ratio. A few examples are shown in Fig.1. Among four carbon stars of Fig.1 it should be remarked HD 19557 and T Lyr are of J-type.

ii) Identification of polyatomic molecular HCN lines. The possibility of the presence of HCN in carbon stars from the identification of five sequences $4\nu_3\nu_2+4\nu_3-\nu_2$, $3\nu_1+2\nu_3$, $\nu_1+3\nu_3$, and $\nu_1+\nu_2+3\nu_3-\nu_2$ was discussed and finally thirty six lines of them as are shown in Table II were selected as suitable lines for identifying HCN(Fujita 1977, 1983). So far in the present investigation HD 182040 is only a star in

M. Jaschek and P. C. Keenan (eds.), Cool Stars with Excesses of Heavy Elements, 31–35.

TABLE I

λ	Identification
7917.70	CN (2,0) Q_1 (10) C_2 (4,1) Q (4)
17.75	CN (8,5) P_2 (18)
17.82	CN (7,4) P_1 (37)
18.05	$C^{13}N$ (2,0) R_2 (1)
18.10	CN (2,0) P_1 (4)
18.22	CN (2,0) Q_2 (15)
18.72	CN (2,0) P_2 (10)
18.86	CN (8,5) R_2 (31)
18.90	$C^{13}N$ (2,0) R_2 (0)
18.97	C_2 (4,1) R (14)
20.06	CN (2,0) R_1 (22)
20.23	CN (2,0) Q_1 (11)
20.67	Atm H_2O
20.78	CN (2,0) P_{12} (3) P_1 (5)
20.83	C_2 (3,0) R (42)
21.13	$C^{13}N$ (2,0) Q_2 (1)
21.32	$C^{13}N$ (2,0) R_2 (11)
7933.15	CN (2,0) Q_2 (18) R_1 (25)
33.45	CN (2,0) R_2 (27)
33.56	CN (2,0) Q_1 (15)
34.12	CN (7,4) R_2 (52)
34.20	$C^{13}N$ (2,0) R_2 (17)
34.52	CN (8,5) R_2 (33)
34.84	CN (2,0) P_1 (9)
35.01	CN (2,0) P_{12} (6)
35.63	$C^{13}N$ (2,0) Q_2 (8)
36.38	CN (2,0) P_2 (13) (7,4) P_2 (39)

TABLE II

HCN

$4\nu_3$		
J	R	P
3	7904.95	
5		7921.62
16	7888.83	
18		7963.68
$\nu_2+4\nu_3-\nu_2$		
J	R	P
3	7954.33	
5		7971.23
6	7949.67	
9		7980.18
13		7990.02
14	7939.66	
17	7936.84	
$3\nu_3+2\nu_2$		
J	R	P
8		7914.59
11		7921.83

$\nu_1+3\nu_3$		
J	R	P
1	8559.19	
2		8567.75
4	8553.43	8572.41
9	8545.06	
14		8599.38
17		8608.70
19		8615.23
21	8531.29	
22		8625.50
25		8636.35
$\nu_1+\nu_2+3\nu_3-\nu_2$		
J	R	P
3	8600.09	
5		8619.88
6	8594.60	8622.45
10		8633.11
11	8586.73	8635.83
12		8638.73
14		8644.71
15		8647.76
16		8650.88
17	8579.33	

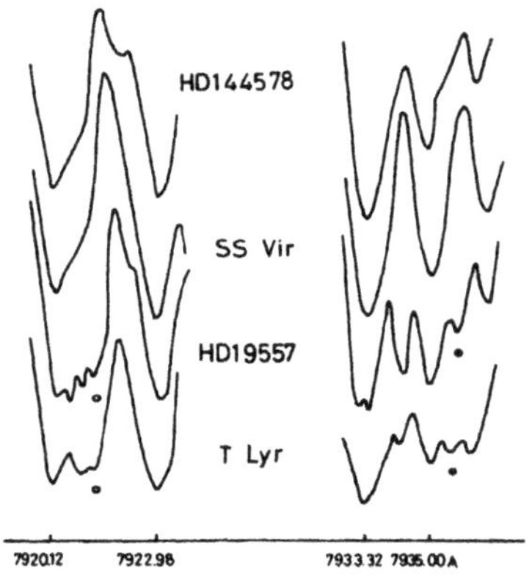

Figure 1. Well recognized profiles of λ7921.13-.32 and λ7934.20 lines indicate $^{12}C/^{13}C < 10$.

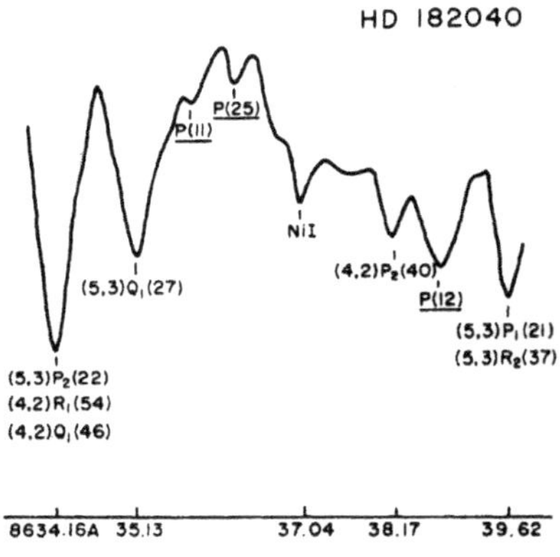

Figure 2. An example of possible presence of HCN in HD 182040.

which HCN is possibly present. Fig 2 shows an example.

iii) Identification of [CI] λ8727 line. Lambert and Swings(1967) assigned λ8727.13 line in the solar spectrum to a [CI] forbidden line $^1D_2-^1S_0$ with a large value of transition probability. In some carbon stars its presence is clearly recognized and the correlation between the central depth of this line and the spectral sub-type of carbon stars was indicated as is shown in Fig.3(Fujita 1973).

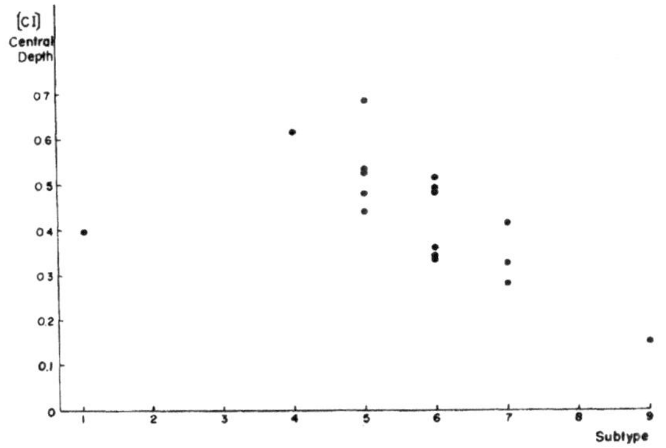

Figure 3. The correlation between [CI] λ8727 intensity and sub-type of C-classification.

iv) Identification of telluric lines. As many telluric lines of H_2O are generally blended with stellar lines in this spectral region, it seems very difficult to estimate true line intensities of ^{12}CN and ^{13}CN lines due to the blending effect. Therefore,it is quite necessary to know exactly the locations of these telluric lines. For this purpose spectra of some carbon stars and an M-type star μCep were traced by the PDS of the Tokyo Astronomical Observatory. In this spectral region of μCep TiO molecular lines are present together with atomic lines of metallic elements(cf: Davis 1947). It became evident, however,that the comparative study of the tracing of μCep with those of the same spectral region of carbon stars is very useful for the identification of ^{12}CN and ^{13}CN. Fig.4 shows two tracings of Y CVn, a carbon star(top) and μCep, an M-type star (bottom). Identified CN lines are marked by (4,2), (3,1), and (2,0) and identified H_2O lines by P2 and Q3. The spectrograms of carbon stars used for our discussion were obtained with the 188 cm reflector of Okayama Astrophysical Observatory and with the 5 m reflector of Palomar Observatory. The author expresses his hearty thanks to staff members of these Observatories for their help in obtaining these spectrograms. His thanks are also due to staff members of Tokyo Astronomical Observatory particularly to Dr. T. Tsuji for obtaining PDS tracings.

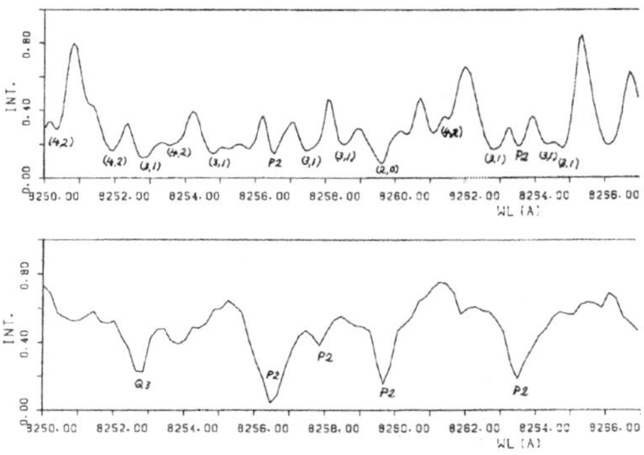

Figure 4. Identified CN Red System (4,2),(3,1),(2,0) and telluric H_2O lines in Y CVn(top), and identified telluric H_2O lines Q3 and P2 in μ Cep (bottom).

REFERENCES

Davis,D.N.:1947, Astrophys.J. 106, 28.
Fujita,Y.:1973, Proc. Japan Acad. 49, 530.
Fujita,Y.:1977, Journ. Roy. Astron. Soc. Canada, 71, 115.
Fujita,Y.:1978, Proc. Japan Acad. 54,Ser.B, 207.
Fujita,Y.:1980, Proc. Japan Acad. 56,Ser.B, 399.
Fujita,Y.:1981, Proc. Japan Acad. 57,Ser.B, 169.
Fujita,Y.:1983, Proc. Japan Acad. 59,Ser.B, 145.
Lambert,D.L. and Swings,J.P.:1967, Solar Phys. 2, 34.

DISCUSSION

JØRGENSEN - Have you been looking for the HCN lines, that you identified in HD 182040, in any carbon stars where you can now definitely conclude that the lines are not present ?

FUJITA - Twenty-three carbon stars (C2 - C9, including J-type) were studied without any evidence of the presence of HCN lines.

RICHER - I am surprised that you would find HCN on HD 182040 because
a) it is hydrogen deficient and
b) hot (∿ 4 000° K). Would you please comment ?

FUJITA - We could not find any CHN lines in HD 137613 which is of the same type as HD 182040. It should be desirable to look for them in other early C-type stars.

QUERCI - Do you plan to publish the identifications of carbon stars in atlases ?

FUJITA - It will be nice to do so. However, it will depend upon the number of requests, and the necessity of publication, and also the economic situation.

PECULIAR COOL STARS IN PLANETARY NEBULAE - THE SPECTRUM OF FG Sge

A. ACKER
Observatoire de Strasbourg
11, rue de l'Université
67000 Strasbourg - France

1\. - Some planetary nebulae (PN) have central stars which appear to be too cool to account for the ionization of the nebular envelope. The most peculiar central stars have a spectral type between A and K. Lutz (1977, 1978), Lutz and Kaler (1983) have made a special study of these stars. Using their work, and the "Catalogue of Central Stars of Planetary Nebulae" (Acker, Gleizes et al, 1982), the caracteristics of these peculiar stars have been compared. The intensive spectrographic study undertaken by F. Gleizes (1984) for about 40 nuclei has also been used.

Table 1 provides the available information for the eleven stars which present a peculiar cool spectrum. Six normal cool nuclei recognised as spectroscopic binaries have not been included in this study.

In some cases, the spectra are similar to those of known peculiar stars (carbon stars ; metal deficient stars, marginal Ba stars that show only enhanced strontium and CH, and a lack of barium enhancement due perhaps to a low - neutron exposure s-process). However, most objects have been variously interpreted as new-born PN, or also as possible symbiotic stars, slow novae, or binary systems undergoing mass exchange.

2\. - The most remarkable object is FG Sge, intensively observed by many authors since 1955, since it exhibits spectacular variations. Let us note that the spectral type of this supergiant has varied from B4 (1955) to A5 (1967) and F5 (1970) ; at present, FG Sge appears as a luminous G-K star surrounded by a low-density, low-excitation nebulosity.

The spectrum of FG Sge was observed in 1979/1980 in the blue and visible regions (Acker, Jaschek, Gleizes, 1982), and, in 1979/1982, in the red region (Acker, 1983).

2.1. - The general appearance of the spectrum (continuum, absorption lines) is that of a G8 - K0 supergiant star. But the spectrum, rich in absorption lines, exhibits many emission features. First, we find the weak [0 II] and [0 III] emission lines due to the old planetary nebula ejected six thousand years ago. In addition, since

M. Jaschek and P. C. Keenan (eds.), Cool Stars with Excesses of Heavy Elements, 37–41.

about a decade ago, the stellar spectrum also presents a series of emission features : about 500 lines were measured in the λλ 3700 - 500 Å region, where they showed up against the faint stellar continuum ; in the red region, no emission features were observed, since the stellar continuum is intense and may possible hide emission lines. In the blue region, the atomic species identified in emission are [Fe II] and O II, well represented ; Fe III, N III, S II, rather well represented ; and weak lines of A II, [F IV], [Ne IV]. Many intense lines remain unidentified (for example λ 4504 A).

2.2. - About 600 absorption lines were measured from λλ 3700 to 5000 Å, and about 50 from λλ 6250 to 6800 A . The lines of light elements are the same as for a normal star (like Arcturus). The lines of the neutral iron peak elements are absent or very weak. The lines of heavy elements : Y, Zr, Ba, La, are enhanced. The lines of the rare earths are very numerous and strong ; the most intense lines are all present for Ce II, Pr II, Nd II, Sm II, Eu I and II, Gd I and II, Tb II, Dy I, Ho I, Lu II, and perhaps Tm I (6460 A) ; we did not find Pm, Er, Yb. The G-band of CH is well observed. A contribution of C_2 is possible at 4737, 6442 and 6480 Å. This spectrum resembles that of a barium star, but it is only a superficial likeness, since in barium stars all elements up to A = 38 are seen as in normal stars, and only heavier elements (later than Sr) appear stronger ; in contrast, in FG Sge, the lines of iron peak species are weaker, and those of the rare earths are enhanced. Thus, FG Sge does not seem to evolve towards a Ba-type star.

It is a red giant with a mass near one solar mass, "pulsating" mass determined from the radial velocity -and photometric light- curves (Mayor and Acker , 1980). If this star experienced a helium shell flash associated with a post-planetary ejection episode, the entire material of the helium-rich zone between the hydrogen and the helium-burning shells is convectively mixed with the outer H-rich shell.As about 90 % of the iron had been converted into s-process species, the observed iron to hydrogen ratio should show no enhancement with the time, but the ratio of s-process elements to hydrogen would have been increased, model consistent with our observations.

At present, the spectrum of FG Sge no longer seems to be evolving ; in 1980, the star may have abandoned the instability strip of the temperature - luminosity diagram

REFERENCES

ACKER A. : 1983, Astron. Astrophys. Suppl. 54, 293
ACKER A. - Jaschek M. - Gleizes F. : 1982, Astron. Astrophys. Suppl. 48, 363
ACKER A. - GLEIZES F. - CHOPINET M. - MARCOUT J. - OCHSENBEIN F. - ROQUES J.M. : 1982, Catalogue of the Central Stars of True and Possible Planetary Nebulae, Publication Speciale du C.D.S. N° 3, Strasbourg and Complem. I, II, 1982 ; III, 1984 : in preparation.
GLEIZES F. : 1984, Thesis : in preparation.
LUTZ J.H. : 1977, Astron. Astrophys. 60, 93
LUTZ J.H. : 1978, IAU Symp. 76 "Planetary Nebulae", Reidel - 185
LUTZ J.H. - KALER J.B. : 1983, Publications of the Astron. Soc. of the Pacific 95, 739.
MAYOR M. - ACKER A. : 1980, Astron. Astrophys. 92, 1

TABLE I : Nuclei of PN with a peculiar cool spectrum

Name		Spectral type of the star*	Magnitude of the star*	Comments on the nature of the object*
Galactic number	Usual name			
80-6°1	CRL 2688	F5 Ia + anomalous lines (molec. C2 emiss. + C3 abs. ; CO emiss.)	V 12.2 B 13.3 U 14.5	Bipolar-, proto PN Carbon star ?
133-8°1	M1 - 2	G2 Ib ? G-type giant with heavy element enhancements	V 13.3 B 14.3 U 14.4	Similar to a marginal Barium star ?
169-0°1	IC 2120	G ; metal deficient	V 15.0 B 16.9 U 17.1	Like HD 221 170
249+6°1	AS 201	G abs. lines	B∿13	Symbiotic star ; proto - PN ?
45+24°1	K1 - 14	G abs. lines	V 13.2 B 13.8 U 13.8	Binary ?
63-12°1	He2-467	G5 II	V 13.3 B 15.1	Binary ? Symbiotic star ?
358-5°2	H1 - 49	Late-type continuum, weak G-band and K-line	p∿13	P Cyg ? Slow nova ? Bep ?
60-7°1	FG Sge	G9 - K0 Ia (1982) ; s-process elements abundants ; neutral iron-peak elements very weak	V 9.1 B 10.6 U 10.9	Post-planetary ejection episode associated with the helium-shell flash (not Ba-star)
53-3°2	HM Sge	M giant + WR ? Abs. CO + H_2O variations : P > 1 year	V 11.5 B 11.6 U 11.2	Association of the spectrum of a late type star, with a rich emission line-spectrum similar to that of P.N.
75+5°1	V1016 Cyg	Late type + emiss.lines variations : P ∿ 450 d.	V 11.2 B 11.2 U 10.3	Interpreted as proto P.N. or symbiotic stars, or slow novae, or binary systems undergoing mass exchange.

Name: Galactic number	Name: Usual name	Spectral type of the star*	Magnitude of the star*	Comments on the nature of the object*
77-5°1	HBV 475	M4 - 5 II + WN. Abs. Ti O (I.R.) variations : P ∿ 950 d.	v 12.5 b 13.3	

* All references can be found in the "Catalogue of central stars of planetary nebulae" (Acker et al, 1982)

DISCUSSION

KEENAN - Is FG Sge similar to Przbylski's famous Ho- star ?

ACKER - Like Przbylski's star, FG Sge presents a weakness of the iron peak elements ; but the Ho was only surely detected in the red range (λλ 6605, 6629).

KAUFMANN - What is the time scale of a mixing up on the s-process elements and is it consistent with the change of a spectral type from B4 to K0 ?

ACKER - The lines in the spectrum indicating s-process elements started to appear in 1968, when the star showed the spectral type A5.

RICHER - You said that in 1955 the spectral type of FG Sge was B4 Ia. Was the spectrum that of a normal B4 Ia star at that time ?

ACKER - The spectrum seems never have been entirely normal. For a summary see Acker et al, AA Suppl. **48**, 363 (1982).

TUTUKOV - FG Sge experienced a spectacular evolution in the last thirty years. Could you say something about variation of the spectrum of the nebulae which surrounds FG Sge ? The recombination time must be rather low.

ACKER - Yes, as the star cools, it brings about a definite weakening of the nebular lines ; at the moment the low excitation spectrum shows only the weak doublets of [O II] and of [O III] hardly visible.

MALANEY - Your statement that FG Sge is no longer evolving is very significant. One proposed mechanism for FG Sge is that it is rapidly undergoing a "loop" in the HR diagram after a final helium shell flash. This loop back to the supergiant region takes about hundred years. Therefore if the star is no longer evolving, it should be placed on the HR diagram in a region where this loop is complete and the subsequent evolution is on a longer time scale. You may then have an observational test of the loop mechanism for FG Sge. However do you think that the period which you have observed this star (1979 - 1982) is long enough to state that its rapid evolution is ended ?

ACKER - Yes, subsequent observations (1983 - 1984) confirm this. To determine its future evolution, Mercedes Jaschek and myself -and certainly future generations in the century to come- would like to monitor FG Sge, the "Rosetta stone" for nucleosynthesis.

THE DISCOVERY AND FREQUENCY OF BARIUM STARS

WILLIAM P. BIDELMAN
Warner and Swasey Observatory
Case Western Reserve University
Cleveland, Ohio 44106 USA

ABSTRACT. The frequency of barium stars as compared to normal giants and intermediate supergiants of classes G and K is assessed through a study of the characteristics of H.D. barium-star suspects south of $\delta = -26^\circ$ found on moderate-dispersion objective-prism plates. For stars of the type that can be found from such plates, the frequency appears to lie between one-half and one percent.

1. INTRODUCTION

The relative numbers of barium stars and normal G and K giants is a matter of considerable interest. Previous estimates range from one-half to one percent which, as we will see, seems to be about the right answer. The present approach is to consider the objective-prism discoveries made on moderate-dispersion spectrograms in the part of the sky south of $\delta = -26^\circ$, in which all of the stars of the Henry Draper Catalogue have recently been classified by Dr. N. Houk of the University of Michigan (1975, 1978, 1982). Counts of normal G and K stars in this part of the sky have been made using data kindly supplied to the writer by Dr. Houk; this is the easy part of the problem. The difficult part is to decide what should be considered as a barium star. South of $\delta = -26^\circ$ Dr. Houk has classified 43 H.D. objects as barium stars. However it is clear from other data that she has been quite conservative in her classifications, and it appears that this number is substantially too low.

2. THE CHARACTERISTICS OF THE STARS

Early in the history of the Michigan Southern Spectral Survey Dr. D. J. MacConnell and I, aided by several skilled graduate students, began preparing lists of peculiar objects of various sorts whose spectra appeared on the 10° objective-prism plates being taken at Cerro Tololo with the Curtis Schmidt (see Bidelman, MacConnell, and Frye 1973). Among the stars noted were many barium stars; in fact south of $\delta = -26^\circ$ this "early-result" program picked up 126 Henry Draper Catalogue stars that were considered as either quite definite barium stars or as reason-

M. Jaschek and P. C. Keenan (eds.), Cool Stars with Excesses of Heavy Elements, 43–46.

ably good prospects. Most of the discoveries of the early-result program were eventually published by Bidelman and MacConnell (1973), but this was not true for the barium stars since most of these--and many others as well--were published in a complete listing of the known barium stars of the entire sky by MacConnell, Frye, and Upgren in 1972. In preparing this list those authors reexamined the previous barium star discoveries, deleting a considerable number (no doubt with good reason) and adding a few that had not been so classified before. Thus the MacConnell, Frye, and Upgren list contains a total of 117 H.D. stars south of $\delta = -26°$. Of these 74 were termed "certain" barium stars and 43 "marginal," the discrimination being based on the visibility and strength of the Ba II λ4554 line. Actually, the writer feels that this distinction was not an entirely happy one, since there are a considerable number of other factors that enter into the confidence with which one can classify such an object on an objective-prism plate. After all, plate quality and exposure level are not always optimal, and the discrimination of barium stars from supergiants of similar type is often quite tricky. It is also true that a fair number of the new discoveries exhibited weaker metal lines than the classical barium stars and thus did not look quite the same. I believe that it would have been preferable to simply have termed some of the objects less likely to be true barium stars than others.

In any case such a list should, of course, be regarded as only an invitation to further study, and, fortunately, two recent slit spectroscopic studies by Catchpole, Robertson, and Warren (1977) and Lü, Dawson, Upgren, and Weis (1983) have given estimates of the Ba II-line strengths for a large number of the stars of the MacConnell, Frye, and Upgren list and for some additional stars that have been subsequently classified as barium stars by Dr. Houk. From a rather careful study of the data it appears reasonable to me to insist that a star must have a λ4554 intensity of more than 1 according to the former investigators or of 1 or more according to the latter to be considered a true barium star, provided, of course, that other criteria as well agree with that assessment. For example, in my own work I have concluded that the ratio of the blend of the Sr II line and the CN band head at λ4215 to the Ca I λ4226 line is a rather sensitive discriminant: this is almost always near to or greater than unity in barium stars but never so in normal objects.

In the above-mentioned papers there are a total of 93 H. D. stars south of $\delta = -26°$ that have sufficiently strong Ba II lines to be considered true barium stars. There are another 29 stars for which slit data are either nonexistent or conflicting. Adding 50% of the latter number to the former gives 107 as a reasonable estimate of the truth.

CONCLUSIONS

The more pronounced barium stars are indeed quite rare. The actual percentages depend, of course, on the spectral range of normal stars considered. Table I gives the numbers of H. D. stars south of $\delta = -26°$ of luminosity classes II, II/III, and III among the G and K stars. It would appear that the frequency of barium stars that can be detected on

TABLE I

Stars of Classes II, II/III, and III

types	no. of stars
G3 and G4	138
G5 and G5/6	341
G6 and G7	1109
G8	2789
G9	1704
KO	4206
KO/K1 and K1	4528
K1/K2 and K2	4003
	18818

Totals:	Class II	198
	Class II/III	400
	Class III	18220

objective-prism plates as compared to that of normal objects of similar luminosity and spectral type is, as previously estimated, between 0.5 and 1.0 percent. It should be emphasized that we are not concerned here with the very weak-Ba or semibarium stars with λ4554 intensity less than 1 discussed by Keenan and Wilson (1977) and others. The frequency of these cannot be determined at all from objective-prism observations but must come from slit work.

REFERENCES

Bidelman, W. P. and MacConnell, D. J.: 1973, Astron. J. 78, 687.

Bidelman, W. P., MacConnell, D. J., and Frye, R. L.: 1973, I.A.U. Symposium No. 50, Ch. Fehrenbach and B. E. Westerlund, eds., p. 77.

Catchpole, R. M., Robertson, B. S. C., and Warren, P. R.: 1977, Monthly Notices Roy. Astron. Soc. 181, 391.

Houk, N. 1978, Michigan Catalogue of Two-Dimensional Spectral Types for the HD Stars, Vol. 2 (Ann Arbor: U. of Michigan Dep't. of Astronomy).

Houk, N. 1982, Michigan Catalogue of Two-Dimensional Spectral Types for the HD Stars, Vol. 3 (Ann Arbor: U. of Michigan Dep't. of Astronomy).

Houk, N. and Cowley, A. P. 1975, U. of Michigan Catalogue of Two-Dimensional Spectral Types for the HD Stars, Vol. 1 (Ann Arbor: U. of Michigan Dep't. of Astronomy).

Keenan, P. C. and Wilson, O.C.: 1977, Astrophys. J. 214, 399.

Lü, P. K., Dawson, D. W., Upgren, A. R., and Weis, E. W.: 1983, Astrophys. J. Suppl. 52, 169.

MacConnell, D. J., Frye, R. L., and Upgren, A. R.: 1972, Astron. J. 77, 384.

DISCUSSION

JASCHEK - Do you think Ba stars can be subdivised ?

BIDELMAN - I am sure that useful subdivisions could be devised by temperature, metallicity, s-process enhancement, etc... But this is a job for slit spectroscopy.

WING - Has the chemical composition of the G-type companion of α Her been analyzed ? I wonder if it's a barium star ...

BIDELMAN - It has not been analyzed but I think it is not a Ba star.

LU - 1) Quantitative Ba/Fe, Sr/F ... ratios will be analyzed for all image-tube spectra at 43 A/mm for classification of heavy elements. Thus strong and weak barium stars can be differentiated more easily than in objective prism spectra.

2) We also noticed the line ratio 4215/4226 shows agreement with Ba intensity.

THE IUE LOW-DISPERSION SPECTRA REFERENCE ATLAS

D. Egret[1], B. J. M. Hassall[2], A. Heck[1,2], C. Jaschek[1],
M. Jaschek[1], A. Talavera[2]
[1]Astronomical Observatory, 11 rue de l'Université
67000 Strasbourg, France
[2]IUE Observatory, Astronomy Division, ESA
Apartado 54065, Madrid, Spain

ABSTRACT. Part one of this Atlas, essentially devoted to normal stars, has just been published by ESA. 229 stellar spectra are presented, together with the corresponding fluxes and an ultraviolet classification. The preparation of this publication has confirmed that MK classification cannot simply be transferred to the Ultraviolet range.

Work on a second part concentrating on peculiar stars is at an advanced stage. More than 650 stars are presently candidates for this volume, and the relative frequency of the individual spectral groups is given in Table I with reference to the Catalogue of Stellar Groups.

1. INTRODUCTION

The ultraviolet stellar classification programme, based on spectra collected by the International Ultraviolet Explorer (IUE) satellite, was initiated in 1978 with the specific goal of building up a set of representative low-dispersion spectra, from which a classification scheme could be constructed in the UV.

First results, concerning O- and B-type stars were presented by Jaschek and Jaschek (1984) and the first volume of the *IUE Low-Dispersion Spectra Reference Atlas* devoted to normal stars has now been published by ESA (Heck et al., 1984).

2. THE MATERIAL

IUE is too well known to be described here. Let us just recall that the wavelength range is 1152 to 3200 Å, in the ultraviolet, with a mean resolution of ∿ 7 Å. Most of the spectra we used have been retrieved from the low-dispersion IUE data base. Some of the existing gaps in the data were filled as far as possible by observations performed during shifts allocated to this programme. The resulting sample remains biased in favour of early-type stars. The absolute calibration applied has been that of Bohlin and Holm (1980).

We tried to produce natural sequences of standard stars in which the features transit smoothly from one standard to the next, staying

M. Jaschek and P. C. Keenan (eds.), Cool Stars with Excesses of Heavy Elements, 47–52.

however as far as possible within the general MK frame. This work has confirmed the conclusions previously drawn on the basis of TD1 data: namely, that one cannot simply transfer MK spectral classifications to the ultraviolet, the order in the MK system not being identical to that which can be established in the ultraviolet spectral range.

3. ATLAS PART 1 : NORMAL STARS

The body of this volume (Heck et al., 1984) consists of the presentation of flux tables (left-hand pages) and the corresponding composite graphs (right-hand pages) for 229 stars, most of them exhibiting normal behavior in the ultraviolet range. A few peculiar stars have, however been included to illustrate typical abnormalities. A set of 34 transparencies are also provided for the most representative standard stars (but only a few later than G5) to allow direct comparison with the spectra.

4. ATLAS PART 2 : PECULIAR STARS

Our present aim is to produce a volume of similar format to the first one, devoted to stars which are peculiar in the ultraviolet range. As before, we draw most of our data from the IUE archive. Up to date, low-dispersion spectra of more than 650 stars have been retained and the relative frequency of the individual stellar categories is illustrated in the table I with reference to the Catalogue of Stellar Groups (Egret and Jaschek, 1984, in this volume)(CGS).

The 14175 entries of the CGS have been used as a starting point in selecting the images. We recognise the existence of some bias in this sample, the most important being that the CGS is constructed on the basis of spectral peculiarities in the optical region. From our experience of producing the normal-star volume, we do not necessarily expect all optically peculiar stars to be peculiar in the UV and vice-versa. Therefore we are also perusing the whole low-dispersion IUE data base with the help of the literature to point out possible new members of the existing groups and possible other UV peculiar groups.

The classification process (in the ultraviolet) of the dearchived images is in progress and an illustration of the available material is given in Figure 1 where two Barium stars are displayed together with K0 and K3 standards, and in Figure 2 (three C stars and two standards).

Acknowledgement. Part of this work is supported by a grant of the Centre National de la Recherche Scientifique (ATP Sol-Espace).

REFERENCES

Bohlin, R.C. and Holm, A.V.: 1980, NASA IUE Newsl. 10, 37

Egret, D. and Jaschek, M.: 1985, this volume

Heck, A., Egret, D., Jaschek, M., Jaschek, C.: 1984, *IUE Low-Dispersion Spectra Reference Atlas- Part 1. Normal stars*, ESA SP-1052

Jaschek, M. and Jaschek, C.: 1984, in *MK Spectral Classification: Criteria and Applications*, in press

TABLE I. Distribution of the stars presently candidates for the peculiar-star part of the atlas, with reference to the Catalogue of Stellar Groups.

Type	n	%
CNO	24	46
OB subdwarfs	46	36
Helium abnormal	41	41
Early-type H-poor	15	63
β Cephei	10	18
Be and shell	48	4
B{e}	12	14
Ap	27	1
Am	19	1
λ Boo	8	73
δ Scuti	10	7
Ae	7	14
Herbig Ae, Be	18	27
Horizontal Branch	15	12
Blue stragglers	8	10
P Cygni	6	50
T Tauri	30	6
Metal-weak	11	1
Halo	7	3
Cool subdwarfs	4	0.5
Me	8	5
Mg II in emission	22	20
Luminous H-deficient	5	16
YY Ori	1	3
C	3	0.1
CN	2	0.4
CH	2	2
CH subgiants	3	25
Barium	7	2
J	1	2
S	4	0.5
VV Cep	16	31
Symbiotic	26	14
Composite	28	2
White dwarfs	72	-
Wolf-Rayet	91	41
Oe, Of	63	37
M dwarfs	5	1

n : number of stars for which IUE low-dispersion spectra are available.

% : percentage relative to the lists of known peculiar stars, from the Catalogue of Stellar Groups (Egret and Jaschek, 1985).

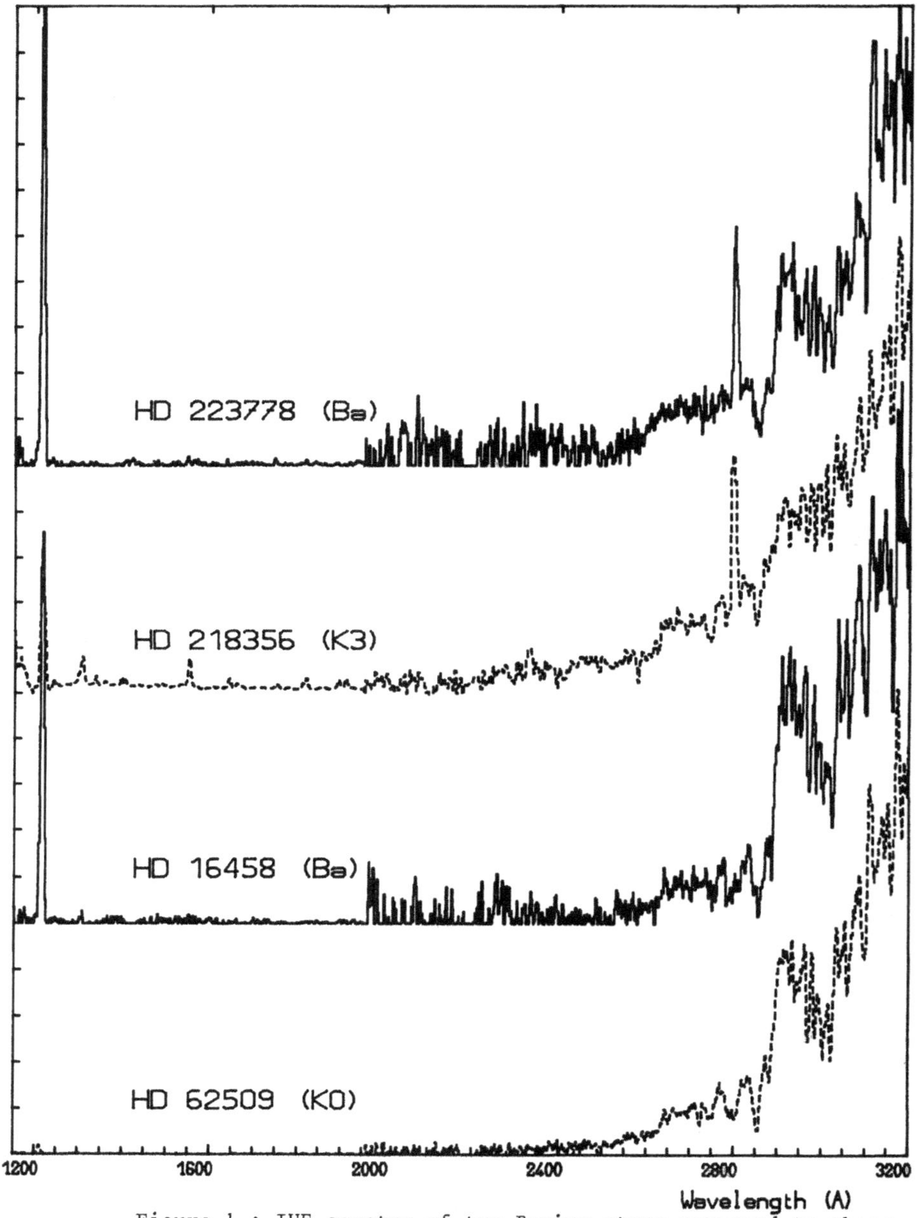

Figure 1 : IUE spectra of two Barium stars compared to those of two K standards (normalized fluxes).

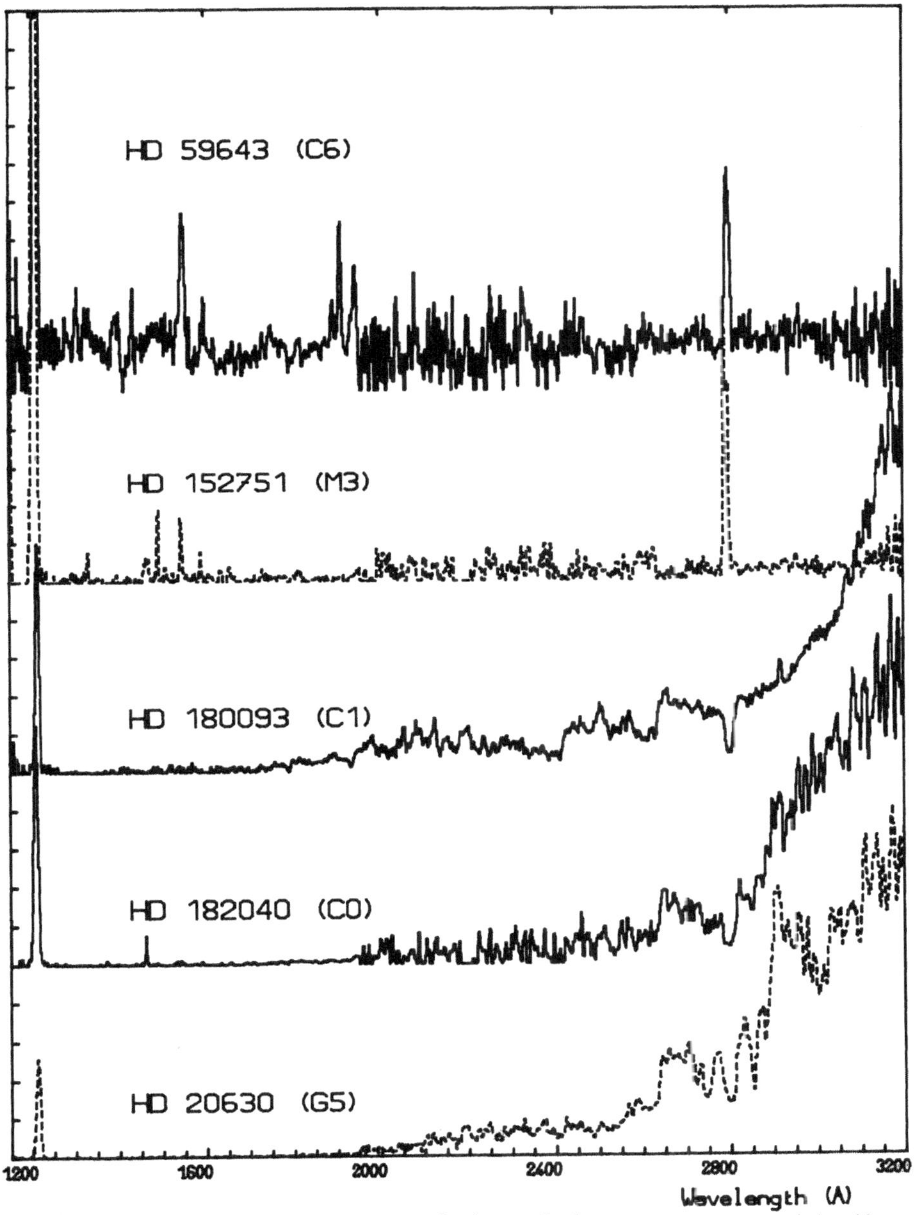

Figure 2 : IUE spectra of three Carbon stars compared to those of two K and M-type standards (normalized fluxes).

DISCUSSION

LU - Do you have any more information of chromospheric Mg II for Ba stars ? High resolution IUE spectra show that Mg II strengths decrease with the increase of Ba intensity.

EGRET - The low dispersion spectra we are using are not suitable for such work. The two Mg II lines are blended.

WING - The Ba star Zeta Cap is also shown from λλ 2500 - 3230 A in the "Atlas of high dispersion IUE spectra of late type stars", published by Perkins Observatory last year (Wing R.F. - Carpenter K.G. and Wahlgren G.M., 1983).

ULTRAVIOLET SPECTRA OF N, R, AND S STARS

[1,3]Hollis R. Johnson, [2,3]Thomas B. Ake, and [1]Joel A. Eaton.
[1] Astronomy Dept., Indiana University, Bloomington, IN 47405
and [2]CSC, Code 685, NASA Goddard Space Flt. Ctr., Greenbelt,
MD 20771, [3]Guest observer with the Int'l Ultraviolet Explorer.

ABSTRACT. Ultraviolet IUE spectra of N, R, and S stars are displayed and discussed. Carbon stars of type N have IUE spectra similar to M stars except that the Mg II lines are much weaker and low-excitation lines from neutral metals are stronger. Spectra of early R stars (R0-R3) closely resemble those of late G or early K in both lines and continuum with no significant emission lines. Late R stars (R5-R8) show spectra like late K or early M stars except that lines from low levels of excitation are stronger. Ultraviolet spectra of S stars show a wide variation in the strength of Mg II lines.

I. INTRODUCTION AND OBSERVATIONS

Because the peculiar red giants cover a fair range in chemical composition, with resultant significant differences in photospheric thermal and density structure, one must consider the following question. Which aspects of chromospheric heating or cooling are affected by changes in chemical composition? How do ultraviolet spectra of the various types of the peculiar red giants differ?

The first attempts to obtain IUE spectra of N-type carbon stars are described by Querci et al. (1982). Low-dispersion IUE spectra of 3 N-type carbon stars have been presented and discussed (Johnson and O'Brien 1983). Spectra for several R stars have recently been obtained and discussed (Eaton et al. 1984). Spectra of several barium stars have been studied generally with an eye toward possible white-dwarf companions. No S-star spectra have been presented except for HD 35155, which is of interest in that it has an interactive companion (Johnson and Ake 1983).

For discussion here, we have selected spectra for a relatively warm and a relatively cool S star. The pertinent data are given in the following table.

Star name	Spectral Class	IUE exposure #
HR 8714	S4, 1	LWR 13317
TV Aur	S5, 6	LWR 15062

M. Jaschek and P. C. Keenan (eds.), Cool Stars with Excesses of Heavy Elements, 53–57.

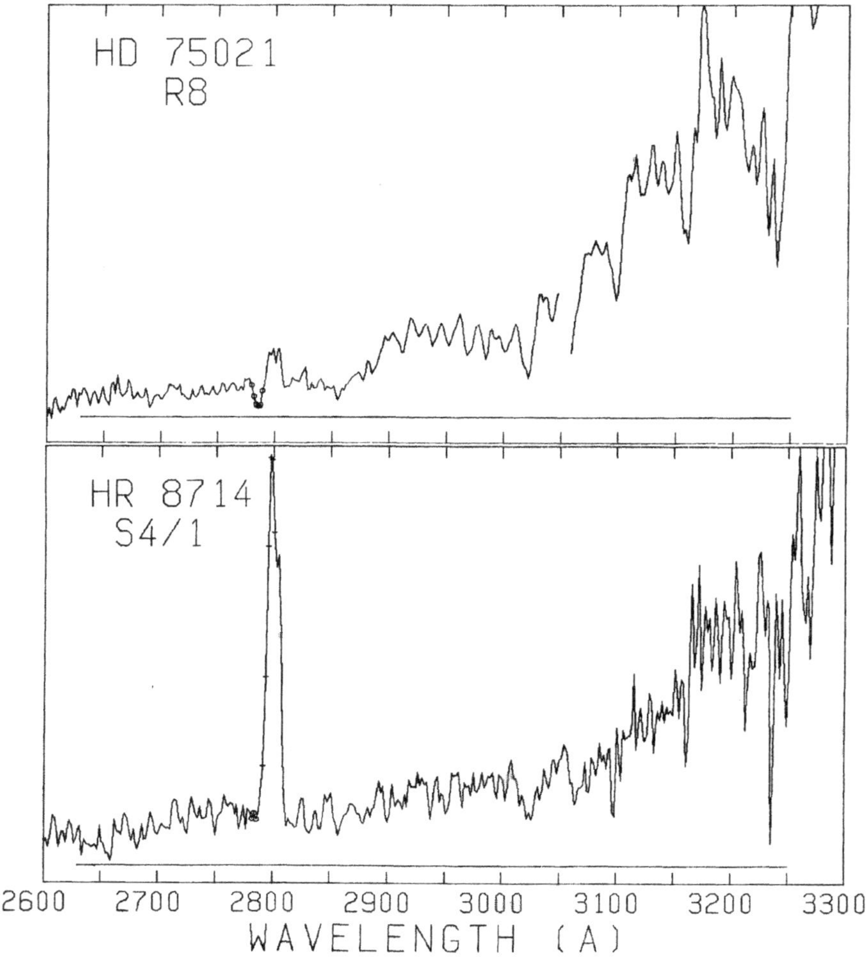

Figure 1. Ultraviolet spectra of late R star (top) and an S star. Horizontal line marks the zero-flux level.

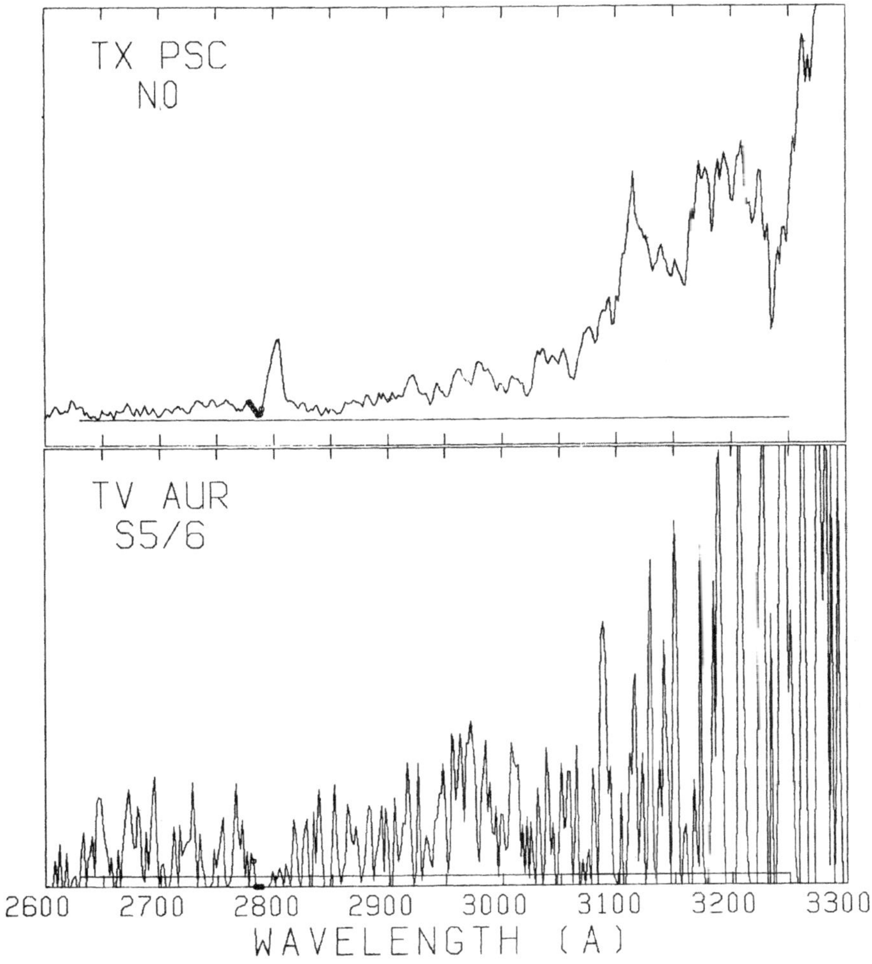

Figure 2. Ultraviolet spectra of N-type carbon star (top) and S star. Horizontal line marks the zero-flux level.

Spectra of these stars are shown in Figures 1 and 2. For comparison we also display spectra of the R8 star HD 25408 (Eaton et al. 1984) and HD 223075 = TX Psc (Johnson and O'Brien 1983).

II. RESULTS AND CONCLUSIONS

A discussion of the IUE spectra of R stars has recently been given by Eaton et al. (1984). Spectra of early R stars are remarkably similar to those of late G stars in both continuum and lines. At the resolution of low-dispersion IUE spectra, no emission features are seen in stars as late as $\sim$ R5. In fact, no indication whatever of a chromosphere is seen. Weak Mg II emission is detected in R8 stars, but only because of the extreme weakness of the continuum at 2800 A.

The warmer N stars have Mg II emission even weaker in terms of bolometric luminosity than do the R stars. They, like the late R stars, have enhanced absorption in low excitation multiplets of neutral metals relative to K and M giants. The N stars are much redder than M giants in terms of f(UV)/f(Bol) as might be expected from their optical colors and likely effective temperatures.

Ultraviolet spectra of S stars (Figs. 1 and 2) show some surprises. Although we have found S stars that resemble M giants, there are others resembling the cool carbon stars. The warmer S stars (e.g., HR 8714, HR 1105, and HD 49368) have strong Mg II λ2800 and relatively featureless continuum emission extending to shortward of 2800 A that are characteristic of M giants. In fact, these stars are comparable to M giants in both f(Mg II)/f(Bol) and f(UV)/f(Bol). There are other cooler S stars, however, that show much lower chromospheric emission. (Our spectra are too lightly exposed to see if they also have ultraviolet absorption line spectra like late R and N stars.) One of these, TV Aur, which is illustrated in Figure 2, shows less Mg II emission than any other cool carbon star observed. Several other cool S stars, (namely, HD 58881, HD 121447, SU Mon, and ER Del), show weak Mg II emission comparable to, or slightly higher than, the warmer N stars.

REFERENCES

Eaton, J. A., Johnson, H. R., O'Brien, G. T., and Baumert, J. H. 1984, Ap. J. (submitted)

Johnson, H. R. and Ake, T. B. 1983, Proc. Conf. on Cool Stars, Stellar Systems, and the Sun (Springer Verlag: New York).

Johnson, H. R. and O'Brien, G. T. 1983, Ap. J. 265, 952.

Querci, F., Querci, M., Wing, R. F., Cassatella, A., and Heck, A. 1982, Astr. Ap. 111, 120.

DISCUSSION

HECK - I noticed the wavelength scales of your IUE spectra go up to 3300 A, while the IUE standard processing stops around 3200 A. So I assume you reprocessed your data trying to get something out of this noisy end ?

JOHNSON - That is correct. The data extend well longward of 3200 A -the normal cutoff wavelength- although the sensitivity is dropping fast there.

LU - Do you have barium stars plotted on your diagram ?

JOHNSON - No, but it would be interesting to try. I suggest there may be a fairly large range.

RICHER - Do you anything at 2200 A in the spectra of these objects, particularly the carbon stars ?

JOHNSON - There is essentially no signal shortward of 2600 A, and there is a hot pixel at exactly 2200 A in the LWR spectrum.

BIDELMAN - TV Aurigae shows a composite spectrum with the companion an A or F stars, so I believe the far UV spectrum (below λ 3100, say) is entirely due to the companion ; this would explain the lack of Mg II emissions. Kondo found a similar thing in the case of α UMa some years ago.

JOHNSON - It is difficult to tell on this rather noisy spectrum. The strong Fe I (M.9) lines characteristic of a cool star seem to be present in the region $\lambda\lambda$ 3000-3100. I suppose a contribution from an F star companion cannot be excluded. The point is, in any case, that S stars apparently show a large range in Mg II although most have a strong emission line.

II. - PHOTOMETRIC PROPERTIES OF THE PECULIAR STARS. LIGHT CURVES OF VARIABLES

PHOTOMETRIC PROPERTIES OF PECULIAR RED GIANTS

Robert F. Wing
Astronomy Department
Ohio State University
Columbus, OH 43210
U.S.A.

ABSTRACT. Four quite different kinds of information are commonly obtained through photometric observations: (1) magnitudes and colors on various systems; (2) energy distributions, leading to estimates of effective temperature, tests of model atmospheres, and the detection of companions or circumstellar shells; (3) spectroscopic data from narrow-band photometry of strong spectral features; and (4) light curves of variable stars. The photometric properties of the peculiar red giants — the cool carbon and S-type stars, and the warmer R-type carbon stars, CH stars, barium stars, and related objects — are reviewed with emphasis on what can be learned from each of these kinds of photometric data.

1. INTRODUCTION

The great diversity of types of peculiar red giant stars — surely to be one of the themes of this conference — is matched by the diversity of photometric systems and techniques used to study them. Normally, the word 'photometry' first calls to mind the measurement of wideband magnitudes and color indices on standard (or at any rate general-purpose) photometric systems. This in itself is a large subject, especially when we consider the many applications of infrared photometry of cool stars. But in addition, cool stars offer numerous possibilities for the measurement of strong spectral features by narrow-band photometry. Furthermore, the spectra of cool stars are often studied with photoelectric spectrum scanners, so that the distinction between photometry and spectroscopy becomes rather arbitrary. Each of these techniques has a variety of applications, each of which can be used to study individual stars or to intercompare classes of stars.

The point of these remarks is that my topic is a large one; a full review could easily fill a book. Indeed, large parts of Golay's (1974) text on photometry and of the recent book by Alksne, Alksnis, and Dzērvītis (1983) on carbon stars, for example, are directly

M. Jaschek and P. C. Keenan (eds.), Cool Stars with Excesses of Heavy Elements, 61–85.

relevant to the topic under review. It became clear in the early stages of preparing this report that I could do no more than present some examples of work in this field. I will try to choose examples that are currently interesting and suggestive of further work, but I must apologize for omitting many important studies and for showing my own biases.

The material to be discussed is grouped under four headings: (a) magnitudes and colors; (b) energy distributions; (c) narrow-band photometry; and (d) variability of red giants. In each section we will see how that particular line of investigation has contributed recently — and should continue to contribute — to our understanding of the peculiar red giants.

It is worth noting at the outset that the types of information on peculiar red giants that can be derived from photometric observations are rather different from the more familiar objectives of photometry of hotter stars. The immediate objective of multicolor photometry of early-type stars is usually to separate the effects of temperature and interstellar reddening on the color of the star, and sometimes to determine the luminosity as well; for intermediate spectral types we can determine the general metallicity also. But for the later-type stars of peculiar (and hence in general unknown) composition, the presence of abundance-sensitive molecular bands affecting all wideband colors makes it virtually impossible to obtain either the luminosity or the reddening, and only in limited cases can the metallicity be obtained with confidence from photometry. Effective temperatures can be studied photometrically for classes of stars whose nearest members are unreddened, but we must keep in mind that in general it is not possible to correct the photometry of individual red giants of peculiar composition for interstellar reddening.

But if we cannot determine luminosity or reddening for peculiar red giants because their spectra are so strongly affected by composition differences, we can at least measure parameters of chemical composition. Recall that in the two-dimensional classification of carbon stars (Keenan and Morgan 1941) and S stars (Keenan 1954), the second dimension is not luminosity but composition. Nearly all molecular band strengths are sensitive to the O/C ratio, and by measuring two or more molecules one can often obtain other composition parameters as well. Narrow-band photometry has been a rich source of information on the chemical compositions, and even isotopic abundances, of the peculiar red giants.

2. MAGNITUDES AND COLORS

Here we consider the measurement of magnitudes and color indices on established systems of photometry, regardless of bandwidth. However, multicolor work aimed at recording energy distributions is discussed in Section 3, and measurements of specific spectral features are covered in Section 4.

2.1. UBV, UBVr, and UBVRI Photometry

The wideband UBV system is most effective in studies of early-type stars, since then the absorption by neutral hydrogen in the U filter is very sensitive to temperature and permits a determination of the star's intrinsic color in the presence of interstellar reddening. At lower temperatures, however, and especially in the stars of peculiar composition, the two color indices of the UBV system are simply too few to separate the effects of luminosity, metallicity, and absorption in various molecular bands, as well as temperature and reddening, on the wide bandpasses. In work on cool stars, the UBV measurements are often supplemented with data taken in wideband red and/or infrared filters; this improves the determination of temperature but does little to elucidate the other parameters.

In the visible region, the reddest of all stars are the cool (N-type) carbon stars by virtue of the rapid increase of their atmospheric opacity toward shorter wavelengths. Stars redder than B-V = 3 nearly always turn out to be carbon stars, although a few are reddened stars of other types. Substantial sets of UBV data for carbon stars have been published by Mendoza and Johnson (1965), Eggen (1972a), and Walker (1976); they are useful primarily in providing quantitative measures of the violet opacity. It should be noted that the problems of transforming such red colors to the "standard system" are so large — in fact, they are insuperable — that each observer or photometer should be considered to define a different photometric system. Therefore, UBV data on cool carbon stars by different observers should not be mixed, and isolated observations of this kind are of very limited value.

The stars of type S (and the intermediate types MS and SC) have not been observed as extensively as the cool carbon stars, but UBV data have been given by Wing (1967), Smak (1968), and Eggen (1972b). These stars are quite red, the majority having B-V colors between 2 and 3 mag, and this has sometimes been taken as an indication that the source of violet opacity in the carbon stars is also present in the S stars. It is therefore worth pointing out that the B-V colors of most S stars are *not* abnormally red for their effective temperatures; they are redder than the colors of M stars because the M stars are *bluer* than blackbodies of the same temperature. The effect of TiO absorption, which is greater in V than in B, is to prevent the B-V color from growing with decreasing temperature in the normal-composition giants. The TiO blanketing effect in the B, V, R, and I bandpasses has been measured with the help of calibrated spectral scans by Smak and Wing (1979), who present their results both for individual stars and as a function of spectral type.

Among the warmer late-type stars, UBVRI photometry has proved useful in recognizing stars of peculiar composition, e.g. the barium stars and the various kinds of warm carbon stars (Eggen 1972c, 1975;

Catchpole, Robertson, and Warren 1977; Lü, Dawson, Upgren, and Weis 1983). Some two-thirds of the barium stars identified spectroscopically have distinctive colors on the UBVRI system, but the remainder do not. Mannery and Wallerstein (1970) used a modified red filter and showed that barium stars, R stars, and CH stars can all be separated from normal stars (although not from each other) by UBVr photometry. Actually the same distinctions can be made from UBV data alone, since the distinguishing characteristic of all these peculiar stars is excessive absorption in the B filter relative to normal stars of the same temperature (or the same V-r color, or U-V color). Because of this excess absorption, which is at least partly due to a strengthening of the CN and CH bands, it is difficult to draw any conclusions about the general metallicity of these stars from wideband photometry.

2.2. JHKL... Photometry

Wideband infrared photometry of late-type stars is given in a very large number of papers, many of which include observations of peculiar stars. A comprehensive catalogue of papers containing the results of infrared photometry, giving the number of stars observed and their range in spectral type, has been compiled by Straižys (1978). Only a few examples will be mentioned here; other papers on infrared photometry are cited in the discussion of energy distributions (Section 3). Papers specifically dealing with JHKL... photometry of chemically-peculiar red giants include the study of carbon stars by Mendoza and Johnson (1965) and the observations of four S stars by Feast, Catchpole, and Glass (1976). The catalogue of South African photometry of late-type stars by Catchpole et al. (1979) includes carbon and S-type stars as well as normal Me Mira variables; Feast et al. (1982) have discussed this material and have shown the characteristic areas occupied by the various types of stars in two-color diagrams. The JHK photometry of Magellanic Cloud variables done in Australia by Wood, Bessell, and Fox (1983) also includes some stars of types S and C, as is shown by the low-dispersion spectra presented in the same paper.

Many of the cooler carbon stars have large excesses in the 8 - 14 μm window, attributed to thermal emission from grains in a large circumstellar envelope. Most of the carbon Miras, for example, show such excesses. But even larger excesses are found in a number of carbon stars that are extremely faint visually and were unknown prior to the IRC 2-μm survey (Neugebauer and Leighton 1969) or the AFGL rocket survey at 4.2, 11.0, 19.8, and 27.4 μm (Price and Walker 1976). Perhaps surprisingly, the colors in the JHK region are strongly correlated with the 10 μm dust emission; a comprehensive review of these properties is given by Alksne et al. (1983). Although spectral classifications are now available for nearly all stars in the IRC (Bidelman 1980), this is not yet the case for the redder AFGL stars or, particularly, the sources recorded by the IRAS satellite, many of which will undoubtedly prove to be chemically-peculiar red giants.

Relatively few stars of types MS, S, or SC have been measured by multicolor infrared photometry, but a much larger number have been recorded in the course of the surveys cited above. Wing and Yorka (1977) identified 65 S-type stars in the IRC (although only 9 of these are listed as of type S in the IRC itself) so that K magnitudes, at least, are available for these stars. The IRAS catalogue, when available, should provide far-infrared photometry for a much larger number of red giants of all kinds.

2.3. Photometry on Intermediate- and Narrow-Band Systems

Several other photometric systems, used primarily for studies of normal stars, have occasionally been applied to the peculiar red giants and, by virtue of their narrower bandpasses, have provided information not available from the UBVRIJHKL... photometry. Here we discuss magnitudes and colors measured on such systems; discussion of indices measuring discrete spectral features by filters specifically designed for that purpose — the primary objective of narrow-band photometry — is deferred to Section 4.

The Strömgren four-color system has seldom been applied to peculiar red giants, but Bond and Neff (1969) presented measurements of several barium stars and showed that their energy distributions differ from those of normal stars in the violet region. The Vilnius seven-color system, a valuable classification tool for stars of all types, has been applied to carbon and S stars by Dzērvītis (1981).

The DDO system, which employs (usually) five filters in the blue spectral region, is primarily useful as an indicator of metallicity and CN strength in G and K stars (McClure 1979; McClure and Forrester 1981). Most of the known barium stars, including the 200 or so discovered by MacConnell, Frye, and Upgren (1972), have been observed on the DDO system (Lü and Upgren 1979; Lü et al. 1983), and they show interesting differences with respect to normal stars. Most barium stars have enhanced CN absorption, and their photometric luminosities are mostly higher than those of ordinary giants. But here the interpretation of the DDO photometry — especially for peculiar stars — is complicated by the fact that the DDO "gravity index" is actually a second CN index, due to the presence of the (0,2) CN band in the '45' filter. I'm not sure that this has ever been stated in print, although it has been known for more than ten years. Because of the abnormal CNO abundances in at least some barium stars, their luminosities cannot be determined by DDO photometry.

A photometric system that avoids some of the problems of the blue spectral region and takes advantage of the brightness of red giants in the near infrared was introduced by Wing (1971). It uses eight interference filters (width 40 - 70 Å) placed between 0.7 and 1.1 μm. The system provides a clean measurement of CN strength, but, despite the great luminosity sensitivity of this molecule, it cannot be used as a luminosity indicator in the chemically-peculiar stars because of its sensitivity to composition. Since there are no atomic lines of any consequence in any of the bandpasses, it is not possible to determine metallicities with this system either, but on the other hand the measures of continuum color are free from atomic line-blanketing effects. The most pervasive absorber in the near infrared is CN, and in strong-CN stars it is simply not possible to avoid CN lines with 50 Å bandpasses; however, the continuum points on the eight-color system are located in regions of minimum CN opacity, and synthetic spectrum calculations are being carried out in collaboration with B. Gustafsson and K. Eriksson in Uppsala to determine this residual absorption.

An early application of the eight-color photometry to peculiar red giants was Baumert's (1972) study of carbon stars, and additional data on the warmer carbon stars have been published by Yorka (1983). A modification of the eight-color system (two of the filters were changed to provide measures of ZrO and LaO) was used by Piccirillo (1977) in an investigation of the S-type stars. Thus there exists a considerable body of data on the magnitudes, colors, and molecular band strengths of peculiar red giants (see also Section 4).

2.4. Application to Determinations of Absolute Magnitude

Photometric determinations of absolute magnitude usually involve the measurement of a narrow-band index which is sensitive to luminosity, e.g. a CN index in G and K stars or the Crawford Hβ index in B stars. As already mentioned, this approach may not be reliable for stars of peculiar composition. But there also are non-spectroscopic methods of determining absolute magnitude which should be mentioned here because they also involve photometry. One is to locate stars of a given type in cluster color-magnitude diagrams, and another is to analyze the space motions of a class of stars to determine their absolute magnitudes statistically.

Many studies have been made of the color-magnitude diagrams of clusters containing one or more red giants, and I will mention only a few good examples. Eggen (1972a) has collected data on carbon stars suspected of membership in open clusters and has presented UBVRI photometry and absolute magnitudes for them. McClure and his co-workers have shown that the star at the tip of the giant branch in NGC 2420 is a classical barium star, while another probable member of the same cluster is a mild barium star (McClure, Forrester, and Gibson 1974; McClure 1984a). The absolute magnitudes of these barium stars are around $M_V = 0.0$, in agreement with the values found statistically for large numbers of field barium stars (Upgren, Lü, and MacConnell 1977).

The discovery in the Magellanic Clouds of numerous carbon stars (Westerlund 1964; Westerlund et al. 1978; Blanco, McCarthy, and Blanco 1980) and, more recently, S stars (Bessell, Wood, and Lloyd Evans 1983; Wood, Bessell, and Fox 1983; Lloyd Evans 1983b) has given us much more precise information on their absolute magnitudes than had been available from studies of galactic stars (see, for example, Richer 1981 and

Frogel et al. 1981). The literature on surveys for, and photometry of, red giants in the Magellanic Clouds has been reviewed recently by Aaronson (1984). We must remember, of course, that the C and S stars found in the Clouds do not represent all the varieties that are known in the Galaxy. In particular, the varieties of relatively low luminosity, such as the early R stars, have not yet been found in the Clouds. Many of the carbon and S stars in the Clouds can be associated with specific star clusters of known ages — largely as a result of a spectroscopic survey of red stars in Magellanic Cloud clusters by Frogel and Blanco (1984) — so that the ages and masses of these chemically-peculiar red giants may be considered rather well known. An interesting recent development has been the identification of S-type stars in galactic globular clusters by Lloyd Evans (1983a, 1984).

Statistical studies of the mean absolute magnitudes of classes of stars are primarily concerned with the measurement of space motions and the elimination of systematic errors, and it is sometimes forgotten that the accuracy of the results also depends upon the quality of the photometry available. Reliance upon existing photometry may necessitate using magnitudes measured in heavily-blanketed spectral regions far from the peak of the star's energy distribution, and perhaps in different filters for different stars, requiring the application of uncertain color indices, or at different (or unknown) phases in the case of variable stars. Such problems are particularly severe in the cooler chemically-peculiar red giants with their very red colors and high incidence of variability. I would therefore like to cite the statistical determination of the absolute magnitudes of carbon stars by Baumert (1974), since it is based upon photometry in the near-infrared continuum at 1.04 μm where the effects of blanketing, variability, and interstellar reddening are small and where the magnitudes bear a simple relation to the bolometric magnitude.

3. ENERGY DISTRIBUTIONS

In addition to the photometry on specific systems, a large quantity of photometry of peculiar red giants has been obtained for the purpose of delineating the shapes of their energy distributions. These observations may consist of calibrated wideband photometry, measurements at selected wavelengths, or continuous spectral scans of either low or high resolution. The result of the study may be the classification of the stellar spectrum, the measurement of molecular band strengths, the determination of effective temperature, the test of a model atmosphere, the identification of continuous opacity sources, the detection of far-infrared excess radiation from a dust shell, or the discovery of a hot companion affecting the short-wavelength region. In the short space that I can devote to this subject, I will touch on only a few topics that have been important in the study of peculiar red giants.

3.1. Infrared Energy Distributions

Before photometric or spectrophotometric measurements can be used to study energy distributions, an absolute calibration of some sort is needed. Early calibrations of the wideband UBVRIJHKL... system (Johnson 1966) and most calibrations of spectrophotometric measurements in the infrared have been based on the assumption that certain stars radiate like blackbodies, or like atmospheric models. While this procedure has produced self-consistent and apparently satisfactory results, it should be checked by direct comparisons of stars to laboratory standards. A project with this objective (Hayes et al. 1980) is now nearing completion at Kitt Peak National Observatory, and a recent review of this area has been prepared by Hayes (1984).

From wideband multicolor photometry, Mendoza and Johnson (1965) showed that the gross energy distributions of carbon stars in the 1 - 4 μm region do not differ greatly from those of oxygen-rich stars, and they expressed the opinion that their effective temperatures could be obtained from the same calibrations of color indices. While this procedure must be regarded as questionable, we are still — after nearly 20 years — unable to provide a better alternative. The handful of effective temperatures that are now available for carbon stars from lunar-occultation measurements of angular diameter (Ridgway et al. 1980) are in general agreement with the photometric values, but they are too few — and individually too uncertain — to establish a true scale of temperature. A basic problem in establishing a temperature scale for carbon stars is that there does not seem to exist a reliable spectroscopic temperature indicator to calibrate: the spectroscopic temperature classes are well known to correlate poorly with color temperature, and all of the stronger spectral features are clearly sensitive to the vagaries of composition. It therefore appears to be necessary to base the temperature scale of carbon stars on the calibration of a photometric color index, although the question of the reddening of individual stars is then necessarily involved. The wideband photometry provides several indices that can serve this purpose, and my own preference would be to use I-K or I-L, or at any rate an index that avoids the J filter, which is heavily depressed in a few (albeit exceptional) carbon stars such as T Lyr.

A different — but closely related — approach involves fitting complete stellar energy distributions (which, for red giants, means primarily infrared distributions) to flux distributions calculated from model atmospheres. Here the derived temperatures depend upon both the absolute calibration of the photometry and the ability of the model to represent the cool stellar atmosphere. The choice of which model should be taken to represent which star is usually based on a comparison of observed and computed spectral features, although even the best models for cool, extended atmospheres have trouble calculating the strengths of strong molecular bands formed close to the surface. Nevertheless, useful results for carbon stars have been obtained by this method, e.g. by Goebel et al. (1980, 1983). In work of this kind it is a great help if the stars observed are sufficiently nearby that the interstellar reddening may be considered negligible: if the reddening is treated as a free parameter, as in the work of Scargle and Strecker (1979), errors in the theoretical band strengths can lead to large, spurious reddening values and nonsensical effective temperatures.

Because of the strong absorption by water vapor in the Earth's atmosphere, the most complete infrared energy distributions have been obtained from above the ground, primarily from the Kuiper Airborne Observatory which operates at altitudes up to 14,000 m. KAO spectra obtained with a circular variable filter provide uniform and nearly continuous spectral coverage through the 1 - 5 μm region, and sometimes out as far as 30 μm.

Airborne observations of carbon stars include studies of the semi-regular variable Y CVn (Goebel et al. 1978, 1980), the R star HD 19557 (Goebel et al. 1983), and the cool Mira variable V CrB (Goebel et al. 1981). Comparisons with carbon-star models have helped to clarify the effective temperatures of these stars and have stimulated further theoretical work to improve the models. As an example of this interplay between observation and theory, the recognition of extensive absorption by HCN and C_2H_2 in V CrB provided the stimulus for *ab initio* calculations of the complex spectra of these polyatomic molecules so that they could be included as opacity sources in model-atmosphere computations (see Jørgensen's paper in this volume).

Recently, calibrated 1 - 5 μm spectra of 10 S-type stars have been obtained and compared to model flux distributions (Augason et al. 1984). Difficulties in matching the observed and calculated spectra in the region of the H^- opacity minimum at 1.6 μm led these authors to suggest that the non-Mira S stars are hydrogen-deficient, an idea proposed in connection with the carbon stars some time ago (Vardya 1966).

The "infrared flux method" of Blackwell and Shallis (1977; see also Blackwell, Petford, and Shallis 1980) is another way of obtaining effective temperatures from the comparison of observed and calculated infrared energy distributions, and it has been applied to the carbon stars by Tsuji (1981). In this method, instead of comparing the stars and models point by point through the spectrum, one compares their bolometric magnitude m_{bol} and a monochromatic magnitude at some infrared wavelength that is relatively free from molecular blanketing, such as L (3.5 μm). Since the bolometric flux is very sensitive to the temperature while the infrared monochromatic flux is not, the "color index" $m_{bol} - L$ can be used as a temperature indicator, and its value for any effective temperature can be calculated from the models. The point that I would like to make is that, given an appropriate set of models, the success or failure of the method depends mainly upon one's ability to measure m_{bol}. Some discussions of the infrared flux method have emphasized the measurement of L and said very little about the rest of the energy distribution; but it is m_{bol}, not L, that is sensitive to temperature and conveys the information that one is trying to extract. There is nothing trivial about measuring m_{bol}, and for the stars of peculiar composition there are no short-cuts — no published tables of bolometric corrections — that can be taken.

Still another method of obtaining effective temperatures from a comparison of stars and models involves the use of models to calibrate a simple color index. Observationally, it is much easier to measure two monochromatic fluxes than the bolometric magnitude, and bandpasses can be chosen so that the fluxes are relatively easy to calculate, i.e. are relatively insensitive to errors in the models. Best of all are narrow-band continuum points, provided that they span an appropriate spectral interval. The effective temperatures of normal K giants have recently been obtained by this method (Wing, Gustafsson, and Eriksson 1984), and we are now extending the calculations to the warmer (R-type) carbon stars. The photometry used for these K and R stars is narrow-band near-infrared data on the writer's eight-color system (Wing 1971). For cooler stars the continuum is less well defined in the near infrared, and it is preferable to combine the I(104) magnitude with a measurement at a longer wavelength. A narrow-band magnitude at 4.00 μm, designated L(400), has been defined by Wing and Rinsland (1981a), and the color index I(104) - L(400) is especially well suited for temperature determinations in the cooler stars. Wing and Rinsland (1981b) have shown how this index behaves in stars and models of normal composition, and the method appears promising as an approach to the temperatures of peculiar red giants as well.

In most late-type stars, the radiation out to 4 or 5 μm is produced in the photosphere. At longer wavelengths, however, it is difficult to obtain any information about the photosphere because the spectrum, in many cases, is dominated by continuous thermal radiation from circumstellar dust. Wideband photometry extending to 10 μm or beyond, and including stars of type S or C, has been published by Mendoza (1967), Gillett et al. (1971), Gehrz and Woolf (1971), Hackwell (1972), and Forrest et al. (1975). Differences in the shapes of the 8 - 14 μm dust bumps in M and C stars have been recorded by filter-wheel spectrophotometry (Hackwell 1972; Forrest et al. 1975; Merrill and Stein 1976a,b,c) and have been interpreted in terms of differences in the composition of the circumstellar grains (see, for example, the discussion by Merrill and Ridgway 1979).

3.2. Spectrophotometry in the Visible and Ultraviolet Regions

Very few calibrated spectral scans have been published for late-type stars in the visible region, even if we define this region generously as extending from the near ultraviolet to the near infrared. One reason may be that the entire visible region is so heavily blanketed by atomic lines and molecular bands that the true continuum is seldom seen, and the possibilities for determining temperatures photometrically are very limited. Measurements of strong spectral features are more often made from simple narrow-band indices (Section 4), rather than from complete scans. The primary use of calibrated spectral scans in heavily blanketed regions has been to detect broad, shallow spectral features such as new molecular bands or continuous opacity sources.

The most extensive sets of calibrated visible-region scans of peculiar red giants have been obtained at Indiana University. Scans of carbon stars are presented by Faÿ and Honeycutt (1972), Faÿ, Honeycutt, and Warren (1973), and Faÿ et al. (1974), while scans of S stars in the red and near infrared regions are shown by Piccirillo (1977). Scans of R-type carbon stars, CH stars, and barium stars form the basis of the investigation by Gow (1976), and the molecular indices given by Gow (1977) for cool carbon stars were also obtained from scans.

An interesting application of Gow's (1976) scans of the warmer peculiar red giants was his study of the excess absorption in the violet region shown by the barium stars. This absorption had been first noted by Bond and Neff (1969) on the basis of intermediate-band photometry, and at very low resolution it appears as a broad feature, approximately 1000 Å wide, centered at about 4000 Å. There has been controversy over the identity and even the reality of this feature (Williams 1975; Fix and Neff 1975; Fix 1976; Gow 1976; Lü and Sawyer 1979; Baird 1982; see also the discussions by Straižys 1982 and McClure 1984a). It is now clear that *most* of the excess absorption originally reported by Bond and Neff is due to enhanced bands of CN, CH, and C_2. Gow's maps of the spectrum of the excess absorption, based on careful comparisons of several barium stars and normal giants, show that, through most of the region of the Bond-Neff depression, there is *no* excess absorption in the barium stars that cannot be readily identified with absorbers that are seen spectroscopically and known to be enhanced in barium stars, viz. diatomic carbon-containing molecules, s-process atoms, and rare earths. The question that remains concerns the identity of a weak residual absorption component between 4000 and 4100 Å which is enhanced in barium stars but not in R or CH stars; the extent of this component is hard to judge because it runs into the H and K lines of Ca II (which are often weak in barium stars) shortward of 4000 Å, and into the (0,1) CN band longward of 4100 Å. Although I don't have a definitive answer, I would like to make a comment: it has always surprised me that, in all the discussions of possible opacity sources in barium stars, there is almost no mention of *atomic lines*. After all, the opacity due to metallic lines in G and K stars has a spectrum rather similar to that of the Bond-Neff depression (Wildey et al. 1962), and in normal stars the overlapping wings of numerous weak lines are presumed to be the dominant opacity source in the violet region. Small differences in line strength — or perhaps just in the relative importance of their *wings* — between the barium stars and the particular field stars to which they have been compared would result in imperfect cancellation, especially in the 4000 - 4100 Å region. This possibility must be investigated before photometry of barium stars can be properly interpreted. For example, we would like to know whether the CN strength measured in the '41' filter of the DDO system is affected by anything other than CN and the usual metallic lines (see the discussion by Lü and Upgren in this volume).

In the ultraviolet below the atmospheric cut-off, red giants are very faint and the main motivation for measuring their energy distributions has been to detect hot companions. The discovery of radial velocity variations in many barium stars (McClure, Fletcher, and Nemec 1980; McClure 1983) and CH stars (McClure 1984b) suggests that they may have white dwarf companions detectable with the IUE satellite. Böhm-Vitense (1980) succeeded in detecting a hot companion of the bright barium star ζ Cap, but it is visible only below about 1800 Å. A high-dispersion spectrum of ζ Cap in the 2500 - 3230 Å region (Wing, Carpenter, and Wahlgren 1983) shows no evidence of a companion: the absorption lines do not appear filled in. Since most barium stars and CH stars are much fainter (i.e. more distant) than ζ Cap, detecting similar companions will be difficult. Further observations of barium stars (Dominy and Lambert 1983) and CH stars (Bond 1984) with the IUE short-wavelength camera have produced very little evidence for additional companions. Several possible interpretations of these results are discussed by McClure (1984c).

A few single red giants of peculiar composition have been observed with IUE, but with limited success. The only high-resolution observation of a carbon star is one of the hydrogen-deficient star HD 182040 (Johnson, Ameen, and Eaton 1984). Low-resolution observations have recorded continuum radiation down to approximately 2600 Å in several relatively warm carbon stars (Johnson and O'Brien 1983; Eaton et al. 1984). On the other hand, long exposures on cooler carbon stars have recorded no light at all, showing that the opacity which depresses their spectra in the violet extends to considerably shorter wavelengths (Querci et al. 1982). An ultraviolet observation of the S star HD 35155 has revealed that it has a hot companion close enough to it to cause physical interactions (Johnson and Ake 1984), and IUE observations of several more S stars are reported in this volume by Johnson, Ake, and Eaton.

4. NARROW-BAND PHOTOMETRY

The spectra of cool stars are filled with strong atomic lines and molecular bands. As a consequence, wideband photometry is very difficult to interpret quantitatively: there are too many significant absorbers within each filter, and there is no way to separate out their effects. The use of narrow bandpasses sharpens the spectral information, and by designing the filters for the specific task at hand one can approach the ideal of measuring one feature at a time.

An equally important use of narrow-band photometry is to *avoid* absorption features and measure the continuum as well as possible. I have already mentioned the use of an infrared narrow-band continuum color, I(104) - L(400), as an effective-temperature indicator and the use of narrow-band spectrophotometry in the 8 - 14 μm window to study the emission from circumstellar grains (Section 3). A related application is to measure a series of widely-spaced continuum points to define the non-blackbody shape of the continuum energy distribution in, say, the region of the H^- opacity minimum. Observations of this kind, and comparisons with continuum flux distributions calculated from model atmospheres, are shown by Wing and Rinsland (1981b). These topics will not be considered further.

In this section we consider photometric measurements of spectral features in peculiar red giants. The material is arranged according to the type of star, starting with the warmest. A more general review of narrow-band photometry of molecular bands, which however emphasized the normal M stars, was given by the writer a few years ago (Wing 1979); an older review, restricted to the one-micron region but with more emphasis on the stars of abnormal composition, is also available (Wing 1974).

4.1. Barium, CH, and R Stars

The barium stars have been studied by a wide variety of photometric techniques. Enhancements of the (0,1) violet CN band, relative to normal class III giants, are seen in most barium stars by DDO photometry (Lü and Upgren 1979; Lü et al. 1983). Williams (1975) used the Cambridge spectrometer with a very narrow (3.2 Å) central bandpass to measure the Ba II line λ6141 in 200 evolved stars, including a number of barium stars and several others in which the feature was found to be strengthened. Gow (1976) used continuous photoelectric scans to measure C_2 indices in barium, CH, and R stars.

The near-infrared spectra of the warmer carbon stars — as in the case of the normal G and K giants and supergiants, and barium stars — are nearly pure CN spectra at low resolution. Consequently the writer's eight-color system (Wing 1971) can obtain for them only three independent quantities: I(104) magnitude, continuum color, and CN strength. The near-infrared CN strengths measured in this manner (Baumert 1971, 1972; Yorka 1983) are rather well correlated with measurements of the violet CN bands in the warmer stars (T_e > 4500 K or so), but in the cooler stars the infrared bands appear generally stronger than the violet bands. Eight-color photometry was used to measure the CN strengths and estimate the effective temperatures of the CH stars in the globular cluster ω Cen (Wing and Stock 1973).

In order to determine CNO abundances in late-type stars by narrow-band photometry of molecular bands, several molecules must be measured. A scanner-defined photometric system to measure five bands of four different molecules — NH, CN, CH, and C_2 — has been described by Wing and Lange (1981) and applied to the warmer carbon and barium stars by Yorka (1983). This system is capable of measuring C_2 in normal G and K giants — in much the same manner as was done earlier by Alexander and Branch (1974) — so that the C_2 enhancements of typical barium stars are well shown. The NH measurements at 3366 Å are difficult but allow the nitrogen abundance to be derived without resorting to the use of CN. An interesting use of the CH and NH measurements is to look for signs of hydrogen deficiency.

Carbon isotope ratios can also be measured by narrow-band photometry, although in general the strong (0,0) bands cannot be employed because of their small isotope shifts. The use of CN (rather than C_2) bands makes it possible to study G and K giants and supergiants and barium stars, as well as carbon stars. Wing (1974) has described a six-color photometric system employing 20 Å scanner bandpasses to measure the profile of the (2,0) and (3,1) bands of the red system of CN. The carbon isotope ratio index defined on this system has been calibrated empirically through observations of G and K stars with spectroscopically-determined isotope ratios (Little-Marenin, Little, and Wing 1979), and results obtained in this way for carbon stars have been compared to spectroscopic determinations by Little, Little-Marenin, and Wing (1980).

4.2. Cool Carbon Stars

Several of the techniques mentioned above in connection with the warmer carbon stars have also been applied to the cool, N-type carbon stars. Of course, systems operating in the blue and ultraviolet, such as the DDO system and most of the "NH-CN-CH-C_2" system, cannot usefully be applied to stars that radiate so little in these regions. Griffin and Redman (1960) did make some measurements of the λ4215 CN band in cool carbon stars, but it is evident that its strength is greatly reduced by the violet opacity. Wing (1967) found that the infrared CN bands are strong in all N stars and concluded that the observed strength of the λ4215 band is more a measure of the violet opacity than of the actual CN abundance.

Visual-region photoelectric scans of carbon stars have been used by Fujita (1970, and references therein) for classification purposes and by Gow (1977) to define a color temperature and indices of C_2, ^{12}CN, and ^{13}CN.

The most extensive narrow-band photometry of carbon stars has been carried out by Baumert (1971, 1972), who measured CN indices and color temperatures on the eight-color near-infrared system for about 350 carbon stars and related objects. Approximately 100 of these are southern carbon stars observed by the writer from Cerro Tololo, and the program included all carbon stars with published proper motions in both hemispheres. These color temperatures are correlated with the values obtained from wideband photometry (Mendoza and Johnson 1965), yet there are differences which are still not adequately understood and which presumably indicate imperfect allowance for blanketing effects on one or both systems.

The SC stars are nearly indistinguishable from C stars on the eight-color system. The extremely cool SC Mira variable VX Aql has been studied by eight-color photometry and image-tube spectroscopy by Greene and Wing (1975).

In the two-micron region, medium-resolution spectra have been used to derive CO indices for carbon stars (Frogel and Hyland 1972; Frogel, Dickinson, and Hyland 1975). This first-overtone CO band is strongest in the carbon stars of low carbon abundance class and weakest in the reddest stars. A filter-defined index of the same band has been measured by Faÿ and Ridgway (1976). Faÿ (1974) has examined the relations that exist between the various C_2, CN, and CO indices in carbon stars.

One of the strongest spectral features in cool carbon stars is a broad band centered at 3.2 μm, attributed to a combination of HCN and C_2H_2 bands. A narrow-band index of this band has been measured by Faÿ and Ridgway (1976). A survey of infrared stars in this index would be an easy way to identify the cool carbon stars among them.

A search technique involving direct photography through narrow-band filters, and which identifies carbon stars by their CN absorption near 8000 Å, has been described by Palmer and Wing (1982). The filters are 100 Å wide and are placed in such a way that the index changes sign when TiO is present. This technique has been used to survey the central regions of eight globular clusters for carbon and M stars (Palmer 1980). A similar technique has been used recently to identify carbon stars in the nearby galaxy NGC 6822 (Aaronson et al. 1984) and in the dwarf elliptical companion of the Andromeda galaxy, NGC 205 (Richer et al. 1984).

4.3. S Stars

The spectra of S stars are probably the most difficult to classify of all stellar types. It is now 30 years since a two-dimensional classification scheme for S stars was developed by the instigator of this conference (Keenan 1954), and it is 50 years since another conference participant introduced its one-dimensional forerunner (Davis 1934); yet I am sure that they would both agree that the S stars still present a major challenge to the classifier.

The problem, basically, is that there is so much *variety* in the spectra of S stars. Even within just the one-micron region (Wing 1972), the spectra of S stars show bands of so many molecules — sometimes individually, sometimes in various combinations — that they cannot be adequately described by any small set of parameters. Some S stars have spectra resembling those of M stars, but with the addition of ZrO and the enhancement of CN; others have entirely different spectra, with no TiO at all, strong bands of oxides of s-process elements such as ZrO, LaO, and CeO, enhanced bands of metallic hydrides such as CaH and FeH, and a large number of unidentified bands. Another problem, upon which I have elaborated elsewhere (Wing 1979), lies closer to the heart of the *purpose* of spectral classification: even if we have measurements of all these molecular bands, we do not know well enough which of them should be considered to be indicators of temperature and which to be indicators of gas pressure, and which are better treated as indicative of abundance anomalies. The model-atmosphere calculations of Piccirillo

(1980) have certainly shed light on this subject, but the ideal classification system still eludes us.

Because of these complications, applications of narrow-band photometry to the S stars have been quite limited. A few continuous scans have been published (Wing 1972; Piccirillo 1976, 1977), primarily as a guide to the appearance of the unidentified bands and an aid to the interpretation of the miscellaneous photometric observations that exist on other systems, such as the writer's 27-color data (Wing 1967, 1972).

The only major effort to study the S stars systematically by narrow-band photometry was made by Piccirillo (1976, 1977), who modified the near-infrared eight-color system to include filters for ZrO and LaO. He was able to derive eight independent quantities from the eight filters: magnitude, color temperature, and indices of TiO, VO, CN, ZrO, LaO, and an unidentified band (see also Wing 1979).

At longer wavelengths, the bands of CO near 2.3 μm and SiO near 4.1 μm are prominent in S stars. Of the 74 stars in which Rinsland and Wing (1982) measured SiO strengths, only 5 are S stars but two of them showed the strongest SiO bands of the entire sample.

The strong pair of ZrO bands at 9299 and 9315 Å is badly contaminated by TiO, CN, and atmospheric water, and consequently it has usually been avoided by narrow-band photometrists. However, it is so strong that it never fails to appear in S stars, or even in the marginal MS stars (Spinrad and Newburn 1965; Spinrad et al. 1966; Wing 1967). In fact, a spectral scan of this region was used by Lockwood and Wing (1982) to prove that VX Sgr is *not* an S star. A narrow-band index of the 9300 Å band could be used to identify S stars in distant stellar systems.

5. VARIABILITY OF PECULIAR RED GIANTS

To this point, we have treated the peculiar red giants essentially as nonvariable stars, with emphasis on their colors and spectra. But we would be remiss not to take note of the significant fact that a large percentage of them are variable in light.

Photographic surveys for variable stars and spectral surveys for peculiar red giants often produce the same stars. On the other hand, the normal M stars are often variable, also. Variability clearly is not *caused* by peculiar composition, or by any particular opacity source in the photospheric layers. Rather, it is more closely related to low effective temperature or high luminosity — or both, which is to say, large radius.

Among the peculiar-composition red giants we can, I believe, find every type of variability that occurs among red giants in general; conversely, among the stars of any variability class we can find examples of just about every kind of abundance anomaly that is known to occur in cool stars. Space limitations will not permit a comprehensive review of so many variability types and composition classes, but I will give some examples to support these statements. My remarks will mostly concern the incidence of composition anomalies among late-type variables rather than the details of their light curves.

5.1. Yellow Variables

In the temperature range 4000 - 6000 K corresponding to spectral types G and K, the great majority of normal-composition stars are nonvariable, although irregular fluctuations of up to a few tenths of a magnitude are probably common among the supergiants. Stars of the peculiar-composition groups which fall in this temperature range are also primarily nonvariable, so far as is known, but I don't think that very many of these stars have actually been tested for variability at the level of 0.1 mag or less. Photoelectric monitoring of the stars of these groups would be a useful occupation for small telescopes at good sites.

There are several well-defined classes of variable stars in the 4000 - 6000 K range: the Cepheid variables, the RV Tauri stars, and the R CrB stars (as well as the not-so-well-defined SRd variables). I will limit my remarks on these stars to consideration of the incidence of chemical peculiarities among them.

5.1.1. Cepheid variables. Although a bit warm to be considered at a conference on cool stars, the classical Cepheids are sometimes cool enough at minimum to develop molecular bands and to invite comparisons with nonvariable G and K stars. These Population I Cepheids have compositions close to solar but with altered CNO abundances (Luck and Lambert 1981). Nitrogen is greatly enhanced, indicating that material processed by the CNO cycle has been mixed into their atmospheres, but no cases of C > O are known.

The Population II Cepheids or W Virginis stars are found in globular clusters as well as in the field, and consequently they are known to be low-mass stars. Interestingly, some Population II Cepheids are carbon stars, the best-known example being RU Cam (Sanford 1928; Gauzit 1959; Wallerstein 1968, 1973). The Cepheid V553 Cen also shows C_2 bands and is found to have moderate excesses of both C and N, but it appears to be only slightly metal-deficient and its population type is somewhat ambiguous (Cottrell 1979; Wallerstein, Brown, and Bates 1979).

5.1.2. RV Tauri stars. These often-neglected variables are evolved stars which occupy the temperature regime between the Cepheids and the Miras. A few of them show pronounced variations in CN strength through the cycle; the most extreme case known is AC Her, which can be classified as a carbon star at minimum light (Rosino 1951; Preston et al. 1963; Baird 1981).

5.1.3. R CrB stars. It appears that all stars of this class are markedly hydrogen-deficient and carbon-enriched. A detailed analysis of the elemental abundances of two of them — R CrB itself and XX Cam — has been carried out by Cottrell and Lambert (1982), who give references to earlier work.

5.2. Small-Amplitude Variables of Late Type

Of the cooler (N- and S-type) peculiar red giants which have been adequately monitored, the majority are small-amplitude semiregular or irregular variables with typical cycle lengths between 50 and 200 days. Others appear to be nearly constant in light output, while still others have the large amplitudes, long periods, and phase-dependent hydrogen emission lines that place them in the category of Mira variables. All of these statements are also true of the normal-composition M stars later than about M3 or M4.

Differences between the M, S, and N stars in the character of their light variations, or in the occurrence of different types of variability, may exist but are difficult to establish with confidence. Comparisons should no doubt be made between stars in the same interval of effective temperature, but the temperatures of S and N stars are much more difficult to judge from their colors or spectra than the temperatures of M stars. In compiling relative statistics on stars grouped by light amplitude, it should be kept in mind that the *visual* amplitudes of M and S stars are increased by changes in the blanketing by metallic oxide bands, which strengthen toward minimum light, whereas the bands of C_2 and CN in carbon stars show relatively little variation with phase.

Partial light curves in UBVRI were obtained for a number of representative semiregular and irregular variables as well as Miras in Eggen's (1972a,b) photoelectric studies of carbon stars and S stars. Light curves for many fainter carbon stars have been published over the last decade in *Investigations of the Sun and Red Stars* (Riga); specific references and a general discussion are given by Alksne et al. (1983).

Detailed studies of two semiregular carbon stars — UX Dra and Y CVn — have been reported recently by Vetešnik (1984a,b). These stars were monitored both photoelectrically in BVR and spectroscopically for radial-velocity variations to test the hypothesis that their light variations are due to radial pulsations. Instead, evidence was found that UX Dra is a binary star, while Y CVn was found to have a period nearly twice as long as the catalogued value. These surprises for two of the brightest carbon stars indicate that a great deal of careful monitoring work is still needed to clarify the nature of small-amplitude variations in cool stars.

5.3. Light Curves of Mira Variables

I have saved the Miras for last for one reason only: I was afraid that if I began with them, I would exhaust my allotted time and space before discussing anything else! This is precisely what happened a few years ago when attempting to review the work on red variables for IAU Commission 27: the title of my report (Wing 1982) had to be changed to "Mira Variables". Here my remarks will be restricted to brief answers to three questions pertaining to the visual and infrared light curves of the Miras of abnormal composition. Elsewhere (Wing 1980) I have reviewed the evidence that Miras are pulsating stars and have documented many of the peculiarities of their variations in light and spectrum.

Can the chemical composition of a Mira variable (i.e. whether it is of type M, S, or C) *be deduced from its visual light curve?* Here we are dealing primarily with Miras of relatively long period, since nearly all the S- and C-type Miras have periods longer than one year. Inspection of the mean visual light curves given in Campbell (1955), which are based on many years of visual monitoring by the AAVSO and are arranged in order of period, shows that there are two major light-curve forms at these longer periods: most have sharply-peaked maxima and broad minima, while others have broad maxima and sharp minima and look as if they have been plotted upside-down. These shapes do indeed correlate strongly with composition classes. The M stars and mild S stars (which have strong TiO bands as well as ZrO) have sharply-peaked maxima corresponding to the time of weakest TiO strength, while the carbon and pure S stars (i.e. those without TiO) have the broader maxima. Of course I am oversimplifying a bit, and exceptions can be found, but it is clear that the presence or absence of TiO has a significant effect on the shape and amplitude of the visual light curve. The visual amplitudes are considerably greater, on the average, for the TiO stars.

Can Miras of types M, S, and C be distinguished on the basis of cycle-to-cycle differences? Here the answer appears to be "no". The observed light curves given in AAVSO Report 38 for the period 1974 - 77 show that major differences in brightness at successive maxima are fairly common, but they seem to occur in all spectral types. The differences evidently cannot be attributed to molecular blanketing effects alone. A similar conclusion was reached by Lockwood and Wing (1971) on the basis of photometry in the near-infrared continuum: large cycle-to-cycle differences are seen in the I(104) light curves and clearly are not the result of changes in the spectrum.

Can the chemical composition of a Mira be deduced from the <u>scatter</u> *in its visual light curve?* Here, surprisingly, the answer is a definite "yes". In looking through the light curves of AAVSO Report 38, which shows the individually reported eye-estimates, one notices immediately that the scatter is much worse — as much as a full magnitude — for some Miras, even bright ones, than for others. The wide light curves invariably belong to carbon stars. The problem is that different observers are using different photometric systems (eyes), and these differences in response do affect brightness estimates for red stars made relative to white comparison stars. The *width* of an AAVSO light curve is thus an index of the *color* of the star in the visible region, and these colors are strongly correlated with the type of spectrum.

Of course, there are numerous other backhanded ways to separate the Miras of types M, S, and C. For example, the relative strengths of the Hβ, Hγ, and Hδ emission lines have nothing to do with Balmer decrement theory but depend upon the overlying molecular absorption spectrum. On the other hand I see no way to distinguish the M, S, and C stars on the basis of infrared continuum light curves, say in I(104) or K. Thus the basic phenomenon of Mira-type variability appears to be quite independent of the chemical composition of the surface layers.

REFERENCES

Aaronson, M. 1984, in IAU Symp. No. 108, *Structure and Evolution of the Magellanic Clouds*, ed. S. van den Bergh and K. S. de Boer (Reidel), p. 183.

Aaronson, M., Da Costa, G. S., Hartigan, P., Mould, J. R., Norris, J., and Stockman, H. S. 1984, Astrophys. J. (Letters) 277, L9.

Alexander, J. B., and Branch, D. 1974, Mon. Not. Roy. Astron. Soc. 167, 539.

Alksne, Z., Alksnis, A., and Dzērvītis, U. 1983, *Properties of Carbon Stars of the Galaxy* (Riga: Acad. Sci. Latvian S.S.R.).

Augason, G. C., Bregman, J. D., Johnson, H. R., and Witteborn, F. C. 1984, preprint.

Baird, S. R. 1981, Astrophys. J. 245, 208.

Baird, S. R. 1982, Astrophys. J. 252, 305.

Baumert, J. H. 1971, in *Proc. Conf. on Late-Type Stars*, ed. G. W. Lockwood and H. M. Dyck (Kitt Peak Nat'l Obs. Contrib. No. 554), p. 155.

Baumert, J. H. 1972, Ph.D. dissertation, Ohio State University.

Baumert, J. H. 1974, Astrophys. J. 190, 85.

Bessell, M. S., Wood, P. R., and Lloyd Evans, T. 1983, Mon. Not. Roy. Astron. Soc. 202, 59.

Bidelman, W. P. 1980, Publ. Warner and Swasey Obs. 2, No. 6.

Blackwell, D. E., and Shallis, M. J. 1977, Mon. Not. Roy. Astron. Soc. 180, 177.

Blackwell, D. E., Petford, A. D., and Shallis, M. J. 1980, Astron. Astrophys. 82, 249.

Blanco, V. M., McCarthy, M. F., and Blanco, B. M. 1980, Astrophys. J. 242, 938.

Böhm-Vitense, E. 1980, Astrophys. J. (Letters) 239, L79.

Bond, H, E. 1984, in *Future of Ultraviolet Astronomy Based on Six Years of IUE Research*, NASA (in press).

Bond, H. E., and Neff, J. S. 1969, Astrophys. J. 158, 1235.

Campbell, L. 1955, *Studies of Long Period Variables* (Cambridge: Amer. Assoc. Variable Star Observers).

Catchpole, R. M., Robertson, B. S. C., Lloyd Evans, T. H. H., Feast, M. W., Glass, I. S., and Carter, B. S. 1979, S. African Astron. Obs. Circulars 1, 61.

Catchpole, R. M., Robertson, B. S. C., and Warren, P. R. 1977, Mon. Not. Roy. Astron. Soc. 181, 391.

Cottrell, P. L. 1979, Mon. Not. Roy. Astron. Soc. 189, 13.
Cottrell, P. L., and Lambert, D. L. 1982, Astrophys. J. 261, 595.
Davis, D. N. 1934, Publ. Astron. Soc. Pacific 46, 267.
Dominy, J. F., and Lambert, D. L. 1983, Astrophys. J. 270, 180.
Dzērvītis, U. 1981, Investigations of the Sun and Red Stars, No. 13, p. 22.
Eaton, J. A., Johnson, H. R., O'Brien, G. T., and Baumert, J. H. 1984, Astrophys. J. (in press).
Eggen, O. J. 1972a, Astrophys. J. 174, 45.
Eggen, O. J. 1972b, Astrophys. J. 177, 489.
Eggen, O. J. 1972c, Mon. Not. Roy. Astron. Soc. 159, 403.
Eggen, O. J. 1975, Publ. Astron. Soc. Pacific 87, 111.
Faÿ, T. D. 1974, Astrophys. J. 190, 597.
Faÿ, T., and Honeycutt, R. K. 1972, Astron. J. 77, 29.
Faÿ, T., Honeycutt, R. K., and Warren, W. H., Jr. 1973, in *Conference on Red Giant Stars*, ed. H. R. Johnson, J. P. Mutschlecner, and B. F. Peery, Jr. (Bloomington: Indiana Univ.), p. 130.
Faÿ, T. D., and Ridgway, S. T. 1976, Astrophys. J. 203, 600.
Faÿ, T. D., Warren, W. H. Jr., Johnson, H. R., and Honeycutt, R. K. 1974, Astron. J. 79, 634.
Feast, M. W., Catchpole, R. M., and Glass, I. S. 1976, Mon. Not. Roy. Astron. Soc. 174, 81P.
Feast, M. W., Robertson, B. S. C., Catchpole, R. M., Lloyd Evans, T., Glass, I. S., and Carter, B. S. 1982, Mon. Not. Roy. Astron. Soc. 201, 439.
Fix, J. D. 1976, Astrophys. J. 203, 463.
Fix, J. D., and Neff, J. S. 1975, Mon. Not. Roy. Astron. Soc. 173, 83P.
Forrest, W. J., Gillett, F. C., and Stein, W. A. 1975, Astrophys. J. 195, 423.
Frogel, J. A., and Blanco, V. M. 1984, in IAU Symp. No. 105, *Observational Tests of the Stellar Evolution Theory*, ed. A. Maeder and A. Renzini (Reidel), p. 175.
Frogel, J. A., Cohen, J. G., Persson, S. E., and Elias, J. H. 1981, in *Physical Processes in Red Giants*, ed. I. Iben, Jr., and A. Renzini (Reidel), p. 159.
Frogel, J. A., Dickinson, D. F., and Hyland, A. R. 1975, Astrophys. J. 201, 392.
Frogel, J. A., and Hyland, A. R. 1972, Mém. Soc. Roy. Sci. Liège, 6th Ser., 3, 111.
Fujita, Y. 1970, *Interpretation of Spectra and Atmospheric Structure in Cool Stars* (Tokyo: Univ. of Tokyo Press).
Gauzit, M. J. 1959, Ann. d'Astrophys. 22, 781.
Gehrz, R. D., and Woolf, N. J. 1971, Astrophys. J. 165, 285.
Gillett, F. C., Merrill, K. M., and Stein, W. A. 1971, Astrophys. J. 164, 83.
Goebel, J. H., Bregman, J. D., Cooper, D. M., Goorvitch, D., Langhoff, S. R., and Witteborn, F. C. 1983, Astrophys. J. 270, 190.
Goebel, J. H., Bregman, J. D., Goorvitch, D., Strecker, D. W., Puetter, R. C., Russell, R. W., Soifer, B. T., Willner, S. P., Forrest, W. J., Houck, J. R., and McCarthy, J. F. 1980, Astrophys. J. 235, 104.

Goebel, J. H., Bregman, J. D., Strecker, D. W., Witteborn, F. C., and Erickson, E. F. 1978, Astrophys. J. (Letters) 222, L129.
Goebel, J. H., Bregman, J. D., Witteborn, F. C., Taylor, B. J., and Willner, S. P. 1981, Astrophys. J. 246, 455.
Golay, M. 1974, *Introduction to Astronomical Photometry* (Reidel).
Gow, C. E. 1976, Astron. J. 81, 993.
Gow, C. E. 1977, Publ. Astron. Soc. Pacific 89, 510.
Greene, A. E., and Wing, R. F. 1975, Astrophys. J. 200, 688.
Griffin, R. F., and Redman, R. O. 1960, Mon. Not. Roy. Astron. Soc. 120, 287.
Hackwell, J. A. 1972, Astron. Astrophys. 21, 239.
Hayes, D. S. 1984, in IAU Symp. No. 111, *Calibration of Fundamental Stellar Quantities* (Reidel), in press.
Hayes, D. S., Joyce, R. R., Ridgway, S. T., Rinsland, C. P., and Wing, R. F. 1980, Bull. Amer. Astron. Soc. 12, 837.
Johnson, H. L. 1966, Ann. Rev. Astron. Astrophys. 4, 193.
Johnson, H. R., and Ake, T. B. 1984, in *Cool Stars, Stellar Systems, and the Sun*, ed. S. L. Baliunas and L. Hartmann (Springer-Verlag), p. 362.
Johnson, H. R., Ameen, M. M., and Eaton, J. A. 1984, Astrophys. J. (in press).
Johnson, H. R., and O'Brien, G. T. 1983, Astrophys. J. 265, 952.
Keenan, P. C. 1954, Astrophys. J. 120, 484.
Keenan, P. C., and Morgan, W. W. 1941, Astrophys. J. 94, 501.
Little, S. J., Little-Marenin, I. R., and Wing, R. F. 1980, Bull. Amer. Astron. Soc. 12, 828.
Little-Marenin, I. R., Little, S. J., and Wing, R. F. 1979, Bull. Amer. Astron. Soc. 11, 646.
Lloyd Evans, T. 1983a, Mon. Not. Roy. Astron. Soc. 204, 975.
Lloyd Evans, T. 1983b, Mon. Not. Roy. Astron. Soc. 204, 985.
Lloyd Evans, T. 1984, Mon. Not. Roy. Astron. Soc. 209, 825.
Lockwood, G. W., and Wing, R. F. 1971, Astrophys. J. 169, 63.
Lockwood, G. W., and Wing, R. F. 1982, Mon. Not. Roy. Astron. Soc. 198, 385.
Lü, P. K., Dawson, D. W., Upgren, A. R., and Weis, E. W. 1983, Astrophys. J. Suppl. 52, 169.
Lü, P. K., and Sawyer, D. 1979, Astrophys. J. 231, 144.
Lü, P. K., and Upgren, A. R. 1979, Astron. J. 84, 101.
Luck, R. E., and Lambert, D. L. 1981, Astrophys. J. 245, 1018.
MacConnell, D. J., Frye, R. L., and Upgren, A. R. 1972, Astron. J. 77, 384.
Mannery, E. J., and Wallerstein, G. 1970, Astron. J. 75, 169.
McClure, R. D. 1979, in *Problems of Calibration of Multicolor Photometric Systems*, ed. A. G. D. Philip, Dudley Obs. Report No. 14, p. 83.
McClure, R. D. 1983, Astrophys. J. 268, 264.
McClure, R. D. 1984a, Publ. Astron. Soc. Pacific 96, 117.
McClure, R. D. 1984b, Astrophys. J. (Letters) 280, L31.
McClure, R. D. 1984c, in IAU Symp. No. 105, *Observational Tests of the Stellar Evolution Theory*, ed. A. Maeder and A. Renzini (Reidel), p. 187.

McClure, R. D., Fletcher, J. M., and Nemec, J. M. 1980, Astrophys. J. (Letters) 238, L35.
McClure, R. D., and Forrester, W. T. 1981, Publ. Dominion Astrophys. Obs. 15, No. 14.
McClure, R. D., Forrester, W. T., and Gibson, J. 1974, Astrophys. J. 189, 409.
Mendoza V., E. E. 1967, Bol. Tonantzintla y Tacubaya 4, 114.
Mendoza V., E. E., and Johnson, H. L. 1965, Astrophys. J. 141, 161.
Merrill, K. M., and Ridgway, S. T. 1979, Ann. Rev. Astron. Astrophys. 17, 9.
Merrill, K. M., and Stein, W. A. 1976a, Publ. Astron. Soc. Pacific 88, 285.
Merrill, K. M., and Stein, W. A. 1976b, Publ. Astron. Soc. Pacific 88, 294.
Merrill, K. M., and Stein, W. A. 1976c, Publ. Astron. Soc. Pacific 88, 874.
Neugebauer, G., and Leighton, R. B. 1969, *Two-Micron Sky Survey - A Preliminary Catalog* (NASA SP-3047).
Palmer, L. G. 1980, Ph.D. dissertation, Ohio State University.
Palmer, L. G., and Wing, R. F. 1982, Astron. J. 87, 1739.
Piccirillo, J. 1976, Publ. Astron. Soc. Pacific 88, 680.
Piccirillo, J. 1977, Ph.D. dissertation, Indiana University.
Piccirillo, J. 1980, Mon. Not. Roy. Astron. Soc. 190, 441.
Preston, G. W., Krzeminski, W., Smak, J., and Williams, J. A. 1963, Astrophys. J. 137, 401.
Price, S. D., and Walker, R. G. 1976, *The AFGL Four Color Infrared Sky Survey: Catalog of Observations at 4.2, 11.0, 19.8, and 27.4 μm*, AFGL-TR-76-0208.
Querci, F., Querci, M., Wing, R. F., Cassatella, A., and Heck, A. 1982, Astron. Astrophys. 111, 120.
Richer, H. B. 1981, in *Physical Processes in Red Giants*, ed. I. Iben, Jr., and A. Renzini (Reidel), p. 153.
Richer, H. B., Crabtree, D. R., and Pritchet, C. J. 1984, Astrophys. J. (in press).
Ridgway, S. T., Jacoby, G. H., Joyce, R. R., and Wells, D. C. 1980, Astron. J. 85, 1496.
Rinsland, C. P., and Wing, R. F., 1982, Astrophys. J. 262, 201.
Rosino, L. 1951, Astrophys. J. 113, 60.
Sanford, R. F. 1928, Astrophys. J. 68, 408.
Scargle, J. D., and Strecker, D. W. 1979, Astrophys. J. 228, 838.
Smak, J. 1968, Acta Astron. 18, 317.
Smak, J., and Wing, R. F. 1979, Acta Astron. 29, 187.
Spinrad, H., and Newburn, R. L.,Jr. 1965, Astrophys. J. 141, 965.
Spinrad, H., Pyper, D. M., Newburn, R. L. Jr., and Younkin, R. L. 1966, Astrophys. J. 143, 291.
Straižys, V. 1978, Bull. Vilnius Astron. Obs., No. 47, p. 7.
Straižys, V. 1983, *Metal-Deficient Stars* (Vilnius: Institute of Physics of the Lithuanian S.S.R.).
Tsuji, T. 1981, J. Astrophys. Astron. 2, 95.
Upgren, A. R., Lü, P. K., and MacConnell, D. J. 1977, Publ. Astron. Soc. Pacific 89, 552.

Vardya, M. S. 1966, Observatory 86, 162.
Vetešnik, M. 1984a, Bull. Astron. Inst. Czech. 35, 65.
Vetešnik, M. 1984b, Bull. Astron. Inst. Czech. 35, 74.
Walker, A. R. 1976, Mon. Not. Roy. Astron. Soc. 174, 609.
Wallerstein, G. 1968, Astrophys. J. 151, 1011.
Wallerstein, G. 1973, Ann. Rev. Astron. Astrophys. 11, 115.
Wallerstein, G., Brown, J. A., and Bates, B. A. 1979, Publ. Astron. Soc. Pacific 91, 47.
Westerlund, B. E. 1964, in IAU Symp. No. 20, *The Galaxy and the Magellanic Clouds*, ed. F. J. Kerr and A. W. Rodgers (Canberra: Australian Academy of Sciences), p. 239.
Westerlund, B. E., Olander, N., Richer, H. B., and Crabtree, D. R. 1978, Astron. Astrophys. Suppl. 31, 61.
Wildey, R. L., Burbidge, E. M., Sandage, A. R., and Burbidge, G. R. 1962, Astrophys. J. 135, 94.
Williams, P. M. 1975, Mon. Not. Roy. Astron. Soc. 170, 343.
Wing, R. F. 1967, Ph.D. dissertation, Univ. of California, Berkeley.
Wing, R. F. 1971, in *Proc. Conf. on Late-Type Stars*, ed. G. W. Lockwood and H. M. Dyck (Kitt Peak Nat'l Obs. Contrib. No. 554), p. 145.
Wing, R. F. 1972, Mém. Soc. Roy. Sci. Liège, 6th Ser., 3, 123.
Wing, R. F. 1974, Highlights of Astronomy 3, 285.
Wing, R. F. 1979, in IAU Colloq. No. 47, *Spectral Classification of the Future*, ed. M. F. McCarthy, A. G. D. Philip, and G. V. Coyne (Vatican Obs.), p. 347.
Wing, R. F. 1980, in *Current Problems in Stellar Pulsation Instabilities*, ed. D. Fischel, J. Rountree Lesh, and W. M. Sparks (NASA: TM-80625), p. 533.
Wing, R. F. 1982, Trans. I.A.U. 18A, 275.
Wing, R. F., Carpenter, K. G., and Wahlgren, G. M. 1983, *Atlas of High Resolution IUE Spectra of Late-Type Stars, 2500 - 3230 Å*, Perkins Obs. Special Publ. No. 1.
Wing, R. F., Gustafsson, B., and Eriksson, K. 1984, in IAU Symp. No. 111, *Calibration of Fundamental Stellar Quantities* (Reidel), in press.
Wing, R. F., and Lange, G. L. 1981, Bull. Amer. Astron. Soc. 13, 546.
Wing, R. F., and Rinsland, C. P. 1981a, Bull. Amer. Astron. Soc. 13, 840.
Wing, R. F., and Rinsland, C. P. 1981b, Revista Méxicana Astron. Astrof. 6, 145.
Wing, R. F., and Stock, J. 1973, Astrophys. J. 186, 979.
Wing, R. F., and Yorka, S. B. 1977, Mon. Not. Roy. Astron. Soc. 178, 383.
Wood, P. R., Bessell, M. S., and Fox, M. W. 1983, Astrophys. J. 272, 99.
Yorka, S. B. 1983, Astron. J. 88, 1816.

DISCUSSION

McCARTHY - How wide a passband do your filters have for discerning different types of cool stars ? Is there overlap ? What would be the wavelengths of maximum transmission for these filters ?

WING - The survey filters used by Palmer and Wing (Astron. J., 1982) are 100 A wide, centered at λ 7810 (Ti 0) and λ 8120 (CN). If you are merely interested in detecting late M stars and cool C stars, the filters could probably be as wide as ∿ 200 A. In that case they would overlap somewhat.

McCLURE - If the Bond - Neff depression is partially due to atomic lines, does this mean that the Ba stars are more metal rich than normal K giants ? Why would CN strong stars not show the same effect ?

WING - I suppose it means that some Ba stars are more metal rich than some of the giants that have been used for comparison. Not all Ba stars show a Bond - Neff depression. For example the energy distribution of HR 774 is compared to K2 giant in Straizys's book on "Metal deficient stars" and there is no sign of a depression. As for the CN strong stars, I agree that they should show the same effect if they are really metal rich. I am not sure that they do not.

BIDELMAN - Can not the Bond - Neff depression be at least partially due to CH ? I thought there was appreciable absorption in the λ 4000 region due to that.

ABSOLUTE SPECTROPHOTOMETRY OF F-, G-, K-, AND M-TYPE STARS

R. Kiehling, J. Dachs
Astronomisches Institut
Ruhr-Universität Bochum
Postfach 102148
D-4630 Bochum 1
West Germany

ABSTRACT. Absolute spectral energy distributions were measured photoelectrically from 3600 Å to 8600 Å with a resolution of 10 Å for 62 bright southern and equatorial stars of intermediate and late spectral types of all luminosity classes in order to provide flux calibration and a grid of secondary spectrophotometric standards for the MK system. Most of the stars are MK standards, several are slightly peculiar (CN, Ba). The results are used to compare measured with theoretical energy distributions. Quantitative classification criteria are derived from measured line and band strengths. At the resolution of our measurements, mildly peculiar stars cannot be easily distinguished from normal stars.

1. OBSERVATIONS

Program stars were selected from the lists of MK standards defined by Morgan et al. (1978) and by Keenan and Pitts (1980). Energy distributions were obtained using the spectrum scanner at the 61-cm telescope of the University of Bochum at the European Southern Observatory, La Silla (Chile). They cover a wavelength range from 3600 (or 3300) Å up to 8600 Å with a resolution of 10 Å in equidistant steps at well defined channel positions. Scans are calibrated using spectrophotometric standard stars of Tüg (1980) measured in the same nights as the program stars. After correction for atmospheric extinction and calibration, energy distributions measured in two or three different nights were averaged for each star. The standard deviation of fluxes obtained in different nights is $0\overset{m}{.}02$ to $0\overset{m}{.}03$ ($\hat{=}$ 2-3 %) in the spectral range from 4000 Å to 8600 Å and rises up to $0\overset{m}{.}05$ to $0\overset{m}{.}07$ for shorter wavelengths because of the smaller fluxes from the stars and lower sensitivity of the scanner in that part of the spectrum. The error may also be larger in strong lines. The precision of wavelength scale is about ±0.5 Å.

M. Jaschek and P. C. Keenan (eds.), Cool Stars with Excesses of Heavy Elements, 87–91.

2. RESULTS

Examples for the results are given in the Figures. Figure 1 shows a sequence of spectra for giants of different temperatures, while in Figure 2 spectra are plotted for K4-K5 stars of different luminosity classes. The strongest spectral features are indicated. All fine details displayed by the energy distributions are real and have been identified.

Figure 3 shows a comparison of the spectral energy distribution for α Boo (K1-IIIb CN-1) with theoretical fluxes taken from the work of Bell and Gustafsson (1978). By interpolation within their grid the model for T_e=4270±80K and log g=2.0±0.3 was found to give the best fit. This result is in good agreement with determinations by other authors, e.g. Martin (1977) who obtained T_e=4300±90K and log g=1.74±0.2.

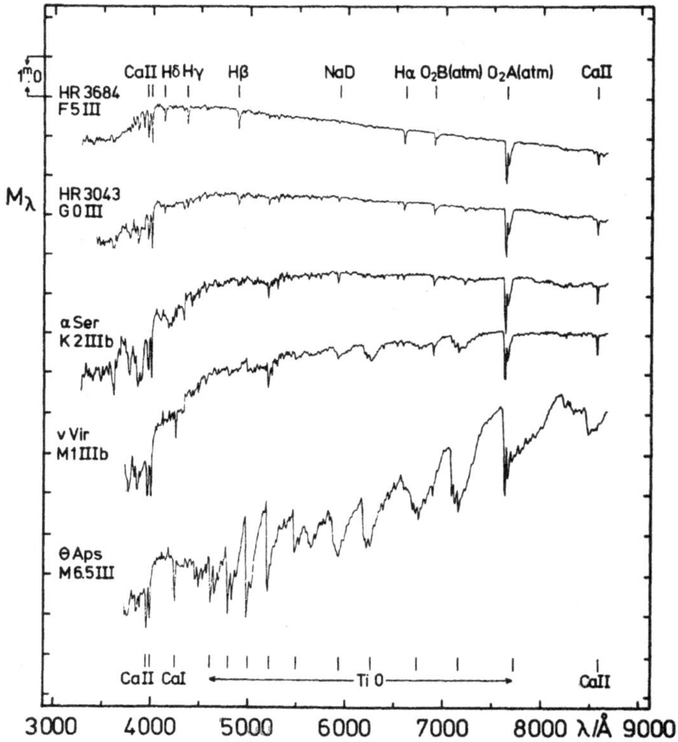

Figure 1. Sequence of spectral distributions of giant stars.

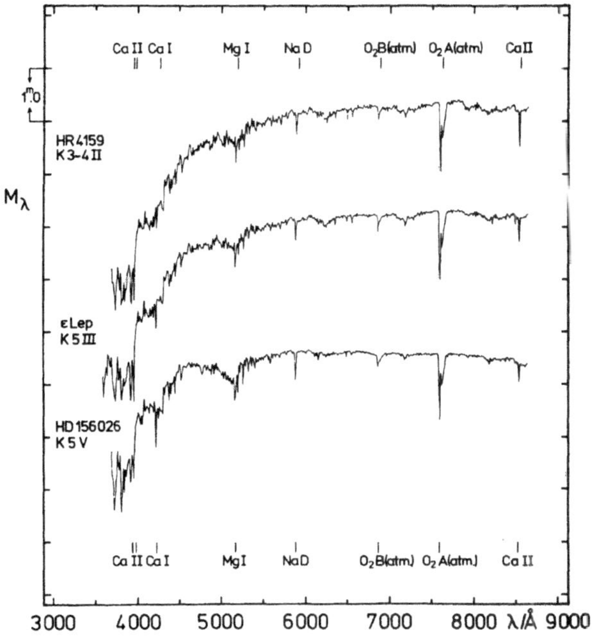

Figure 2. Spectral energy distributions of K4-K5 stars of different luminosity classes.

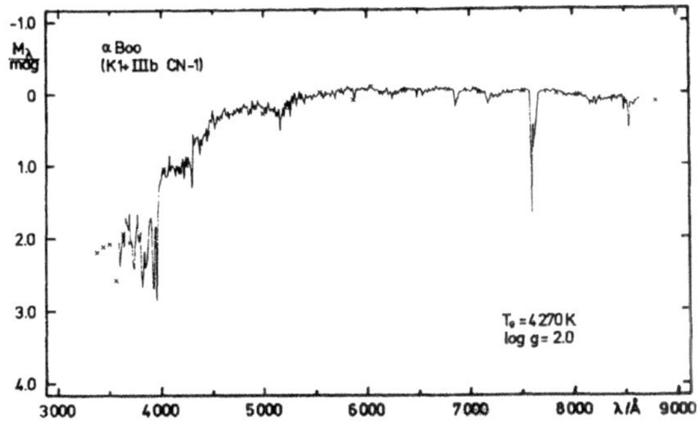

Figure 3. Comparison of the spectral energy distribution of α Boo (solid line) with an interpolated model (marked by crosses) from Bell and Gustafsson (1978).

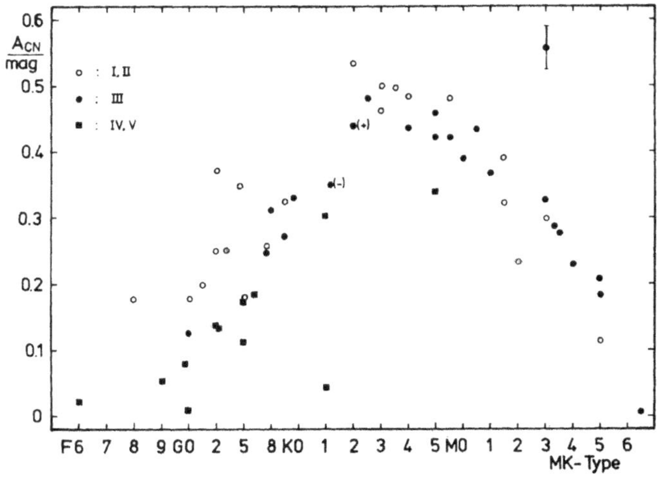

Figure 4. Absorption strength of CN λ 4216 Å versus spectral type. No dependence on luminosity class is found. Stars mildly overabundant or underabundant in CN like α Ser (K2IIIb CN1 Fe4143-1, marked by "+") and α Boo (K1+IIIb CN-1, marked by "-") cannot be distinguished from the other stars at the resolution of our spectrophotometry.

In addition, absorption strengths were measured for the strongest features. Typical results are shown in Figure 4. Mild overabundances or underabundances of CN or Ba appear to escape detection at the resolution of 10 Å used in our spectrophotometry.

Support from the Deutsche Forschungsgemeinschaft for this work is gratefully acknowledged.

REFERENCES

Bell, R.A. and Gustafsson, B. 1978, Astron.Astrophys.Suppl. 34, 229.
Keenan, P.C. and Pitts, R.E. 1980, Ap.J.Suppl. 42, 541.
Martin, P. 1977, Astron.Astrophys. 61, 591.
Morgan, W.W., Abt, H.A., Tapscott, J.W. 1978, "Revised MK Spectral Atlas for Stars Earlier than the Sun", Yerkes and Kitt Peak Observatories.
Tüg, H. 1980, Astron.Astrophys.Suppl. 39, 67.

DISCUSSION

JASCHEK: I propose that you include some typical peculiar stars in your program; this would be very important for calibration purposes for further photometric and spectrophotometric work.

KIEHLING: Our first intention was to establish a grid of spectrophotometric measurements of normal stars. Indeed, for the future we are planning to include many more peculiar stars in our program.

CALIBRATIONS OF PHOTOMETRIC INDICES ON NEW TEMPERATURE SCALES AND PROBLEM OF SPECTRAL CLASSIFICATION IN M AND C STARS

Takashi Tsuji
Tokyo Astronomical Observatory
University of Tokyo
Mitaka, Tokyo, 181 Japan

ABSTRACT. C-classification of carbon stars shows poor correlation with the effective temperature based on infrared photometry. This is due to a peculiarity of the atmospheric structure of carbon stars: Because of the increased line-blanketing(backwarming) effect due to increased molecular absorption in cooler stars, kinetic temperature of the line formation layers shows little decrease for large decrease of the effective temperature(e.g., ΔT_{ex}(FeI)=250K for ΔT_{eff}=3600-2600=1000K!). For this reason, usual spectral classification may be very difficult for carbon stars, and C-classification too cannot be free from this fundamental difficulty.

1. Most of our knowledges on stars, especially on faint stars, depend on photometry and spectral classification. Therefore, calibrations of photometric indices and spectral types by means of the fundamental stellar parameters are of basic importance in our studies of stars and of stellar systems. These two basic tools, photometry and spectral classification, are complementary each other in that they reflect the properties of stellar photospheres in somewhat different ways. For this reason, it is important to understand photometric and spectroscopic characteristics of stars on the basis of the unified picture on stellar atmosphere before we could be confident of proposed calibrations. We examine this problem for the case of red giant stars(M and C stars).

2. Recently, effective temperatures for large sample of red giant stars have been determined(Tsuji 1981a,b) on the basis of the infrared flux method(IRFM) of Blackwell et al(1980). The resulting effective temperatures show good agreement with those based on measured angular diameters (Ridgway et al. 1980). Then, calibrations of photometric indices in terms of effective temperatures are straightforward. Some examples are shown in Figs.1a-b. In these new calibrations, all the colour indices show positive and tight correlations with effective temperatures(Tsuji 1981c). On the contrary, correlations of the spectral types with temperatures are not so simple. For K-M giant stars, recent calibration based on the infrared flux method(curve D in Fig.1c) and that based on angular diameters(curve C) show agreement within ± 100K. Thus upward revision of the

M. Jaschek and P. C. Keenan (eds.), Cool Stars with Excesses of Heavy Elements, 93–97.

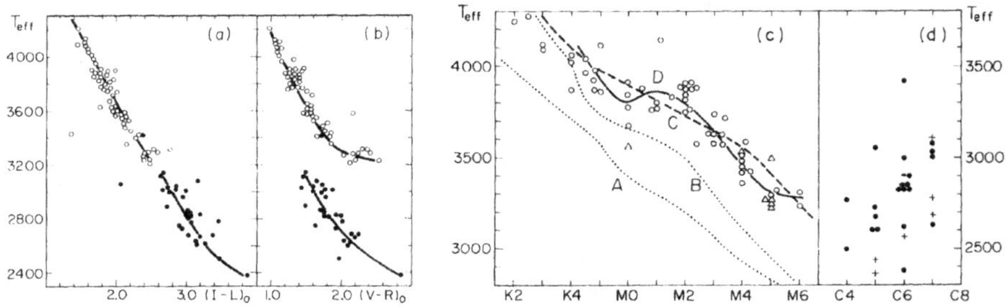

Figure 1. (a)(b) Photometric colour-T_{eff} calibrations for M giants(open circles) and C stars(filled circles). (c) Spectral type-T_{eff} calibrations for M giants: A(Kuiper 1938), B(Johnson 1966), C(Ridgway et al. 1980), and D(Tsuji 1981). T_{eff} by IRFM for M stars of luminisity classes II(triangle) and III(circle) are shown. (d) T_{eff} for carbon stars based on IRFM(filled circle) and on angular diameters(plus).

---

stellar temperature scale against previous calibrations(curves A and B) is well confirmed. However, spectral types and effective temperatures show poor correlation for some individual stars(filled circles in Fig. 1c), and something is still not well understood(see also; Keenan 1982).

The situation for carbon stars is more serious; C-classification proposed by Keenan & Morgan(1941) shows poor(or almost negative) correlation with effective temperatures based on the infrared flux method(filled circles in Fig.1d) as well as on angular diameters(plus signs in Fig.1d) which include the latest one by Ridgway et al(1981). Further, almost all the observations are difficult to understand on the basis of C-classification. For example: (1) Infrared colours show poor correlation with C-subtype(e.g., Richer 1971,Noguchi et al. 1981). (2) The 3 μm absorption shows no correlation with C-subtypes(e.g.,Noguchi et al.1981) while this should be highly sensitive to temperature if it originates from polyatomic molecules. In fact, the 3 μm absorption shows expected dependence on the effective temperature(Tsuji 1981c, see also Fig.3a). (3) Major spectroscopic features such as C_2 band, SiC_2 bands, and UV absorption show large scatters if plotted against C-subtypes(Yamashita 1972). Some of these scatters were attributed to abundance variations(but not yet confirmed), and second parameter was introduced in C-classification(Keenan & Morgan 1941). It is to be noted, however, that such large scatters do not appear if the above noted spectroscopic features are plotted against effective temperatures(Tsuji 1981c). For all these reasons, physical basis of C-classification should once be questioned.

3. A new series of model atmospheres of carbon stars are constructed with polyatomic molecular opacities included(Tsuji 1984). We assumed log g=0.0 ,v_{tur}=3.0Km/s, and chemical composition of C/O=2 and N/O=3. The effect of spherical extension is known to be not very important in C stars(Scholz & Tsuji 1984) and not considered here for simplicity. In resulting thermal structures shown in Fig. 2a, open and filled circles indicate the layers of $\tau_{0.8}$= 0.01 and 1, respectively($\tau_{0.8}$is the

optical depth in continuum at 0.8 μm). Note that temperatures at line formation layers for cooler stars remain relatively higher because of the increased backwarming effect due to increased molecular absorption.

In Fig.3a, predicted 3 μm-index=$-\mathrm{In}(F_{3.1\mu m}/F_{cont})$ based on emergent fluxes(Fig.2a) shows reasonable agreement with observation(Noguchi et al. 1981). In Fig.3b, predicted line intensities based on the weighting function method(Cayrel & Jugaku 1963) are shown for selected spectral lines. Excitation temperatures determined from the line intensities are shown in Fig.3c. It is to be noted that the excitation temperature is much higher than the effective temperature in carbon stars and, moreover, the excitation temperature decreases only slightly for large decrease of the effective temperature; these results can readily be explained by the characteristics of the thermal structure in carbon stars noted above. Because of this slow decrease of the kinetic temperature in line formation layers, increase of Na D line intensity is only modest for large

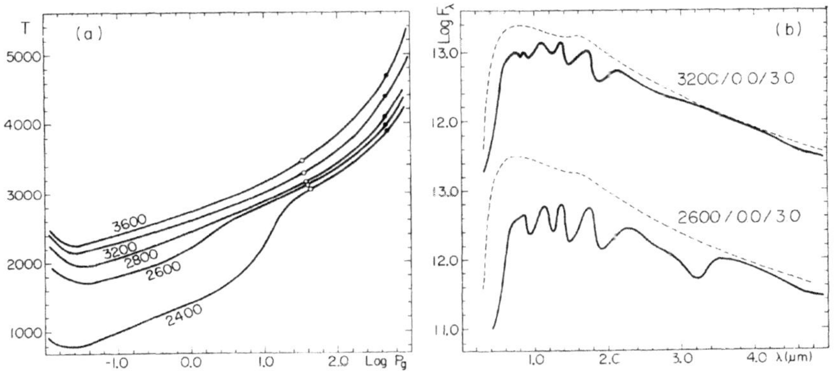

Figure 2. (a) Model atmospheres of carbon stars with T_{eff} indicated. (b) Line-blanketed(solid lines) and line-blocking free (dashed lines) fluxes predicted from models of $T_{eff}/\log g/v_{tur}$ indicated.

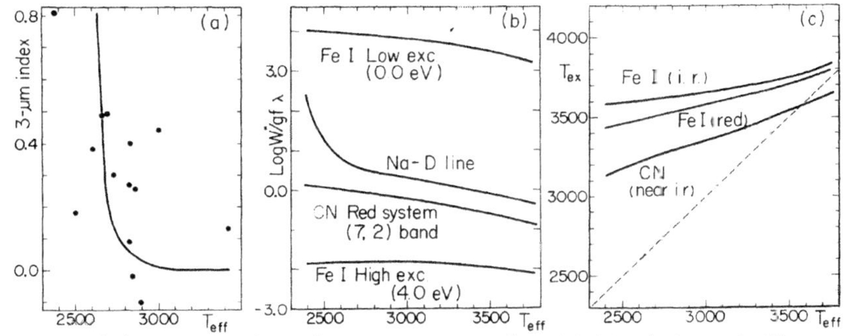

Figure 3. (a)Observed (dots) and predicted(solid line) 3 μm-indices. (b)Logarithm of the predicted line intensity. (c)The relation between excitation temperature and effective tempeature. All for C stars.

decrease of effective temperatures(T_{eff}=2600-3800K). Furthermore, (7,2) band of CN red system, which acts as a pseudo-continuum to Na D lines, also increases in parallel with D lines(Fig.3b), thus compensating for the slight increase of D line intensity with decreasing temperature.

A simple analysis outlined above explains why spectral classification of cool carbon stars is so difficult. For example, if carbon stars in the effective temperature range between 2600 and 3600K, which corresponds to the excitation temperature difference of 250K, are to be classified into 5 spectral subtypes, we must detect 50K difference in excitation temperatures on low dispersion spectrograms. Moreover, Na D line which is used as the major temperature indicator in C-classification should show little dependence on temperature by the reasons just noted. If D lines show some variations among carbon stars, it may reflect some parameters other than temperature, and this explains why most of the temperature-sensitive observed quantities plotted against C-subtypes show so large scatters as noted before. Thus, it seems to be difficult to justify C-classification as a temperature classification.

4. In general, the effect of abundance anomaly in a star, such as excess of heavy element, on spectral classification is to enhance the relevant spectral lines, from which we recognize the peculiarity of the star. The excess of carbon over oxygen in cool stars, however, produces more serious effect upon the spectral classification, and it offers an interesting example in which heavy line-blanketing effect by strong molecular absorption changes the atmospheric structure in such a way that even the most fundamental spectral features such as used for spectral classification are seriously disturbed. For this reason, some considerations on atmospheric structure should be necessary before we could establish the physical basis of the spectral classification for carbon stars.

REFERENCES

Blackwell,D.E.,Petford,A.D.,Shallis,M.J.:1980, Astron. Astrophys. 82,249
Cayrel,R.,Jugaku,J.:1963, Ann. Astrophys. 26, 495
Johnson,H.L.:1966, Ann. Rev. Astron. Astrophys. 4, 193
Kuiper,G.P.:1938, Astrophys. J. 88, 429
Keenan,P.C.:1982, Publ. Astron. Soc. Pacific 94, 299
Keenan,P.C.,Morgan,W.W,:1941, Astrophys. J. 94, 501
Noguchi,K.,Kawara,K.,Kobayashi,Y.,Okuda,H.,Sato,S,Oishi,M.:1981, Publ. Astron. Soc. Japan 33, 373
Richer,H.B.:1971, Astrophys. J. 167, 521
Ridgway,S.T.,Jacoby,G.H.,Joyce,P.R.,Wells,D.C.:1981, Physical Processes in Red Giants ed. I.Iben,Jr. & A.Renzini, p.47
Ridgway,S.T.,Joyce,R.R.,White,N.M.,Wing,R.F.:1980, Astrophys. J. 235,126
Scholz,M.,Tsuji,T.:1984, Astron. Astrophys. 130, 11
Tsuji,T.:1981a, Astron. Astrophys, 99, 48
Tsuji,T.:1981b, J.Astrophys. Astron. 2, 95
Tsuji,T.:1981c, J.Astrophys. Astron. 2, 253
Tsuji,T.:1984, Astron. Astrophys. 134, 24
Yamashita,Y.:1972, Ann. Tokyo Astron. Obs. 2nd Ser. 13, 169

DISCUSSION

WOOD - Have you compared T_e values derived from the infrared flux method with T_e values derived directly (for both C and M stars) from lunar occultation angular diameters ?

TSUJI - For K - M giant stars, the T_e scale based on the infrared flux method and that based on angular diameters (Ridgway et al, 1980) show good agreement (see Fig. 1c). For carbon stars, T_e values based on IRFM and those based on angular diameters (6 stars at present) also show reasonable agreement, and these two sets of T_e values anyhow show positive correlation (Fig. 1d).

TSUJI (additional remark) - This morning, Dr Jaschek presented us an interesting diagram showing a correspondence between the temperature scale or K - M giants and that of C stars based on Dr Keenan's earlier work. That diagram shows that the excitation temperature of cool carbon stars correspond to that of middle M - giants. Although Dr Keenan is too modest in saying that this still needs confirmation, it is interesting to notice that this is just what I have shown in Fig. 3c ; $T_{ex} \sim 3500°$ K is predicted for cool carbon stars and close to that of middle M - giants for which $T_{ex} \sim 3500°$ K. Then, our prediction on excitation temperature of cool carbon stars may already have an observational confirmation, and we hope it will be confirmed. Also, if I remember correctly, similar correspondence between T_{ex} (M stars) and T_{ex} (C stars) was down by Fujita et al (Publ. Dom. Astrophys. Obs., 1965) based on higher dispersion spectroscopic analysis.

A NEW PHOTOMETRIC SYSTEM FOR MONITORING CARBON STAR VARIABILITY

F.and M. Querci
Observatoire du Pic du Midi et de Toulouse
14 Avenue Edouard Belin
31400 Toulouse, France

1. INTRODUCTION

Usually, photoelectric photometry gives data regarding such quantities as spectral type, gravity, absolute magnitude, chemical composition parameters, CN and CH anomalies, metal deficiencies and interstellar reddening. It also turns out to be a very powerful tool for monitoring the variability of the cool stars. The identification of the observed spectra and the computation of the partial pressure versus optical depth via model atmospheres demonstrate that the variable flux obtained with the wide-band and intermediate-band photometries are due to the superposing of spectral features coming from different layers of the star atmosphere. Consequently, the inordinately wide filters cannot be employed to follow the physical parameter variations of different atmospheric layers, and a new photometric system for the monitoring of carbon stars (miras, SRabc, Lbc, RCB, HdC, CH, strong-CN,...) is proposed. The flux variations recorded by well-suited filters are the ground-based material for hydrodynamical investigations of these stars.

2. WIDE-BAND AND INTERMEDIATE-BAND PHOTOMETRIES

Among the wide-band photometries, the most used are the well-known UBV system (e.g. Mendoza and Johnson, 1965; Catalogue of Nicolet, 1978), and its extention into the infrared region (e.g. Bergeat et al.,1976,1980). Table 1 gives the molecular absorptions which contribute to the visual and near-infrared filters of these systems.

Table 1

Filters	Molecules	Filters	Molecules
U	CN-violet+C_3+SiC_2 +ε C_2 Swan	B	CN-violet+C_2 Swan +ε C_3
V	C_2 Swan+CN-violet +ε SiC_2 (Merrill-Sanford)	R	CN-red+ε C_2 Swan
I	CN-red+C_2 Phillips	J	CN-red+C_2 Ballik-Ramsay+C_2 Phillips+CO-third overt.
H	CN-red+CO-sec.overt. +ε C_2 Ballik-Ramsay	K	CN-red+CO-first overt. +ε HCN+C_2H_2

M. Jaschek and P. C. Keenan (eds.), Cool Stars with Excesses of Heavy Elements, 99–104.

The intermediate-band photometries (with filter bandpasses in the range 200Å-500Å) are particularly developed for investigations on the intermediate-temperature stars (A to G types), and their filters are adapted to the spectral features and continuum region of these stars. The Vilnius system (Straizys and Zdanavicius, 1970) was used for the analysis of color excess ratio for barium and CH stars by Bartkevicius and Sviderskiené (1981). The Geneva system (Rufener, 1976), the VILGEN system (Straizys et al., 1982) and the DDO system (McClure, 1977; Hesser et al., 1977) especially used for CH and CN anomalies have occasionally monitored a few carbon stars. Upon to request, the Centre de Données Stellaires de Strasbourg (CDS data bank) gives the whole bibliography about the required star and the Catalogues on which the star is noted; it is a powerful tool in star variability studies.

It is not the aim of this paper to review all the works done in cool star research with these photometric systems. However, regarding the scope of variability, we point out the work of Payne-Gaposchkin (1975) on M stars, which demonstrates that the indices of the M miras are variable during their cycle and different from the indices of the non-variable stars of same spectral type and luminosity class. As for the carbon stars, the results obtained with the wide and intermediate-band photometries are linked to their most striking observational behaviour (Querci, 1985).

3. NARROW-BAND PHOTOMETRIES

It is well accepted that the larger the filters, the greater is the penetrating power of a photoelectric system. So, the narrow-band photometries do not intend to prospect stars as far as possible into space.

3.1. Wing system and Lockwood and McMillan system

The first measurements of miras, SRa, SRb, SRc, Lb and Lc carbon stars with narrow-band photometry were made by Baumert (1971,1972) with the Wing's eight color system (Wing, 1971). Multiple observations for a few variables indicate a wide range in temperature (filters: λ=7806Å, $\Delta\lambda$=42Å; λ=10392Å, $\Delta\lambda$=55Å) and in CN-index (filters: λ=8122Å, $\Delta\lambda$=43Å; λ=10868Å, $\Delta\lambda$=75Å), exhibited by these stars. Another narrow-band system (Lockwood and McMillan, 1971) has been used for carbon stars, but the width of its filters is larger than in the Wing system. The use of these two systems in carbon star research only consists to measure the importance of the CN-absorption with respect to the continuum in the 0.7μm-1.1μm range.

3.2. A proposed photometric system for C stars

We would study the behaviour of a set of selected stars by a permanent monitoring in the largest spectral range. The width of the filters defined as described below, limits the monitoring of the faint stars (m $\geqslant$11). Having observationally analysed the spectrum of some carbon stars from spatial UV (Querci and Querci, 1983) ,through the IR region (Connes et al., 1968) to the millimetric-region (Querci et al.,

1979), as well as theoretically by model-atmosphere synthetic-spectra (Querci and Querci, 1975, 1976), we propose a set of filters (Table 2) adapted to the current receivers: photomultipliers, photodiodes, photovoltaic detectors and bolometers used in astronomical research.

Table 2

Involved feature	$\lambda_{Å}$	$\Delta\lambda_{Å}$	Star prototype	Remarks (depicted stellar region...)
			Spatial UV	
C IV	1550	80	HD 59643	Chromosphere or PN (fig.1)
MgII h & k	2800	80	TX Psc	Envelopes for miras? Chromospheres for SRa? (reference filter)
FeII V1 + VII V7	3280	80	TW Hor	strongly variable; large (month) and short (days) time-scales
			Near-UV and Visible	
C_3	4058	110	Y CVn	photospheric variable feature
CH	4287	102	CH stars	
H_γ	4343	22	miras	together with H_α, they give the
H_β	4866	20	miras	evolution of the shock-front
SiC_2	5005	54	UX Dra	to be connected with SiC particle emission at 11μm
C_2 Swan	5600	60	all stars	(reference filter)
			Visible and Near-infrared	
Na-D	5892	55	WZ Cas	Such as Li I, very large in some stars
H_α	6563	13	miras	in emission and absorption for miras and SRa
Li I	6708	13	UX Dra	Electron-donor, NLTE effects, hot or cool chromosphere
Red-CN	8126	53	all stars	(2,0) band and (3,1) bandhead
I104	10405	48	all stars	continuum, (reference filter) from Wing's system (1971)
			Infrared	
CN	2.13μm	0.28μm	all stars	anticorrelation CO-CN
CN+CO	2.40μm	0.30μm	"	from Bergeat et al.'s system(1976)
$HCN+C_2H_2$ $+\varepsilon$ C_3	3.10μm	0.7μm	"	very abundant in the outer layers of cool stars
continuum	3.80μm	0.7μm	"	pseudo-stellar continuum, perturb. by $HCN+C_2H_2$,(reference filter)
11.3 μm feature	11μm	2.6μm	"	SiC particles, outer layers
			Radio	
Continuum	1.5mm			Flare in V Cyg
CO	2.6mm			Appearence and disappearence of CO lines Temporal variations of CO lines
HCN	3.4mm			

The first tests with the near-UV, visible and near-IR filters were made by us and Gillet (1981) on the 61cm and 100cm ESO telescopes and on the 80cm and 100cm OHP telescopes. An accuracy of 0.01 to 0.02 mag. was obtained with the 80cm-telescope on stars of $m_v \approx 9$ to 10. These filters are manufactured by Infrared Industries Inc., Waltham, Mass., USA. (The transmission curve of each filter is available upon request to one of us). The standard stars which define the photometric system and the absolute calibration stars are from Hayes (1970). The observational technique and the data reduction procedure are from Rufener (1964).

Let us note that the spectrophotometry is also a good technique for investigating the color changes during the light variations. For example, besides many M stars, Maehara and Yamashita (1978) follow variations with time on the carbon star SS Vir.

4. PROPOSITION FOR A COOPERATIVE PHOTOMETRIC PROGRAM

As an evidence, the proposed filters are too numerous for one photometer! To cover the prospected spectral range, a cooperative program has to be developed. The monitoring of a selected set of carbon stars during a few months or years must be organized. The time-scale of the observations must be adapted to the suspected phenomena that we investigate: hours for some phenomena at a given phase, days or weeks for other phenomena in other stars, etc. The list of stars, of their suspected phenomena and time-scale are available upon request. The expected IAU Working Group "Chemically peculiar late-type stars" could act as a close link between the interested astronomers.

This cooperative photometric program should be all more efficient that the prospected range is the greatest and the observers the most numerous.

4.1. The Amateurs

The photometric photometry is striding forward among several Amateur Societies. At present, their telescopes reach diameters up to 40cm and even to 50cm; the used photometers like the Starlight with a photomultiplier or the SSP-3 with a photodiode are extremely powerful. A few amateurs use photovoltaic detectors. The international collaboration is growing: for example let us quote IAPPP, GEOS, AFOEV.

4.2. The ground-based and space middle-size telescopes

Many are the middle-size telescopes (60cm to 100cm) around the world. However, the selection comittees do not attribute the necessary observation time for investigations in cool star variability research because it is highly telescope-time consuming. The observations with an adapted set of filters or CVF wheels from ground-based and space telescopes (Maran et al., 1977) could be organised through the IAU Working Group.

4.3. The large facilities

Why this amount of efforts and collaborations? To prepare the observations with larger instruments from the ground (e.g. with Fourier transform spectroscopy etc.) and in space (KAO, ISO, ST, ...). The procedure of Target opportunity has to be developed when the small or middle-size telescopes predict an important event. It seems to be necessary to prepare large cooperative observational programs for chosen phases.

5. CONCLUSION

What is the ultimate aim of all these very expensive observations? To understand the hydrodynamical behaviour of these stars, which are in one of the most important place of the HR diagram.

The 1970's decade saw the development of static model-atmosphere calculations. However, numerous are the observations which demonstrate that the carbon stars have variability on many time scale and rapid evolution. The proposed observing program will lay the observational foundation for understanding this variability.

References

-Bartkevicius,A. and Sviderskiené,Z.: 1981,Bull. Vilnius Obs. 58, 63

-Baumert,J.H.: 1971, Proceedings of the conference on late-type stars, Ed. G.W.Lockwood and H.M.Dyck, Kitt Peak Nat.Obs.Contrib. N°554, 155

-Baumert,J.H.: 1972, Ph.D. Dissertation, Ohio State University

-Bergeat,J.,Sibille,F.,Lunel,M. and Lefevre,J.: 1976, A.& A. 52, 227

-Bergeat,J. and Lunel,M.: 1980, A. & A. 87, 139

-Connes,P.,Connes,J,Bouigue,R,Querci,M.,Chauville,J. and Querci,F.: 1968, Ann.d'Ap. 312, 483

-Gillet,D.: 1981, Thèse de 3ième cycle, Université de Paris VII

-Hayes,D.: 1970, Ap.J. 159, 165

-Hesser,J.E.,Hartvick,F.D.A. and McClure,R.: 1977, Ap.J.suppl.ser. 33, 471

-Lockwood,G.W. and McMillan,R.S.: 1971, Proceedings of the conference on late-type stars, Ed. G.W.Lockwood and H.M.Dyck, Kitt Peak Nat.Obs.contrib. N°554, 171

-Maehara,H. and Yamashita,Y.: 1978, Annals of Tokyo National Obs. 17, 2d serie, nb.2, 93

-Maran,S.P.,Heinsheimer,T.F.,Stecker,T.L.,Chapman,R.B.,Hobbs,R.W., Anand,S.P.S.,Michalitsianos,A.G.,Wright,F.M. and Kipp,S.L.: 1977, Infrared Physics 17, 565

-McClure,R.: 1977,Ap.J. 216, L101

-Mendoza,V.E.E. and Johnson,H.L.: 1965, Ap.J. 141, 161

-Nicolet,B.: 1978, A. & A.suppl.ser. 34, 1

-Payne-Gaposchkin,C.: 1975, Proc. IAU Symposium 72, "Abundance Effects in Classification" (B.Hauck and P.C.Keenan, eds.)

-Querci,F.: 1985, "M,S and C Stars" in Monograph series on Nonthermal Phenomena in Stellar Atmospheres (H.R.Johnson and F.Querci, eds.), CNRS-NASA, chap.1

-Querci,F. and Querci,M.: 1975, A. & A. 39, 13

-Querci,F. and Querci,M.: 1983, Proceedings of the Japan-France seminar on: Active phenomena in the outer layers of sun and

stars, 3-7 october,1983, Collège de France, Paris
-Querci,M. and Querci,F.: 1976, A. & A. 49, 443
-Querci,M.,Courtin,R.,Querci,F.,Coron,N. and Gispert,R.: 1979, A.& A. 77, 155
-Rufener,F.: 1964, Thèse, Publ.Obs. de Genève,Serie A, 413
-Rufener,F.: 1976, A. & A.suppl.ser. 26, 275
-Straizys,V. and Zdanavicius,K.: 1970, Bull.Vilnius Obs. 29, 15
-Straizys,V.,Jodinskiené,E. and Hauck,B.: 1982, Bull.Vilnius Obs. 30, 50
-Wing,R.F.:R.F.: 1971, Proceedings of the conference on late-type stars, Ed. G.W.Lockwood and H.M.Dyck, Kitt Peak Nat.Obs.contrib. N°554, 145

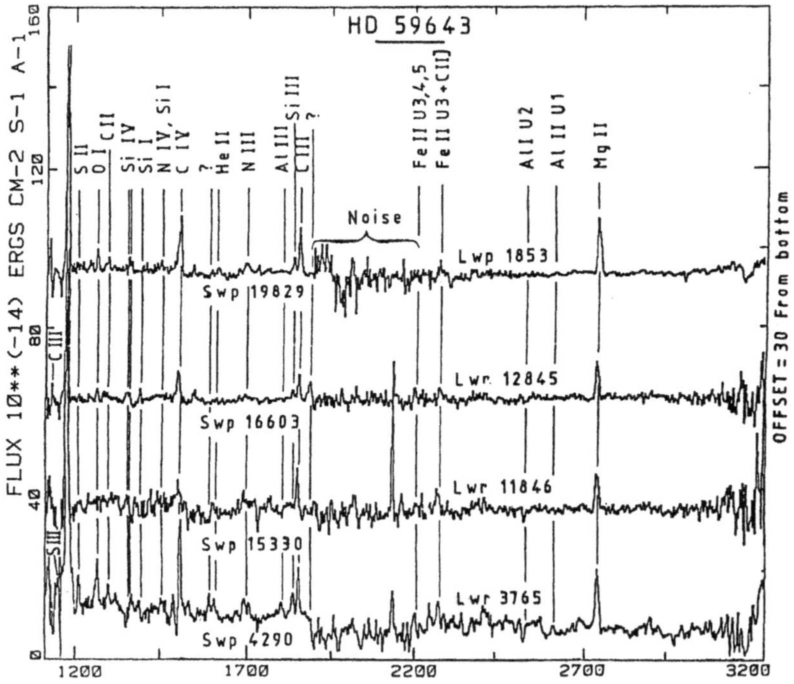

FIGURE 1- IUE spectra of carbon star HE 59643 (R8;C6,3) taken at various dates from 1979 to 1983 with the SWP and LWR or LWP cameras, illustrating emission line variations. Note the particularly high variable emission in the CIV around 1550Å and MgII around 2800Å. The SWP 4290 and LWR 3765 images were recorded on February 17,1979 and February 14,1979 respectively. The SWP 15330 and LWR 11846 images were recorded on October 25,1981, the SWP 16603 and LWR 12845 images on March 23,1982 and SWP 19829 and LWP 1853 images on April 27,1983.

SOME PROPERTIES OF S MIRA VARIABLES

M. S. Vardya
Tata Institute of Fundamental Research
Bombay 400 005
India

ABSTRACT. Some correlations between period of pulsation, spectral type at maximum, mean difference in visual magnitude between maxima and minima, intensity of Tc I line, ratio of the number of days from minimum to the next following maximum to the period, and infrared colour index [11μ] - [3.5μ] are described for S Mira variables.

INTRODUCTION

Mira variables are large amplitude red stars, of spectral types Me, Se, and Ce. S Miras are supposed to be in an evolutionary phase between M and C types. Therefore, the study of their properties should cast light on their evolutionary status and the manner transition takes place between M, S and C. With this in view, we have undertaken to investigate the properties of S Miras, including MS and CS, in the first instance and supplement it later on, if needed, with similar properties of M and C Miras.

We have tried to find correlations between period of pulsation (P), spectral type at maximum, mean range in visual magnitude between maxima and minima ($< m_v >$), intensity of Tc I line (I(Tc I)), ratio of the number of days from minimum to the next following maximum to the period (f), and infrared colours. We will describe here a few of these correlations.

RELATION BETWEEN SPECTRAL TYPE AT MAXIMUM AND P

We have plotted in Fig. 1 spectral type at maximum (Keenan et al 1974) as a function of period for 25 S Miras; the numbers near each point denote the abundance class. The figure shows that the temperature spectral class advances with P, as found for M Miras (Keenan et al 1974). This can

M. Jaschek and P. C. Keenan (eds.), Cool Stars with Excesses of Heavy Elements, 105–109.

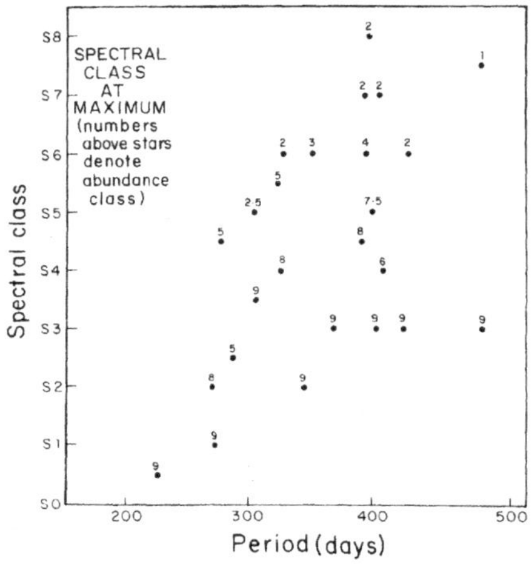

Figure 1. Mean spectral class at maximum visual light versus period for S Mira variables, with the abundance class marked over each point.

be qualitatively understood as follows : Period is roughly equal to 2R/Vs, where R is the radius of the star and Vs the mean velocity of sound. As the star expands, i.e. R increases temperature decreases and therefore spectral class advances. But as R increases, P will also increase as Vs is not very sensitive to temperature.

It is also found that at the same temperature spectral class, period has a tendency to increase with the abundance class. For example, T Cam (S5, 7.5) has a higher period than Z Del (S5, 2.5), and T Sgr (S4.5,8) higher than U Cas (S4.5,5) The period increases from R Gem → R Lyn → R Cyg → W Aql four S3,9 stars from 370 to 490 days but the abundance effect is not noticed perhaps due to saturation effect as all have the highest abundance class. This abundance effect may be indicative of the evolutionary phase of S Miras. However, it will be more instructive to compare relative abundances, when available, rather than abundance class.

CORRELATION BETWEEN I(Tc I) AND f

There are nine S Miras for which intensity of Tc I 4297 is available (Little-Marenin and Little 1979), including one MS star. Correlations of I(Tc I) with f, $<\Delta m_v>$, and P were attempted. The best correlation was found (Fig. 2) with f in the sense that I(Tc I) increases with decrease in f; the correlation coefficient, r = -0.8, which has a probability, p = 0.009, to be accidental. Solid line gives the least square fit.

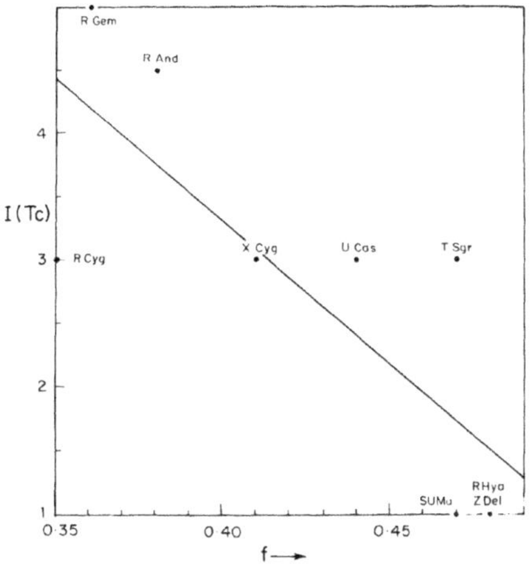

Figure 2. Estimated intensity of Tc I 4297 versus f, the ratio of rise time from visual minimum to next maximum light to the period for S Miras. The straight-line gives the least square fit.

This is perhaps the second example that the parameter f, which can be thought as the asymmetry factor of the visual light curve, has been found to be correlated with activity of the long period variables. Bowers (1975) and Bowers and Kerr (1977) had found that Mira variables showing OH emission tend to have smaller values of f than non-OH Miras.

A small value of f(<0.5) indicates a sharp or steeper rise time to decay time in the light curve. Smaller values of f may therefore be suggestive of deeper dredging of material, which is brought to the surface and ejected out as well. Note that Sanner (1978) has suggested that virtually all Tc stars have massive circumstellar gas envelopes.

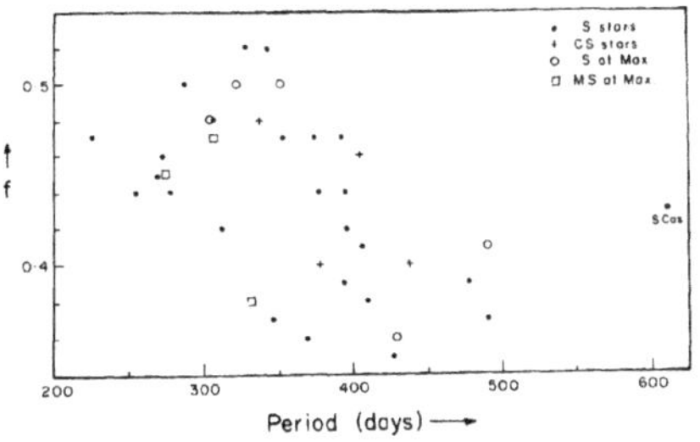

Figure 3. Correlation of f with period for S Miras. The misfit S Cas may have period twice the normal.

CORRELATION BETWEEN f and P

Data for 38 S Miras, including 4 CS, 5 S at maximum only and 8 MS at maximum only are available, as plotted in Fig. 3. There is a very good correlation with f decreasing with increase in P, with $r = -0.46$, $p = 0.003$. In the figure, one star, S Cas, having the largest period of 611.77 days among S Miras, appears to be a misfit. If its period is reduced by half, the correlation improves considerably with $r = -0.53$, $p = 0.0007$. The fact that S Cas appears to have a period equal to 2P in comparison to the rest of the stars, is not a new phenomenon (cf Campbell 1955; Keenan and Landi 1966), and has ramifications vis-a-vis mode of pulsation (cf Wood 1981).

IR COLOUR INDICES VS $<\Delta m_V>$

Infrared colour indices, [4.9μ] - [3.5μ], [8.4μ] - [3.5μ] and [11μ] - [3.5μ] are available for nine S Miras including three MS and one CS star (Gillett et al 1971) with mean values available for three of them (Forrest et al 1975). Using the mean values, where available, we have found the correlation between [11μ] - [3.5μ] and $<\Delta m_V>$ with $r = -0.79$ and $p = 0.01$ as is evident from Fig. 4. Note that the three MS stars form top envelope of the spread.

In Fig. 4, we have also plotted values for M and S Miras. The figure clearly shows that there are three separate

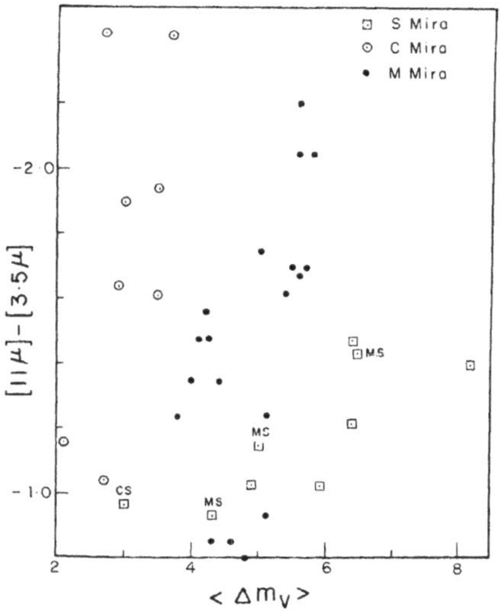

Figure 4. Infrared colour index [11μ] - [3.5μ] versus mean visual magnitude for S, C and M Miras. MS and CS stars are marked.

relations for the three spectral type Miras, such that C Miras show the steepest increase in IR excess with $<\Delta m_V>$ and S Miras the least, with M Miras falling in between. This is not surprising. Carbon stars have O/C<1, whereas M stars have O/C>1, with S stars having O/C larger but very close to unity. The particulate matter, giving rise to IR excess, is comprised mostly of graphite in C stars and silicates in M and S stars; in S stars, as O/C ≈ 1, particulate matter will be the least. Graphite has a far higher emission efficiency than silicates. Hence the result. As MS stars should have O/C larger than S stars, it is not surprising that they are on the border of M and S stars. As there is only one CS star, it is difficult to say much about it.

Though the majority of the M Miras show a good trend, four stars (V Cas, U Cet, S Lac and R Vir), all from Gillett et al (1971) align better with S stars than with M. This may be due to statistical fluctuations. It is likely, however, that these stars are in a transition phase from one class to the other. It will be good to obtain mean colours for these stars, and if even then, they are found to be misfits, a detailed analysis may reveal their true evolutionary phase.

CONCLUSIONS

Some interesting relations have been obtained for S Miras, in spite of the fact that in some cases, the number of stars available were small and the data did not pertain to similar phases in the pulsation cycle. It is food for thought why abundance class advances with period and I(Tc I) increases with decrease in f. Mass loss may have an important role, which we have not even considered here, in clarifying our understanding of these objects.

REFERENCES

Bowers, P.F.: 1975, Astron. Astrophys. 39, 473.
Bowers, P.F., Kerr, F.J.: 1977, Astron. Astrophys. 57, 115.
Campbell, L.: 1955, Studies of Long Period Variables, AAVSO, Cambridge, Mass.
Forrest, W.J., Gillett, F.C., Stein, W.A.: 1975, Ap. J. 195,423.
Gillett, F.C., Merrill, K.M., Stein, W.A.: 1971, Ap.J. 164, 83.
Keenan, P.C., Garrison, R.F., Deutsch,A.J.: 1974, Ap.J. Suppl.28, 271.
Keenan, P.C., Landi, D.J.: 1966, Ap. J. 146, 583.
Little-Marenin, I.R., Little, S.J.: 1979, Astron.J. 84, 1374.
Sanner, F.: 1978, Ap. J. 219, 538.
Wood, P.R.: 1981, in Physical Processes in Red Giants, I. Iben Jr. and A. Renzini (eds.), 205.

III. – DISTRIBUTION AND SPACE MOTIONS OF THE PECULIAR STARS. MEMBERSHIP IN CLUSTERS AND MAGELLANIC CLOUDS

THE DISTRIBUTION AND MOTIONS OF PECULIAR RED GIANTS

R.M. Catchpole and M.W. Feast
South African Astronomical Observatory
P.O. Box 9 Observatory 7935
South Africa

1. INTRODUCTION. For the present conference, the principal interest in studying the distributions and motions of peculiar red giants (in our own and other galaxies) is in the information that this gives us on the population types (ages, masses) and evolutionary phases of these stars.

At the outset it has to be borne in mind that the abundance data, although sparse, indicate that there may well be fundamental differences between different types of P giants. This is shown in Table 1 (Catchpole 1982a, Luck & Bond 1982). It must also be realized that the various subgroups (CH, BaII, S, SC, CS, C, R, N) are not always defined in quite the same way and that it is not always clear into which subgroup a particular star falls.

TABLE 1

Type	Abundance Compared with Sun 5th Period (Rb Sr Y Zr)	Abundance Compared with Sun 6th Period (Cs Ba La Ce)
K, M, R	Normal	Normal
BaII, S, SC, N, CH Super Li rich S	Over Abundant	Over Abundant
^{13}C rich Super Li rich N	Normal	Over Abundant

2. CARBON STARS

As with some other types of peculiar red giants, our understanding of carbon stars has been vastly improved in recent years by the study of these stars in the Magellanic Clouds and other nearby galaxies (mainly dwarf spheroidals).

M. Jaschek and P. C. Keenan (eds.), Cool Stars with Excesses of Heavy Elements, 113–133.

The first detection of C stars in the field of the LMC was made by Westerlund (1960) whilst C stars in Magellanic globular clusters (SMC) were found by Feast and Lloyd Evans (1973). Since such stars are not found in galactic globular clusters (leaving aside the rather different CH stars), but are known in a few old or intermediate age open clusters in our Galaxy, this immediately suggested that these SMC clusters were somewhat younger (say $\sim 1.5 \times 10^9$ years old) and contained somewhat more massive stars (say 1.5 - 2.0 $M_\odot$) than globular clusters in our Galaxy.

Clusters and groups which contain (or possibly contain) carbon stars are listed by Scalo & Miller (1979) and further possible examples have been given by Alksne & Alksnis (1980). The most certain examples of cluster membership are for one star each in the intermediate age clusters NGC 7789, 2477 and 2660. Bessell, Wood and Lloyd Evans (1983 = BWL) consider a second possible C star in NGC 2477 to be a heavily reddened background star. BWL point out that the presence of just a single C star each in NGC 2477 and NGC 2660 as well as the absence of C stars in M67 and NGC 188 is consistent with these clusters being similar to NGC 419 (SMC) in which 13 C stars are known, if the C star population scales with the total luminosity (i.e. mass) of the clusters.

Since Westerlund's objective prism survey of the LMC (1960) there have been a number of additional surveys which are listed in Table 2.

TABLE 2

C Star Objective Prism Surveys in the Magellanic Clouds

Source	Cloud	λ μm
Westerlund (1960)	LMC	0.8
Sanduleak Phillip (1977) (SP)	LMC	0.5
Blanco Blanco McCarthy (1978)	LMC SMC	0.8
Westerlund Olander Richer Crabtree (1978) (WORC)	LMC	0.8
Blanco McCarthy Blanco (1980) (BMB)	LMC SMC	0.8
Blanco McCarthy (1983)	LMC SMC	0.8

Blanco and McCarthy found 11000 C stars in the LMC and 2900 in the SMC. They find that whilst the numbers of C stars in the two Clouds are in the same ratio as the masses of the Clouds, the ratio of C to M type stars is quite different in the two Clouds. This is shown in Table 3 where the C/M ratio is derived using M stars of type M6 and later. The C/M ratio does not change very significantly over the face of the LMC but there is a large change from centre to edge of the SMC.

TABLE 3

Ratio of Carbon to M Stars in Various Systems

	C/M	Refs
SMC centre	19.2 ± 0.8	1
SMC periphery	4.7 ± 0.4	1
LMC	2.2 ± 0.1	1
Solar neighbourhood	0.01	2
Galactic Centre	<0.001	1

1 Blanco McCarthy (1983)
2 Blanco et al. (1978)

The figures for the Clouds are to be compared with the low figure for the solar neighbourhood and the even smaller value at the galactic centre. Recent studies have suggested that the younger population in the SMC has a great depth in the line of sight (e.g. Mathewson & Ford 1984, Feast 1984 and references therein) due to tidal interaction between the two Clouds. It would be of great interest to see whether the variations of C/M in the SMC reflects some aspect of this history.

Infrared photometry of samples of Magellanic Cloud C stars allows us to deduce their bolometric absolute magnitudes. These lie between

TABLE 4

JHKL Photometry of Cool Stars in the Magellanic Cloud Stars

Glass	(1979)
Feast Catchpole Carter Roberts	(1980)
Frogel Persson Cohen	(1980)
Mould Aaronson	(1980)
Catchpole Feast	(1981)
Cohen Frogel Persson Elias	(1981)
Aaronson Mould	(1982)
Frogel Cohen	(1982)
Glass Feast	(1982)
Mould Aaronson	(1982)
Bessell Wood Lloyd Evans	(1983)
Wood Bessell Fox	(1983)
Frogel	(1983)
Frogel Richer	(1983)

−3.5 and −6.5 (Frogel et al. 1980a). A histogram shows that there is a prominent peak at −5.0 (Richer 1980). The problem of this distribution is well known. There is an absence of bright ($M_{Bol} \sim -7.0$) C stars which were expected theoretically and the fainter C stars are less luminous than was theoretically anticipated. C stars in the open clusters NGC 7789 and NGC 2660 in our Galaxy have $M_{Bol} = -5.5$ (Frogel et al.1980a) consistent with the Magellanic Cloud C star luminosities.

C stars in our Galaxy are listed in Stephenson (1973). Later surveys are given in Table 5. Feunmayor (1981) made a deep photo-

TABLE 5

Recent S and C Star Objective Prism Surveys in the Galaxy

Source		Object	λ μm
Alksnis Alksne	(1977)	C	
Daube Duncans	(1977)	C	
Westerlund Olander	(1978)	S	0.8
Nandy et al.	(1978)	C	0.8
MacConnell	(1979)	S MS C SC	0.65
Kurtanizde West	(1980)	C	0.5
Fuenmayor	(1981)	C	0.8

graphic survey of 322 square degrees in both the galactic centre and the anticentre directions. This survey confirms the report of Blanco (1965) that there is a deficiency of C stars towards the galactic centre compared to the anticentre. As shown in Table 3 C stars are very rare in the galactic centre itself. McCarthy (see Blanco et al. 1983) has found only one C star compared to more than 1000 late M giants in the Sgr I and NGC 6522 fields. Catchpole (1982b) has detected a C star (star 11) in the Sgr I field at the centre, from infrared scans.

JHK photometry provides a means of comparing different samples of carbon stars. A useful way to do this is shown in Figures 1 and 2 (Catchpole unpublished). In these figures J-H is plotted against a reddening free index $b_1 = (H-K) - 0.336\ (J-K)$ (Feast et al. 1982).

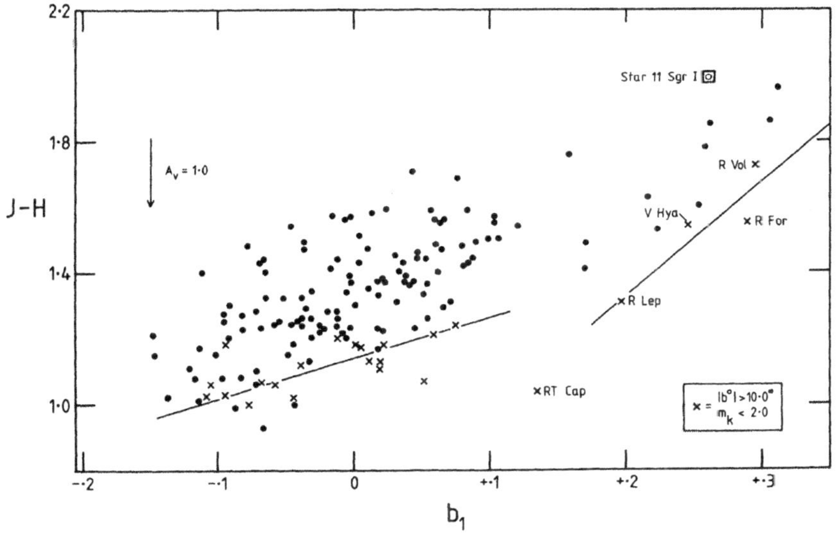

Fig. 1. (J-H) is plotted against the reddening free index b_1 for Westerlund (1971) galactic carbon stars, between ℓ = 235° 254°. The crosses are bright unreddened stars which define the top intrinsic lines.

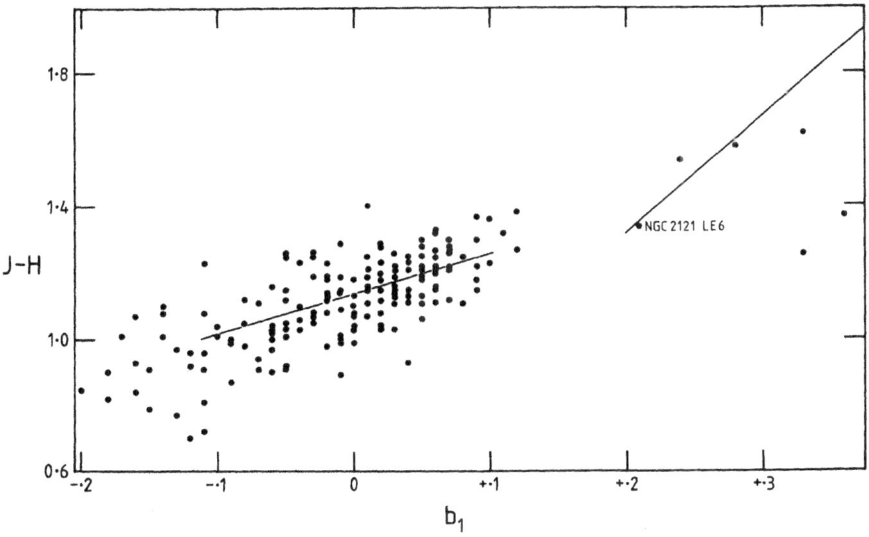

Fig. 2. Plot for Magellanic Cloud carbon stars, from sources listed in Table 4. Axes are as defined in Fig. 1. Note how the stars fall close to the line defined by the bright galactic C stars.

Bright stars away from the galactic plane, for which interstellar reddening will be negligible, form relatively narrow sequences (crosses in Figure 1). The very red Mira variables R Lep, V Hya, R Vol and R For form a group on their own (though there are other Miras at low b_1 values). The dots are results of unpublished photometry (Catchpole) of C stars between longitude 235° and 254° in Westerlund's (1971) galactic survey. They are evidently displaced from the intrinsic lines by reddening. In fact assuming $M_K = -8.0$ (consistent with the mean Magellanic Cloud value) gives individual reddenings which are consistent with Neckel & Klare's (1980) results on the distribution of interstellar absorption. Evidently stars with $b_1 \geq 0.15$ are candidates for cool Mira variables and several of these stars have been found to be variable. It is interesting that the one C star found in Sgr I lies in this region of the diagram. Figure 2 shows the galactic intrinsic lines with the available photometry for Magellanic Cloud stars (suitably transformed to the SAAO system (Glass 1983)). Evidently the intrinsic colours are similar to those in our Galaxy. The extension of the Magellanic Cloud stars to the blue may be a selection effect. Of the few stars with $b_1 \geq 0.15$, NGC 2121 LE6 has Hα emission (Lloyd Evans private communication) indicating, as expected from the above discussion, that it may be a Mira variable.

It is evident that the presence of C stars in some old or intermediate age clusters means that at least some C stars evolve from relatively low mass stars. This is confirmed by the kinematic discussion of Dean (1976) who analysed radial velocities of 427 C stars. He finds from velocity dispersions that the majority are kinematically similar to F5 dwarfs (main sequence mass $\sim 1.2\ M_\odot$, age $\sim 2.8 \times 10^9$ years) corresponding to a young disc population. Eggen (1972a) had earlier concluded that there were N type carbon stars in both the old and young disc populations. However a range of masses and ages is almost certainly present amongst the C stars. UV Aur has a B8.5 V companion (Sanford 1950) from which a main sequence mass of $\sim 4M_\odot$ and an age of $\sim 10^8$ years can be derived (cf. Dean 1976). W CMa seems likely to be a member of the young ($\sim 3 \times 10^6$ years) Canis Major OBI association (Herbst et al. 1977). The status of TW Hor is less clear; Bouchet & The (1983) place it in the open cluster NGC 1252 but Eggen (1984) finds the existence of the cluster dubious and assigns TW Hor to the Hyades Supercluster.

Little has yet been published on the kinematics of C stars in the Magellanic Clouds. Joslin & Richer (1982) report that stars from the WORC (Table 2) sample in the LMC follow the normal kinematics of other stellar populations in the LMC. However C stars from the SP (Table 2) sample do not, showing no velocity gradient across the Clouds. It would be important to follow this up. Due to different methods of detection the SP sample probably contains some stars that are hotter and bluer than the WORC sample. A dependence of population type on colour is possible since Dean (1976) found that earlier C stars (which will be mainly R type stars) in our Galaxy, came from a kinematically older population (old disc population) than the late C stars (young disc population).

3. S TYPE STARS

The most recent review of the distribution of S stars within the Galaxy is by Yorka & Wing (1979). They show that the longitude distribution of S stars is similar to that of cool carbon stars. Out of the galactic plane the S type Miras are brighter than the non-Miras on the average and Yorka and Wing take this to indicate that their absolute magnitudes are brighter by about 1.5 mag. Yorka and Wing argue for $M_V = -1.6$ for S type Miras at maximum light in which case $\bar{z} = 230$ pc. If the absolute magnitude is the same at all periods then $\bar{z}$ decreases from 364 pc for periods less than 300 days to 110 pc for periods over 400 days. Alternatively the absolute magnitude could be a function of period. It is difficult to obtain conclusive results since the total sample is small (43 stars). Yorka and Wing conclude that most S type stars (Mira and non-Mira) belong to a young disc population (older population I). Earlier Eggen (1972b) had argued for S stars belonging to an old disc population.

Despite these results there is evidence that the S star phenomenon covers practically the entire range of stellar populations. The Se variable NT Tel (Catchpole & Feast 1971a, Andrews 1975) has $b = -25.4^{\circ}$ and is 3 kpc from the galactic plane (if $M_{Bol} = -5.4$ as for other variables, see below). Its radial velocity is +325 km s^{-1}. This object thus appears to be a member of the extreme halo population. On the other hand there are four S type stars in the young η Carinae complex, at least one of which is a radial velocity member (Catchpole & Feast 1976). In this case we presumably have S stars that have evolved from massive main sequence stars. Craine et al. (1983) have found two pure S type stars in the region of the cluster and nebulosity NGC 6820/23. If these stars are cluster members they must also be young objects. Such cases may however be rare since Yorka and Wing find no significant correlation between the positions of S type stars and OB associations. At 80 A/mm the η Carinae S stars TT9 and TT12 are indistinguishable from NT Tel (except that TT9 is a lithium star).

There are five S type stars with infrared photometry for which individual bolometric absolute magnitudes can be obtained from membership in clusters, binaries etc. These have M_{Bol} near −5.4. (Feast et al. 1976). The two known extragalactic pure S stars have similar luminosities (SMC −5.1 (Blanco et al.1981), Fornax ~ −5.0 (Westerlund 1983)).

Surveys for S stars since the publication of Stephenson's (1976) catalogue, are listed with the C star surveys in Table 5.

An interesting point is raised by MacConnell's survey (1979). He has searched 4500 square degrees within ±15° of the galactic plane in the southern Milky Way and finds 494 C stars, 370 S/MS stars and 4 SC stars. Only 34 of his C stars are new discoveries as opposed to 204 S stars. As a result he finds a higher S/C ratio (0.75) than previous workers. For instance Wing & Yorka (1977) found that for stars chosen by infrared magnitude S/N = 0.3 (changing to 0.5 when corrections for selection effects are applied). The chief reason for MacConnell's higher ratio is that he observed ZrO 6474A and TiO 6195A (at 420 A/mm) and was therefore able to detect a much more subtle enhancement of

ZrO than was possible in previous surveys. This matter is of some importance because it is connected to estimates of the relative duration of the S and C star phases (assuming the same stars go through both). Such matters are in fact unlikely to be solved satisfactorily until we have rather precise definitions of the boundaries between the S, MS and M stars and how these are related to the physical and chemical properties of the stars.

Westerlund & Olander (1978) used infrared objective prism plates to search for S stars in a region $\ell = 235^\circ$ to 7°, $b = \pm 5^\circ$. The plates had been used earlier to find C stars (Westerlund 1971). An examination of this work shows the remarkable result that whilst the C stars are fairly uniformly distributed in galactic latitude the S stars are strongly concentrated to the plane $|b| = 1.2$ (see histogram Figure 3). (The stars between $\ell = 330^\circ$ and 350° have a mean $|b|$ of only $0^\circ.5$). For $M_I = M_{Bol} -1 = -6.4$ which is probably an overestimate of the luminosity this gives a mean height from the plane of about 90 pc for the whole sample and a mean distance of 4.5 kpc. Further work on this problem is desirable since it may indicate that a significant young Population I has been isolated.

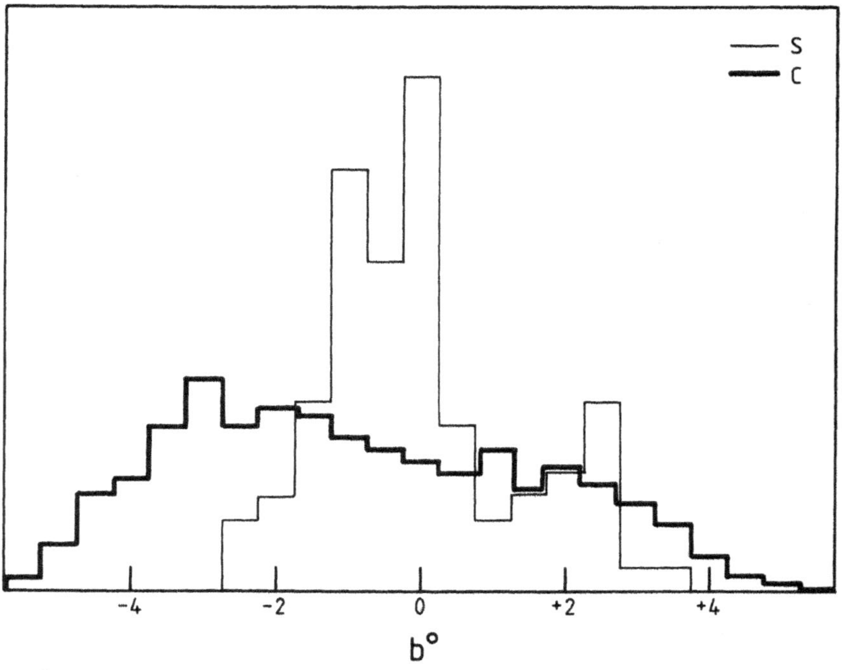

Fig. 3. Normalized histograms showing the distribution of C and S star latitudes for stars found on Westerlund's (1971) survey plates. There are 74 S stars and 1126 C stars. The S stars show a strong concentration to the plane indicating a young population.

In an attempt to improve our understanding of the kinematics of S type stars radial velocities are being determined (Catchpole in progress) for stars in the Henize (1965) survey. This includes both S and Se stars. The work is being combined with estimates of the intensities of spectral features and JHKL photometry. So far velocities are available for 135 stars. These, together with published velocities for 16 additional stars, for which JHKL photometry is available, have been used to derive the preliminary results shown in Table 6. Three points emerge from these results. (1) There are no stars in the sample with radial velocities greater than 100 km s^{-1}, demonstrating that halo S stars like NT Tel must be very rare. (2) The velocity dispersion is consistent with a galactic disc population. (3) If we adopt M_{Bol} = -5.4 for the S stars we find an Oort constant A = 4.5 km s^{-1} kpc^{-1}. The variation of A with stellar population is not yet well defined observationally. Whilst young objects give a value of A near 17, the older Me Mira variables give A $\sim$ 5 (a value in rough agreement with that anticipated from classical ellipsoidal theory (Feast 1963, 1972)). Either the S stars belong to a similar population to the Me Miras or the sample studied contains a significant proportion of S stars of lower luminosity than we have assumed.

4. SC STARS

The first attempt to define the kinematic properties of the SC stars was made by Catchpole & Feast (1971b) who found M_V = -2 and a velocity dispersion of 31 ± 7 km s^{-1}. Catchpole (1982c) considered the problem further and concluded that these stars belong to a population between old disc and old population I with M_{Bol} = -4.5 ± 1 (corresponding to M_V = -1.5). Baumert (1974) used radial velocities, proper motions and 1.04 μm photometry, to deduce $M_{1.04}$ = -5.3 ± 1.5 for 15 SC stars, corresponding to M_{Bol} = -4.3.

Using radial velocities from Catchpole & Feast (1971b), Mikami (1978), Walker (1979), Dean (1976) and Wallerstein (private communication), Catchpole has made a new kinematic solution for 30 SC and CS stars. (see Table 6). Note that these SC and CS stars are collectively described as SC stars in the classification scheme of Boeshaar & Keenan (1979). The SC stars show clear signs of differential galactic rotation.

TABLE 6

Kinematics of S and SC Stars

	M_{bol} = −5.4 Oort Constant A km s^{-1} kpc^{-1}	Deviation from solar motion in V direction km s^{-1}	Velocity Dispersion km s^{-1}	$\bar{z}$ pc	No. Stars
SC*	10.0 ± 2.5	+6 ± 5	28	195	27
Se	11.2 ± 2.5	+6 ± 7	35	600	29
S	4.5 ± 0.8	+7 ± 3	32	680	124

* Omitting GP Ori, −8° 1906 and Case 598.

However the existence in the sample of three stars with radial velocities greater than 80 km s^{-1} (GP Ori, −8° 1906 and Case 598) shows that we cannot be dealing entirely with a young population. Case 598 in particular has a velocity of opposite sign to that expected from differential galactic rotation. The overall velocity dispersion suggests we are dealing with a population similar to that of the bulk of the S stars. It is therefore interesting that if we adopt M_{Bol} = −5.4 which is already slightly fainter than the three known Magellanic Cloud SC stars, which have M_{Bol} from −5.5 to −6.1 (Lloyd Evans 1984c) we obtain a low value of the Oort A. Whilst this may be in accord with ellipsoidal theory (as discussed above), we cannot exclude the possibility that the galactic SC star sample is on the average of lower luminosity than the Magellanic Cloud one. In particular there are significant spectroscopic differences between the galactic SC stars and the three known in the Clouds (BW89 in the LMC (Richer & Frogel 1980) and CV78 and CV84 in the SMC (Lloyd Evans 1984c). The equivalent widths of the D lines for these SC stars, given by Lloyd Evans (3 to 10A), are considerably less than for those in the Galaxy which range from 13 to 56A (Feast 1953). This makes the LMC super lithium rich SC star BW89 quite remarkable with nearly equal equivalent widths of lithium 6707A (5A) and the D lines (6A) suggesting [Li/Na] ~ 5.2 much greater than in any galactic super lithium rich star (Scalo et al. 1975). However the lithium equivalent width is similar to that of galactic super lithium rich stars indicating that the [Li/H] ratio may also be similar. These results could have an important bearing on the understanding of the super lithium rich phenomenon.

5. S AND C TYPE MIRA VARIABLES

The Mira variables with M type spectra (Me variables) form a rather well defined group. They have a good period-luminosity relation and a dependence of population type (i.e. mass/age) on period (Feast 1963), (Glass & Lloyd Evans 1981). The shorter period Me Miras are old low mass objects (e.g. the 200 day Miras in metal rich globular clusters);

the longer period stars (300-400 days) being somewhat younger objects of moderate mass. The Me Mira variables also show how seriously incomplete our understanding of cool objects can be if we depend entirely on visual and photographic surveys. Radio and infrared surveys have shown that the Me sequence is extended to longer periods ($\sim$ 1000 days) by the type II OH/IR sources which are concentrated to the galactic plane and are presumably of higher mass than optically selected Me Miras.

It would be desirable to clarify the relationship of the Ce and Se Miras to the Me variables. Unfortunately this is not at present possible with any certainty. However at least for the Ce variables there are a few interesting hints.

The two carbon Mira variables (C7 and 153) in the S Doradus field of the LMC fit the period-bolometric luminosity relation for Me Miras of Glass & Lloyd Evans (1981). It would obviously be important to obtain much more data for carbon Miras in the LMC to see whether this holds in general. The C type Mira 371/31 in the SMC also falls near the Glass-Lloyd Evans relation (Lloyd Evans 1984c).

At least some carbon Mira variables belong to an old, low mass population. Stephenson's catalogue (1976) contains four Ne variables with $|b| \geqslant 40^{o}$ (Table 7). The distances from the galactic plane are considerable if these stars follow the Mira Period-luminosity relation (in the K form of Glass & Feast 1982). In addition one of the stars, V CrB, has a high radial velocity (-115 km s^{-1}) also indicating membership of an old population. The JHK colours of Mendoza & Johnson (1965) (J-K = 2.1 K-L = 0.9) place V CrB amongst the coolest carbon stars (cf. Figure 2 of Feast et al. 1982).

Table 8 contains four objects which were discovered in infrared surveys at high galactic latitude and are found to contain carbon stars. These objects all have large amplitude, long period, variability and are almost certainly dust enshrouded C type Mira variables. The existence of such objects shows that our statistics of carbon Miras in the galaxy must be incomplete. It is not clear whether these objects extend the C Mira variable sequence to longer periods and higher masses and are thus the carbon rich equivalent of the OH/IR stars or whether they are simply ordinary C type Miras undergoing episodic dust obscurations. Recent work makes this second hypothesis attractive. HV2379 a C type Mira of period 355 days in the LMC suddenly went faint in 1981 probably due to dust obscuration and R For (Period = 388 days) in our Galaxy has recently had a milder obscuration phase (Bessell & Wood 1983, Feast et al. 1984). In the latter case it was shown that the obscuration was due to relatively large particles ($\sim$ 0.15 μm radius).

The existence of heavily obscured carbon Miras shows that there is still some possibility that the missing high luminosity C stars (see section 2) may be heavily enshrouded as originally proposed by Iben (1981). The large dust particles found in the case of R For are efficient at absorbing in the 1-3 μm region and, if typical, will make dust enshrouded Miras difficult to detect even in this wavelength range. Frogel & Richer (1983) have made a 2.2 μm survey, of the LMC Bar West field, searching for luminous stars and conclude, that if such exist,

TABLE 7

Ne Variables with |b| ≥ 40° (Stephenson's Catalogue)

CCCS	Name	Period (days)	b	Type	RV km s^{-1}	z (pc)
103	R For	388	-68°	M	+ 30	775
1999	SS Vir	355	+63°	M	+ 3	378
2293	V CrB	358	+51°	M	-115	772
2306	RR Her	240	+47°	SRb	- 37	1204

TABLE 8

Dust Enshrouded Carbon Mira Variables

Object	Period (Days)	b	Ref
IRC + 10° 216	640	+44°	Whitteborn et al. (1980)
CRL 3099	438	-47°	Preliminary period (SAAO unpublished)
CRL 3068	~ months	-40°	Lebofsky & Rieke (1977)
CIT6 (RW LMi)	635	+56°	Cohen & Schmidt (1982)

they must be fainter than K = 11.0. As just noted three carbon Miras in the Magellanic Clouds fall close to the Glass-Lloyd Evans relation for Me Miras. Some other large amplitude red variables with peculiar spectra in the Magellanic Clouds fall above this relation (cf. Lloyd Evans (1984c). These include two SC stars (CV84 and CV78 in the SMC) and a number of MS stars. In our own Galaxy the super lithium rich Se variable T Sgr (which has a companion of known distance, also falls above the relation whilst the Se variable R And (a member of the Wolf 630 group) lies close to the relation (cf. Feast et al. 1976). As noted by Lloyd Evans (1984c) these more luminous objects may form a sequence within the AGB region.

6. MAGELLANIC CLOUD CLUSTERS CONTAINING P GIANTS

Both C and S type (mainly MS, not pure S type) stars are found in globular clusters in the Magellanic Clouds and they are of great importance in understanding the relation of these stars to the age of the system. Lloyd Evans (1983b, 1984a, 1984c) has recently gone over these matters on the basis of his own and other work. He has adopted as a basis the classification of the clusters by age proposed by Searle et al. (1980). It is found that S stars tend to be found preferentially in the younger clusters. The luminosity of the M to S and S to C transitions become brighter as the cluster's ages decrease.

Lloyd Evans also notes that there are more M and S stars of later spectral type as well as a greater degree of ZrO enhancement in the younger clusters. The S stars in the clusters have M_{Bol} mainly in the range -4.3 to -4.8 whilst for C stars the range is -4.3 to -5.6.

It is of interest to see whether there are S stars in globular clusters in our Galaxy. Lloyd Evans (1983a) found several MS stars in ω Cen but considers that they result from an unusual primordial chemical composition rather than from internal nuclear processing. They do not for instance show the violet and ultraviolet deficiency normally associated with such stars. On the other hand Lloyd Evans (1984b) finds two stars in NGC 6723 which show very red B-V and U-B colours as well as strong D lines and enhanced ZrO and thus appear more similar to field MS stars. These stars are of special interest since they have a very faint M_{Bol} (-3.3). This is nearly a magnitude fainter than the faintest known S star in the Magellanic Clouds.

7. Ba II STARS

Barium stars were first recognized as a distinct group by Bidelman & Keenan (1951). They show overabundances of s process elements and carbon compared with normal G and K giants. MacConnell, Frye & Upgren (1972 = MFU) and Upgren et al. (1977) listed 150 southern stars that they considered certain Ba II stars from objective prism plates and derived M_V = -0.4 from proper motions. Catchpole, Robertson & Warren (1977 = CRW) obtained spectra for about half of the MFU list and photometry for nearly all and divided them into three groups; Ba strong, Ba weak and metal weak. The first two groups represent 2/3 of the sample and are typical examples of "classical" BaII stars. The metal weak group forms a sequence of spectra between CH stars and BaII stars and includes a number of subgiant CH stars similar to those found by Bond (1974). CRW find that the velocity dispersion increases in the sequence Ba weak (17 ± 2.4 km/sec) Ba strong (25 ± 2.8 km/sec) and metal weak (33 ± 4.5 km/sec) whilst the luminosity, as determined from differential galactic rotation solutions, is greatest for the Ba weak group. There is a difference of 3 magnitudes between the luminosities of the Ba strong and weak groups. This is roughly consistent with a range of individual values of M_V from +2.5 to -3.0 from the Wilson-Bappu effect (Warner 1969). Not only do the Ba II stars have a wide range of absolute magnitudes but they have a wide range of ages. The kinematics discussed above indicate different groups with ages ranging from those of early main sequence stars to those of old disc stars. The ages evidently extend to extreme Population II since Griffin et al. (1982) show, the very metal deficient star HD115444 with [M/H] = -2.95, to be a Ba II star.

Current work centres on; the discovery by McClure (1983) that many Ba II stars are binaries; the detection of white dwarf companions to some Ba II stars (Bohm Vitense 1980, Dominy & Lambert 1983); and the conclusion by the latter writers that the abundances observed in Ba II stars cannot be accounted for simply in terms of mass transfer. These aspects will be discussed elsewhere at this meeting.

8. CH STARS

The CH stars which were originally defined by Keenan (1942) are now recognized as extreme Population II carbon stars (on the basis of their generally high velocities). They include the "classical" CH stars, the CH-like stars (Yamashita 1975) with lower radial velocities but probably high space motions, and the subgiant CH stars of Bond (1974), Luck & Bond (1982). CH stars have been found in globular clusters; ω Cen (Harding(1962), Bond (1975), Norris & Bessell (1977)), M22 (McClure & Norris 1977) and M55 (Menzies, unpublished). Richer & Westerlund (1983) and Sanduleak (1980) have suggested a CH star candidate in the Magellanic Stream though this is in some doubt. The luminosity of NGC 121 V8 in the SMC is similar to that of CH stars (Bessell et al. 1983 see also Feast 1974). NGC 121 is older than clusters containing "normal" C stars (Aaronson & Mould 1982).

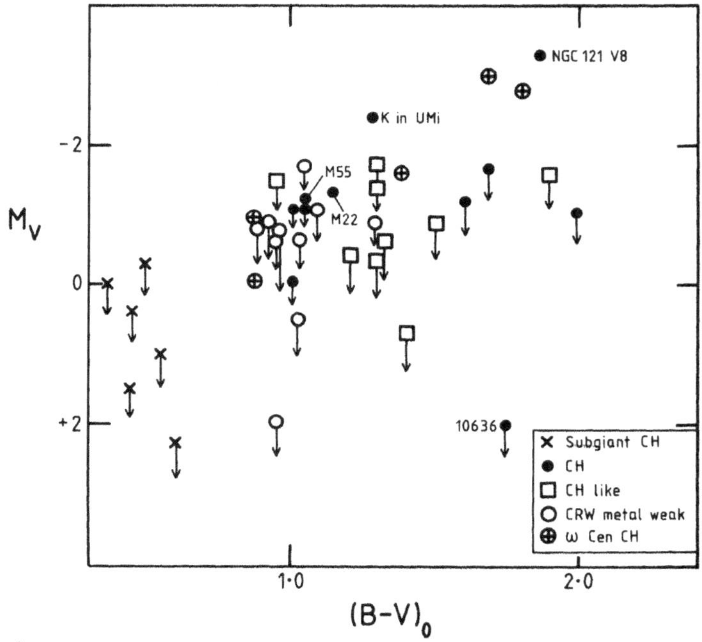

Fig. 4. The arrowed symbols indicate upper limits on the absolute magnitudes of the stars on the suposition that they will not escape from the Galaxy. Absolute magnitudes are also given for CH stars in clusters. Cluster data is drawn from the following sources. M22, Hesser et al. (1977) and Harris & Racine (1979). M55, Menzies private communication. Star K in Ursa Minor, Aaronson et al. (1983). ω Cen, Wing Stock (1973) Bond (1978) Norris & Bessell (1977).

The existence of subgiant CH stars with M_V = +2 (Bond 1974),whilst the ω Cen CH stars have M_V = −2.5, indicates the wide range of absolute magnitudes for these stars. Figure 4 shows a plot against B−V of the maximum absolute magnitudes that various CH stars and metal weak Ba stars can have and still remain bound to the Galaxy. (data from Yamashita (1975) and Feast, Catchpole & Warren (1977) and references therein). Also included are some stars with known distances. The trend of increasing absolute magnitudes with colour is evident. The star HD10636 is of particular interest since it has a large proper motion and has a low upper limit to its luminosity despite being the same colour as the ω Cen CH stars. This suggests a considerable intrinsic spread in luminosities at a given temperature. Luck & Bond (1982) find absolute magnitudes from spectroscopically determined gravities that place the subgiant CH stars on the main sequence, which is about 2 magnitudes fainter than the upper limits indicated by Figure 4. Table 9 lists abundance determinations for CH stars. These also indicate that the group is quite heterogeneous. The recent announcement by McClure (1984) that some subgiant and classical CH stars are binaries is obviously of great importance in understanding these stars.

TABLE 9

Abundances in CH Stars

	[Fe/H]	Refs	
HD201626	−1.48	1	CH
HD26	−0.70	1	
HD209621	−1.30	2	
HD198269	−1.56	3	
HD176021	−0.40	4	Subgiant CH
HD204613	−0.30, −0.4	4, 5	
HD4395	−0.3	5	
HD11377	−0.2	5	
HD88446	−0.4	5	
HD89948	−0.3	5	
HD182274	−0.4	5	
HD216219	−0.5	5	

1 Wallerstein Greenstein (1964)
2 Wallerstein (1969)
3 Lee (1974)
4 Sneden Bond (1976)
5 Luck Bond (1982)

9. C AND S STARS IN GALAXIES BEYOND THE MAGELLANIC CLOUDS

Surveys for carbon stars have recently been extended to galaxies beyond the Magellanic Clouds. Carbon stars have been discovered in all 7 known dwarf spheroidal galaxies in the local group and one carbon star has recently been found in the Magellanic irregular galaxy NGC 6822. The current position is summarized in Table 10. The carbon stars tend to fall into two groups one with M_{Bol} near -3.5 near the top of the globular cluster giant branch and the other with a range of luminosities greater than M_{Bol} = -4.0. Carbon stars of similar colours and luminosities to the CH stars in ω Cen are known in the Sculptor, Draco, Carina and Ursa Minor dwarf spheroidals. However with the exception of star K in Ursa Minor which is a CH star (and the possible exception of the Carina stars which have not been examined in detail, spectroscopically) the stars are not spectroscopically similar to CH stars. So far the only S star found in one of these systems is in Fornax (Westerlund 1983).

TABLE 10

Carbon Stars in Nearby Galaxies

	No found	Est. Total No	Ref for No	M_{Bol} ~-3.5	M_{Bol} ≤ -4	Ref
Leo I	1	>1	1	?	?	1
Leo II	4	>4	1	?	?	1
Ursa Minor	2	2	1	✓		1
Sculptor	3	5	1	✓		2 4
Fornax	49	65	1		✓	2 3 4
Carina	8	8	1	✓	✓	5 6
Draco	3	3	1	✓		7
NGC 6822	1	500-700	6	?	✓	8

1 Aaronson et al. (1983)
2 Richer Westerlund (1983)
3 Aaronson Mould (1980)
4 Frogel et al. (1982)
5 Cannon et al. (1981)
6 Mould et al. (1982)
7 Aaronson et al. (1982)
8 Aaronson et al. (1984)

10. CONCLUSIONS

Much of the work to date on the distribution of peculiar red stars has been of an initial survey nature. The work on the Magellanic Clouds and elsewhere is beginning to allow us to assign reasonably good absolute magnitudes to these stars as a function of their physical characteristics. These data, together with accurate photometry (particularly in the infrared) and radial velocities for galactic stars, should allow us to study the distribution and kinematics of this type of star in our own Galaxy with much higher precision than previously.

11. ACKNOWLEDGEMENTS

We wish to thank Dr T Lloyd Evans, Dr I S Glass and Dr J W Menzies for unpublished information and helpful discussions and Professor G Wallerstein for data on SC stars.

REFERENCES

Aaronson, M., Da Costa, G.S., Hartigan, P., Mould, J.R., Norris, J. & Stockman, H.S, 1984. Astrophys. J., **277**, L9.

Aaronson, M., Liebert, J. & Stocke, J., 1982. Astrophys. J., **254**, 507.

Aaronson, M. & Mould, J., 1980. Astrophys. J., **240**, 804.

Aaronson, M. & Mould, J., 1982. Astrophys. J. Suppl., **48**, 161.

Aaronson, M., Olszewski, E.W. & Hodge, P.W., 1983. Astrophys. J., **267**, 271.

Alksnis, A. & Alksne, Z., 1977. Photometric Investigations of Carbon Stars, p. 7, ed. A. Balklavs = Riga Radioastrofiz. Obs. Tr., **16**.

Alksne, Z. & Alksnis, A., 1980. Inf. Bull. Variable Stars, **No. 1831**.

Andrews, P.J., 1975. Mon. Not. R. astr. Soc. **173**, 701.

Baumert, J.H., 1974. Astrophys. J., **190**, 85.

Bessell, M.S. & Wood, P.R., 1983. Mon. Not. R. astr. Soc., **202**, 31P.

Bessell, M.S., Wood, P.R. & Lloyd Evans, T., 1983. Mon. Not. R. astr. Soc., **202**, 59.

Bidelman, W.P. & Keenan, P.C., 1951. Astrophys. J., **114**, 473.

Blanco, B.M., Blanco, V.M. & McCarthy, M.F., 1978. Nature, **271**, 638.

Blanco, V.M., 1965. Galactic Structure, p. 241, eds. Blaauw, A. & Schmidt, M., Univ. Chicago Press.

Blanco, V.M., Frogel, J.A. & McCarthy, M.F., 1981. Publs. astr. Soc. Pacif., **93**, 532.

Blanco, V.M. & McCarthy, M.F., 1983. Astr. J., **88**, 1442.

Blanco, V.M. McCarthy, M.F., & Blanco, B.M., 1980. Astrophys. J., **242**, 938.

Boeshaar, P.C. & Keenan, P.C., 1979. Rich. Astr., **9**, 39.

Bohm-Vitense, E., 1980. Astrophys. J., **239**, L79.

Bond, H.E., 1974. Astrophys. J., **194**, 95.

Bond, H.E., 1975. Astrophys. J., **202**, L47.

Bouchet, R. & The, P.S., 1983. Publs. astr. Soc. Pacif., **95**, 474.

Cannon, R.D., Niss, B. & Norgaard-Nielsen, H.U., 1981. Mon. Not. R. astr. Soc., **196**, 1P.

Catchpole, R.M., 1982a. Thesis, University of Cape Town.

Catchpole, R.M., 1982b. Mon. Not. R. astr. Soc., **200**, 33P.

Catchpole, R.M., 1982c. Mon. Not. R. astr. Soc. **199**, 1.

Catchpole, R.M. & Feast, M.W., 1971a. Observatory, **91**, 29.

Catchpole, R.M. & Feast, M.W., 1971b. Mon. Not. R. astr. Soc., **154**, 197
Catchpole, R.M. & Feast, M.W., 1976. Mon. Not. R. astr. Soc., **175**, 501.
Catchpole, R.M. & Feast, M.W., 1981. Mon. Not. R. astr. Soc., **197**, 385.
Catchpole, R.M., Robertson, B.S.C. & Warren, P.R., 1977. Mon. Not. R. astr. Soc., **181**, 391.
Cohen, J.G., Frogel, J.A, Persson, S.E. & Elias, J.H., 1981. Astrophys. J., **249**, 481.
Cohen, M. & Schmidt, G.D., 1982. Astrophys. J., **259**, 693.
Craine, E.R., Turnshek, D.E. & Turnshek, D.A., 1983. Publs. astr. Soc. Pacif., **95**, 185.
Daube, I. & Duncans, L., 1977. Photometric investigations of carbon stars, p. 100, ed. A Balklavs = Riga Radioastrofiz. Obs. Tr., **16**.
Dean, C.A., 1976. Astr. J., **81**, 364.
Dominy, J.F. & Lambert, D.L., 1983. Astrophys. J., **270**, 180.
Eggen, O.J., 1972a. Astrophys. J., **174**, 45.
Eggen, O.J., 1972b. Astrophys. J., **177**, 489.
Eggen, O.J., 1984. Publs. astr. Soc. Pacif., **96**, 70.
Feast, M.W., 1953. Les Processus Nucleaires dans les Astres. Inst. d'Astrophys. Univ. Liege Mem. 8°, **No. 357**, 413.
Feast, M.W., 1963. Mon. Not. R. astr. Soc., **125**, 367.
Feast, M.W., 1972. Vistas in Astronomy, **13**, 207.
Feast, M.W., 1974. Stellar Instability and Evolution, (IAU Symposium 59) p 93. Eds P Ledoux, A Noels and A W Rodgers, Reidel, Dordrecht, Holland.
Feast, M.W., 1984. Cepheids: Observation and Theory (IAU Colloquium 82) ed. B F Madore, Cambridge University Press (in press).
Feast, M.W., Catchpole, R.M. & Glass, I.S., 1976. Mon. Not. R. astr. Soc., **174**, 81P.
Feast, M.W., Catchpole, R.M. & Warren, P.R., 1977. Observatory, **97**, 140.
Feast, M.W., Catchpole, R.M., Carter, B.S. & Roberts, G., 1980. Mon. Not. R. astr. Soc., **193**, 377.
Feast, M.W. & Lloyd Evans, T., 1973. Mon. Not. R. astr. Soc., **164**, 15P
Feast, M.W., Robertson, B.S.C., Catchpole, R.M., Lloyd Evans, T., Glass, I.S. & Carter, B.S., 1982. Mon. Not. R. astr. Soc., **201**, 439.
Feast, M.W., Whitelock, P.A., Catchpole, R.M., Roberts, G. & Overbeek, M.D., 1984. Mon. Not. R. astr. Soc., (in press).
Frogel, J.A., 1983. Astrophys. J., **272**, 116.
Frogel, J.A., Blanco, V.M., McCarthy, M.F. & Cohen, J.G., 1982. Astrophys. J., **252**, 133.
Frogel, J.A., Cohen, J.G., Persson, S.E. & Elias, J.H., 1980a. Physical Processes in Red Giants, p. 159, eds Iben, I. & Renzini, A., Reidel, Dordrecht, Holland.
Frogel, J.A. & Cohen, J.G., 1982. Astrophys. J., **253**, 580.
Frogel, J.A., Persson, S.E. & Cohen, J.G., 1980. Astrophys. J., **239**, 495.

Frogel, J.A., & Richer, H.B. 1983. Astrophys. J., **275**, 84.
Fuenmayor, F.J., 1981. Rev. Mex. Astron. Astrof., **6**, 83.
Glass, I.S., 1979. Mon. Not. R. astr. Soc., **186**, 317.
Glass, I.S., 1983. Mon. Notes astr. Soc. Sth. Afr., **42**, 43.
Glass, I.S. & Feast, M.W. 1982. Mon. Not. R. astr. Soc., **199**, 245.
Glass, I.S. & Lloyd Evans, T., 1981. Nature, **291**, 303.
Griffin, R. & R., Gustafsson, B. & Vieira, T., 1982. Mon. Not. R. astr. Soc., **198**, 637.
Harding, G.A., 1962. Observatory, **82**, 205.
Harris, W.E. & Racine, R., 1979. Annual Rev. Astron. Astrophys., **17**, 241.
Henize, K.G., 1965. List of S stars (private circulation).
Herbst, W., Racine, R. & Richer, H.B., 1977. Publs. astr. Soc. Pacif., **89**, 663.
Hesser, J.E., Hartwick, F.D.A. & McClure, R.D., 1977. Astrophys. J. Suppl., **33**, 471.
Hodge, P.W., 1971. Annual Rev. Astr. Astrophys., **9**, 35.
Iben, I., 1981. Astrophys. J., **246**, 278.
Joslin, G. & Richer, H., 1982. J.R. astr. Soc. Can., **76**, 321.
Keenan, P.C., 1942. Astrophys. J., **96**, 101.
Kurtanizde, O.M. & West, R.M., 1980. Astr. Astrophys. Suppl., **39**, 35.
Lebofsky, M.J. & Rieke, G.H., 1977. Astron. J., **82**, 646.
Lee, P., 1974. Astrophys. J., **192**, 133.
Lloyd Evans, T., 1983a. Mon. Not. R. astr. Soc., **204**, 975.
Lloyd Evans, T., 1983b. Mon. Not. R. astr. Soc., **204**, 985.
Lloyd Evans, T., 1984a. Mon. Not. R. astr. Soc., **208**, 447.
Lloyd Evans, T., 1984b. Mon. Not. R. astr. Soc., **209**, in press.
Lloyd Evans, T., 1984c. Bright Red Variables of Large Amplitude in the Magellanic Clouds. Mon. Not. R. astr. Soc., submitted.
Luck, E.R. & Bond, H.E., 1982. Astrophys. J., **259**, 792.
MacConnell, D.J., 1979. Astr. Astrophys. Suppl., **38**, 335.
MacConnell, D.J., Frye, R.L. & Upgren, A.R., 1972. Astr. J., **77**, 384.
Mathewson, D.S. & Ford, V.L., 1984. Structure and Evolution of the Magellanic Clouds., IAU Symp. 108, p. 125, eds van den Bergh, S. & de Boer, K.S., Reidel, Dordrecht, Holland.
McClure, R.D., 1983. Astrophys. J., **268**, 264.
McClure, R.D., 1984. Astrophys. J., **280**, L31.
McClure, R.D. & Norris, J., 1977. Astrophys. J., **217**, L101.
Mendoza, V., E.E. & Johnson, H.L., 1965. Astrophys. J, **141**, 161.
Mikami, T., 1978. Ann. Tokyo astr. Obs., 2nd series, **17**, 1.
Mould, J. & Aaronson, M., 1980. Astrophys. J., **240**, 464.
Mould, J. & Aaronson, M., 1982. Astrophys. J., **263**, 629.
Mould, J.R., Cannon, R.D., Aaronson, M. & Frogel, J.A., 1982. Astrophys. J., **254**, 500.
Nandy, K., Smriglio, F. & Buonanno, R., 1978. R. Obs. Edin. Publs., **9**, , no. 5, p. 125.
Neckel, Th. & Klare, G., 1980. Astr. Astrophys. Suppl., **42**, 251.
Norris, J. & Bessell, M.S., 1977. Astrophys. J., **211**, L91.

Richer, H.B., 1980. Physical Processes in Red Giants, p. 153 eds. Iben, I. and Renzini, A., Reidel, Dordrecht, Holland.
Richer, H.B. & Frogel, J.A., 1980. Astrophys. J., **242,** L9.
Richer, H.B. & Westerlund, B.E., 1983. Astrophys. J., **264,** 114.
Sanduleak, N., 1980. Publs. astr. Soc. Pacif., **92,** 246.
Sanduleak, N, & Phillip, A.G.D., 1977. Publ. Warner Swasey Obs., **2,** no. 5.
Sanford, R.F., 1950. Astrophys. J., **111,** 270.
Scalo, J.M. & Miller, G.E., 1979. Astrophys. J., **233,** 596.
Scalo, J.M., Despain, K.H. & Ulrich, R.K., 1975. Astrophys. J., **196,** 805
Searle, L., Wilkinson, A. & Bagnuolo, W.G., 1980. Astrophys. J., **239,** , 803.
Sneden, C. & Bond, H.E., 1976. Astrophys. J., **204,** 810.
Stephenson, C.B., 1973. Publ. Warner Swasey Obs., **1,** no. 4.
Stephenson, C.B., 1976. Publ. Warner Swasey Obs., **2,** no. 2, 23.
Upgren, A.R., Lu, P.K. & MacConnell, D.J., 1977. Publs. astr. Soc. Pacif., **89,** 552.
Walker, A.R., 1979. SAAO Circular, **1,** no. 4, p. 112.
Wallerstein, G., 1969. Astrophys. J., **158,** 607.
Wallerstein, G. & Greenstein, J.L., 1964. Astrophys. J., **139,** 1163.
Warner, B., 1969. Mon. Not. R. astr. Soc., **144,** 333.
Westerlund, B., 1960. Uppsala astr. Obs. Ann., **4,** no. 7.
Westerlund, B.E., 1971. Astr. Astrophys. Suppl., **4,** 51.
Westerlund, B.E., 1983. Astr. Astrophys., **118,** L5.
Westerlund, B.E., Olander, N., Richer, H.B. & Crabtree, D.R., 1978. Astr. Astrophys. Suppl., **31,** 61.
Westerlund, B.E. & Olander, N., 1978. Astr. Astrophys. Suppl., **32,** 401.
Wing, R.F. & Stock, J., 1973. Astrophys. J., **186,** 979.
Wing, R.F. & Yorka, S.B., 1977. Mon. Not. R. astr. Soc., **178,** 383.
Witteborn, F.C., Strecker, D.W., Erickson, E.F., Smith, S.M., Goebel, J.M. & Taylor, B.J., 1980. Astrophys. J., **238,** 577.
Wood, P.R., Bessell, M.S. & Fox, M.W., 1983. Astrophys. J., **272,** 99.
Yamashita, Y., 1975. Publ. astr. Soc. Japan, **27,** 325.
Yorka, S.B. & Wing, R.F., 1979. Astr. J., **84,** 1010.

DISCUSSION

LLOYD EVANS - 1. MS stars in Magellanic Clouds and in ω Cen span $4^{m}5$ in M_{bol}. So I question the value of kinematic studies of absolute magnitudes in our galaxy.

2. NGC 6723 has two marginal MS stars, $M_{bol} = -2.3$. These are not detectable in Magellanic Clouds as LMC are too faint to obtain a spectra of sufficient signal/noise ratio. Here LMC clusters could well contain similar stars.

CATCHPOLE - I agree. Your first point does raise problems and that is why I think the S stars are best compared with the Me Miras as a whole, even though the Me stars can be shown to cover a range of kinematic properties. Confining the selection of the stars to a range of Zr0 and Ti0 strengths does reduce the observed errors.

DAVIS LOCANTHI - Is an effort being made to search for Li - rich stars ?

CATCHPOLE - I do not know.

DAVIS LOCANTHI - There should be a search for Li- rich stars and their properties studied.

RICHER - In your M_V, B - V diagrams for CH, subgiant CH, CH like and ω Cen CH stars, where do the ω Cen S - type stars lie ? What (if any) do you suggest is their relationship to the ω Cen CH stars ?

CATCHPOLE - I think they will lie below the CH stars in this diagram and have no relationship to them because the S - type stars that Lloyd Evans finds he considers due to peculiar primordial abundances.

THE CATALOGUE OF STELLAR GROUPS. II. COOL PECULIAR STARS

D. Egret and M. Jaschek
Centre de Données Stellaires, Observatoire de Strasbourg
11, rue de l'Université, 67000 Strasbourg, France

ABSTRACT. The "Catalogue of Stellar Groups" is a set of files listing all the kinds of stars with known peculiarities. A first volume covering the earlier types was published two years ago (Jaschek and Egret, 1982). Preliminary lists for about thirty groups of cool peculiar stars are now available, and the corresponding volume is expected to be published soon. Histograms in visual magnitudes and plots of the galactic distribution are presented here for some of the groups under discussion during the colloquium.

1. INTRODUCTION

A large effort has been undertaken at the Stellar Data Center in order to produce a catalogue listing all the kinds of stars with spectral peculiarities: the Catalogue of Stellar Groups. The first volume, covering the earlier types was published two years ago (Jaschek and Egret, 1982). For each star within a group, we gave an identification, the coordinates and visual magnitude, and reference(s) to the paper(s) in wich the peculiarity is discussed. The first volume contained 6350 entries covering seventeen different spectral groups. The same work is now in progress for the later types.

2. THE SECOND VOLUME : THE LATER GROUPS

Preliminary lists for about thirty groups of cool peculiar stars are now available, and the second volume of the "Catalogue of Stellar Groups" is expected to be published soon. A provisional list of the groups considered for inclusion in the volume is given in Table I.

Two remaks are to be made concerning this work: the first one is that the files are the result of a compilation of all the lists published in the field; and thus we do not expect to bring something new to the specialists of a given group. We are, more practically, addressing to the non-specialists for which a bibliographical research of the whole literature would be a very time-consuming task.

The second remark is the following: the CGS is directly connected to the SIMBAD Data Base of the Centre de Données Stellaires. The cross-

M. Jaschek and P. C. Keenan (eds.), Cool Stars with Excesses of Heavy Elements, 135–140.

TABLE I. Number (n) of stars in each group of cool peculiar stars, and number of stars n(HD) within each group also belonging to the Henry Draper Catalog.

Spectral group	n	n(HD)
P Cygni	12	9
FU Ori	6	1
Luminous Hydrogen-deficient	31	15
VV Cep	51	29
Symbiotic	188	25
Barium	321	294
Weak G-band	26	26
Carbon	3484	186
J	55	40
Li	18	15
S	834	127
SC or CS	51	14
MS	298	31
CH	85	56
Subgiant CH	12	10
CN	537	536
Fe or F shell	9	7
Metal-weak	832	630
T Tau	520	23
YY Ori	33	1
M dwarf	991	29
Me	152	10
Cool subdwarf	843	140
Composite spectra	1263	1200
Degenerate	159	-
ZZ Ceti	25	-
BY Dra	41	12
RS CVn	93	84
FK Com	5	4
UV Ceti	75	12

identifications, fundamental data, astrophysical measurements and bibliographical references for the stars can be listed very easily. All these data are kept up-to-date. As a complement to the preceding remark, we may note that this facility can also be very helpful even for the specialist of a given group !

3. STATISTICS ABOUT SOME LATE-TYPE GROUPS

Figure 1 displays the histograms in V magnitudes of the stars known in five of the groups of peculiar stars under discussion during this colloquium. These histograms are clearly affected by observational biases

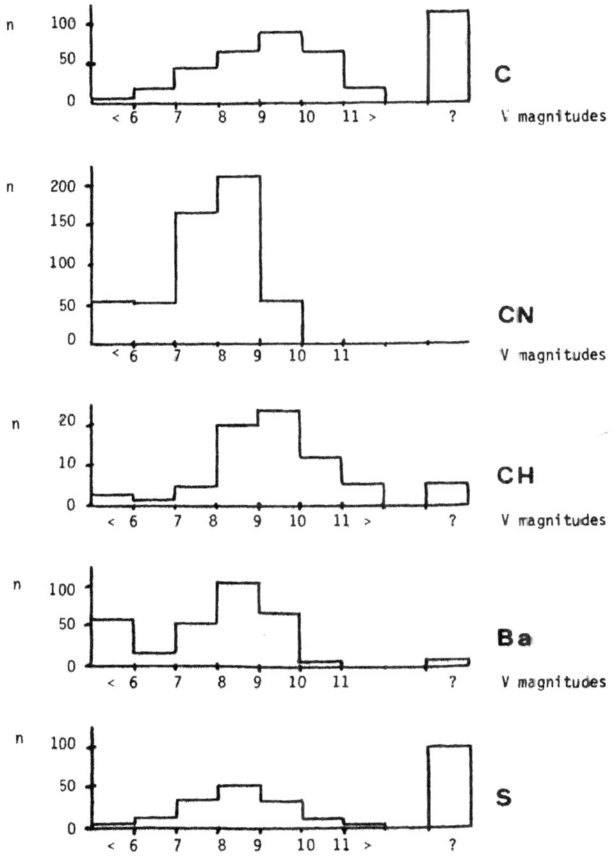

Figure 1. Histograms in visual magnitudes of some spectral groups from Table I.

and illustrate our rather poor knowledge of the precise spectral classification of a common sample such as the sample of the HD stars (in the range of magnitudes 7 to 9).

These biases are also illustrated by the Figures 2 to 4 presenting the galactic distribution of the stars from three different groups : the CH stars, S stars and CN stars. The galactic center is the center of the figure; the equatorial plane is represented by the dashed line. The last group is typically affected by the "Michigan syndrome": a large part of the CN stars presently known were discovered by Nancy Houk during her reclassification of the HD stars for the Michigan Survey of Spectral Classification for the HD stars (now available for -90° to -26°).

4. FINAL REMARK

The second volume of the "Catalogue of Stellar Groups" dedicated to the cool peculiar stars will be published by the Stellar Data Center (Jaschek and Egret, 1985).

Printouts of the preliminary lists can be obtained from the authors.

REFERENCES

Jaschek, M. and Egret, D.: 1982, *Catalog of Stellar Groups. Part one: The earlier Groups*, Publication Spéciale du C.D.S. n°4.

Jaschek, M. and Egret, D.: 1985, *Catalog of Stellar Groups. Part two: The later Groups*, Publication Spéciale du C.D.S. n°5, *in preparation*.

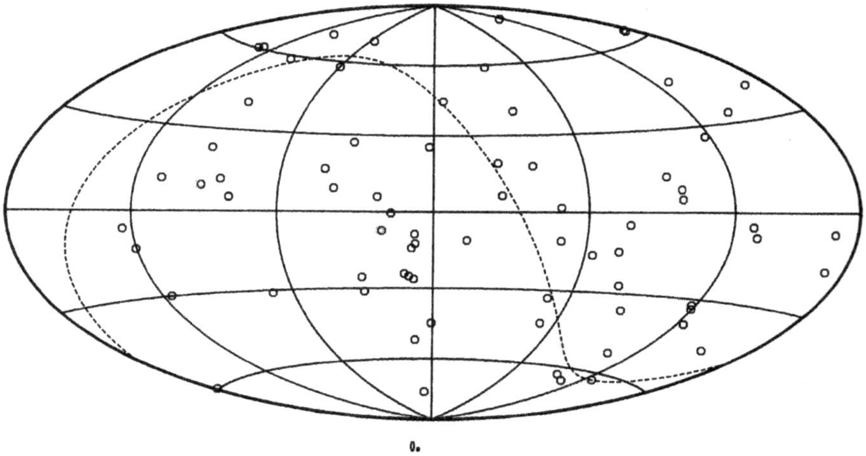

Figure 2. Distribution in galactic coordinates of CH stars. The dashed line if the curve $\delta=0°$. The galactic center is the center of the figure.

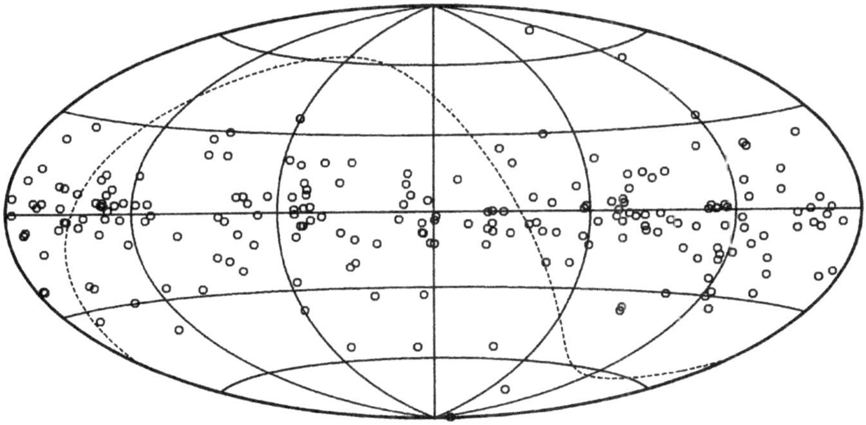

Figure 3. Distribution in galactic coordinates of the S stars (see Figure 2).

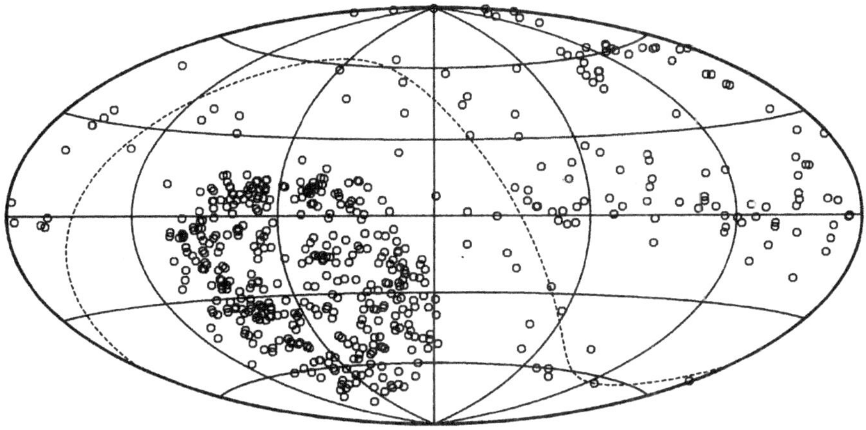

Figure 4. Distribution in galactic coordinates of CN stars (see Figure 2).

DISCUSSION

CANNON - Does the CGS have a magnitude limit, or do you include all known peculiar stars such as faint examplars in globular clusters and the Magellanic Clouds ?

EGRET - There is no magnitude limit and we intent to include all known (and published !) lists of peculiar stars.

BENNETT - Are you planning to include classes of photometric variable stars, e.g. the RS Cvn or BY Dra variables in the CGS in the future ?

EGRET - Both groups are already included in the CGS.

KINEMATICS OF LATE-TYPE GIANTS

F. Ochsenbein - M. Holtzer
C.D.S. - Observatoire de Strasbourg
France

ABSTRACT - The well defined data found in the Bright Star Catalog ($m_V \leqslant 6\overset{m}{.}50$) were used to derive the main kinematical parameters ($V_\odot$, ℓ_A, b_A, σ_{res}) of stars grouped according to spectral classification. Monotic variation of the dispersion of the residual velocities σ_{res} (Fig. 1) among main sequence stars is stressed and the incidence of high-velocity stars in late-type giants is discussed.

The purpose of this communication is to establish a properly-defined basis for kinematical studies of late-type giants in the solar neighborhood. To accomplish this, we used samples from the "Bright Star Catalog" (Hoffleit and Jaschek, 1982), which covers the whole sky, is complete down to a fixed limiting magnitude ($V = 6\overset{m}{.}50$), and is composed almost entirely of stars having a known radial velocity. We retained only normal stars and rejected objects with spectral peculiarities and composite spectra.

The sample were defined by spectral classification, i.e., spectral type and luminosity class, but only those groups composed of a sufficiently large number of stars ($N > 60$) were retained. Using the average radial velocities given in the catalog, we derived the solar motion and the dispersion of residual velocities for each group according to the following formula :

$$V_{obs} = V_{res} + V_\odot (\cos b \;\cos b_A + \sin b \;\sin b_A \cos (\ell - \ell_A)) + K$$

where

V_{obs} = observed radial velocity, corrected from the galactic rotation

$V_\odot$ = solar velocity

$\ell,\ b$ = galactic coordinates of the stars

$\ell_A,\ b_A$ = galactic coordinates of the apex

V_{res} = peculiar radial velocity of the star

M. Jaschek and P. C. Keenan (eds.), Cool Stars with Excesses of Heavy Elements, 141–146.

K = constant term

and

$$\sigma^2_{res} = \frac{1}{n-1} \sum V^2_{res}$$

All stars having a residual radial velocity larger than 60 km/s were eliminated because we considered them to be high-velocity stars and their inclusion would have implied a mixture of stellar population.

The results are given in Table 1.

TABLE 1 – Main kinematical results for the lrgest spectroscopic classes

R_{LIM} : estimate of the limiting distance (in pc).
HV : number of high-velocity stars (residual velocity > 60 km s^{-1})
N : sample size
$V_\odot$: peculiar velocity of the Sun
ℓ_A, b_A : galactic position of the apex
σ_{res} : dispersion of the residual velocities
K : expansion term

GROUP	R_{LIM}	HV	N	$V_\odot$	ℓ_A	b_A	σ_{res}	K
A0V	125	1	193	19.0	55	+23	12.1	-0.1
A1V	110	0	180	19.3	53	+24	12.6	0.1
A2V	95	0	160	16.0	57	+29	14.7	-0.5
A3-5V	80	0	236	12.7	40	+35	14.2	-2.0
A6-F0V	60	0	165	22.5	36	+31	14.6	0.7
Am	100	0	165	21.5	20	+21	14.8	-0.6
F1-4V	50	2	149	17.0	48	+32	19.2	1.8
F5-7V	35	2	226	20.4	41	+25	20.1	0.0
G5-7III	150	4	177	15.0	45	+26	19.0	-1.4
G8-9III	175	9	384	16.8	48	+27	19.0	-0.4
K0III	180	9	512	21.6	59	+13	21.2	-0.4
K1III	185	6	247	16.0	70	+24	21.2	1.0
K2III	190	9	293	20.9	66	+30	22.7	0.5
K3III	200	4	216	21.7	57	+17	22.1	2.7
K4III	220	6	151	20.8	60	+19	22.7	0.2
K5III	240	8	228	21.1	48	+14	22.0	-3.2
M0-2III	300	11	235	25.6	50	+17	22.0	-1.4
M3-6III	350	11	153	20.0	50	+19	22.2	-1.4

As can be see from this table, the parameters of the peculiar motion of the sun ($V_\odot$ ℓ_A b_A) are distributed over a large range of values. We are unable to detect any clear relation of sucessive values of $V_\odot$ ℓ_A and b_A with spectral classification. Furthermore, the K-term shows no clear trend and since its value is close to zero, it could have been omitted. On the other hand, the dispersion of the residual radial velocities (σ_{res}) increases monotonically along the main sequence with an apparent gap at F0. The results are plotted in Fig. 1. For late-type giants the dispersion σ_{res} covers a narrow range between 19 and 23 km/s, and is comparable to the dispersion of late F-type dwarfs.

An examination of the influence of high-velocity stars on the kinematic parameters (Table 1) reveals that the ratio of high-velocity objects is only significant among the late-type giants. The influence of such high-velocity stars is illustrated in Table 2, in which the late-type giants are grouped in samples larger than 500 stars, and kinematical parameters were computed for samples either including or excluding high-velocity stars : the apex and the solar motion is almost constant among the groups, except for the latest types K5 - M6, where the ratio of high-velocity stars is much larger (about 5 %).

Table 2 - Main kinematical parameters for late giants
- without exclusion of high-velocity stars (1st lines)
- with exclusion of high-velocity stars (2nd lines)

	N	$V_\odot$	ℓ_A	b_A	σ_{res}
G5 - 9III	574	16.9	50	+26	21.8
	562	16.0	47	+28	19.1
K0-1III	791	20.5	63	+18	23.2
	772	20.2	62	+16	21.0
K2-4III	668	22.3	63	+21	26.9
	650	20.9	61	+22	22.5
K5-M6III	658	25.7	56	+16	29.2
	625	22.8	50	+17	22.8

The distribution of the residual velocities is illustrated in Fig. 2a - 2d, in which the observed histograms are compared with the expected ones, with high-velocity either excluded (dashed lines) or included (dotted lines). The shape of the observed distribution, at least for low residual radial velocity stars, is more compatible with dashed-line curves : the dispersion of residual velocities derived from samples in which high velocity stars have been removed is quite compatible with the observed distribution.

Our conclusions are in good agreement with previous values listed in the literature (Delhaye, 1965 ; Mihalas, 1981).

We wish to stress that our values are derived from complete samples with well-defined limits in luminosity, spectral type, and apparent magnitude, and refer, therefore, to a sample which can be deliminated in space. Consequently, a relation more direct than was previously possible may now be established between spatial and kinematic parameters.

REFERENCES

Delhaye J. : 1965, "Galactic structure" - Stars and stellar systems - ed. A. Blaauw and Maiten Schmidt - The University of Chicago Press

Hoffleit D., Jaschek C. : 1982, "The bright star catalog" - Yale University Observatory

Mihalas D., Binney J. : 1981, "Galactic astronomy" - Freeman and Co

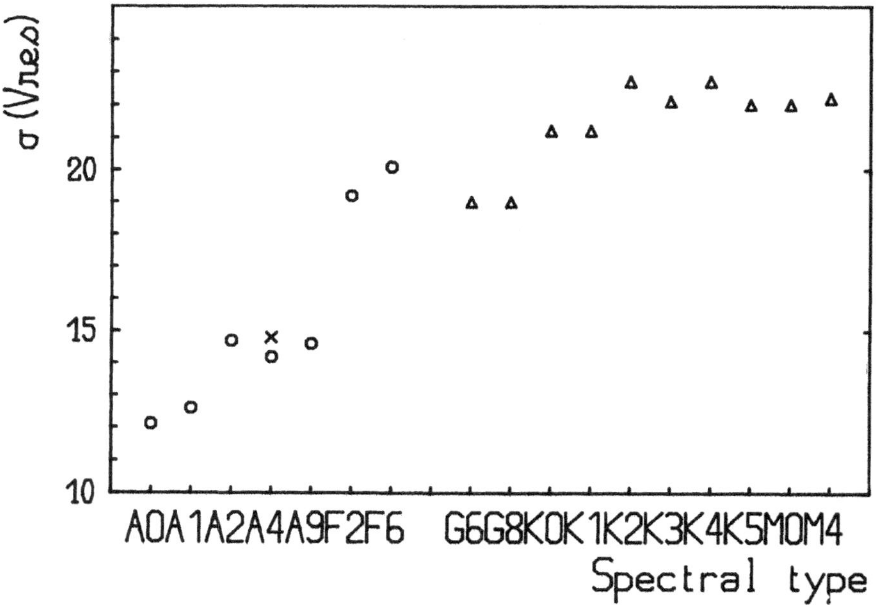

Figure 1. - Dispersion of the residual velocities versus spectral type for main-sequence (circles), giants (triangles) and Am (cross)

Fig. 2a

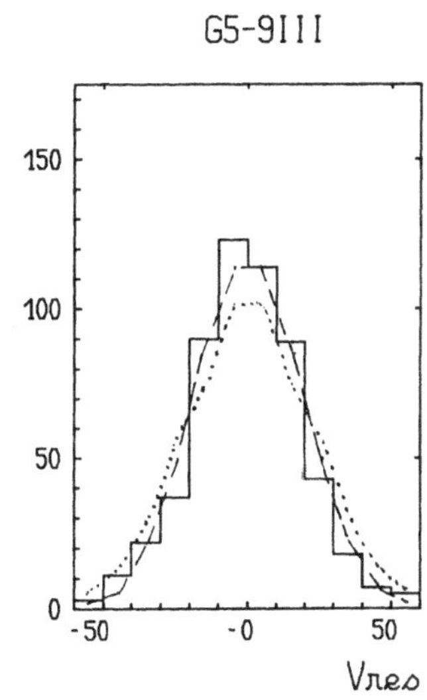

Fig. 2b

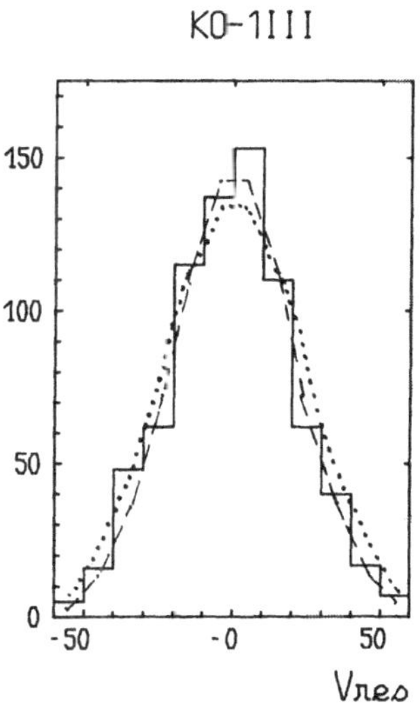

Fig. 2c

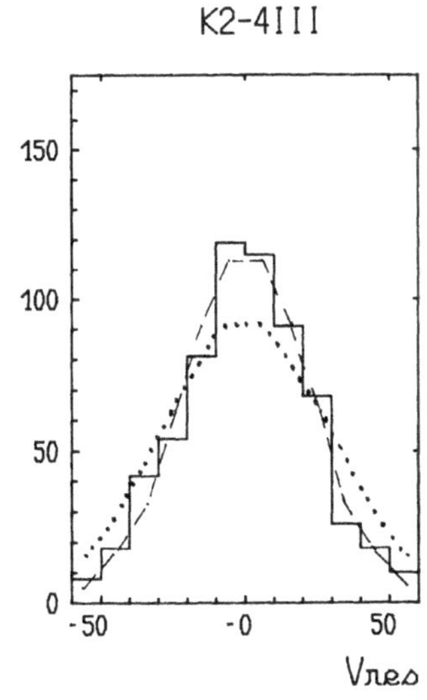

Fig. 2d

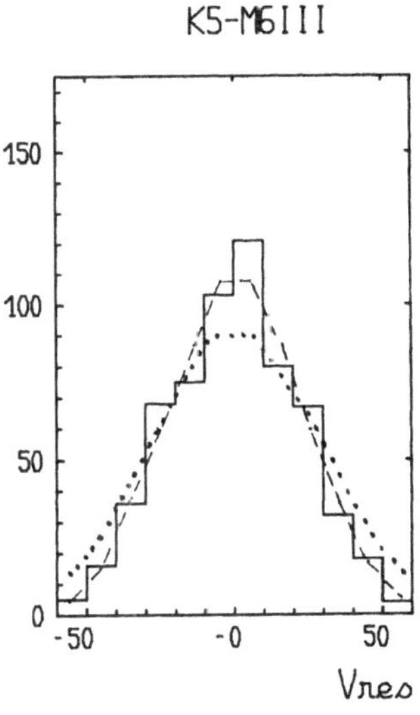

Figures 2a - 2d : Histograms of the residual radial velocities, compared to expected histograms if high-velocity stars are included (dotted lines) and excluded (dashed lines)

DISCUSSION

GRIFFIN R.F. - The rejection of stars with velocities above a certain limit seems rather arbitrary. Its sole advantage is that the distribution that remains is more like a gaussian. But in practice most observational distributions contain more small -and correspondingly, therefore, more large- values than the gaussian distribution. The only real conclusion is that the gaussian is not an accurate model of the observed distributions.

OCHSENBEIN - The gaussian velocity distributions have some physical meaning in a well-mixed population ; the classical way for the interpretation of distribution with more large values is a several-component model (superposition of several gaussians).

CAYREL - Can you suggest an explanation for the gap of 4 km/s^{-1}, between σ 15 km/s^{-1} and 19 km/s^{-1} ?

OCHSENBEIN - I have no explanation, but this gap should be correlated with the mean ages of each group.

BIDELMAN - I think it might have been better to get points at A7 and F0 rather than just at A9. This might tend to fill in the gap between A9 and F2.

OCHSENBEIN - A9 point corresponds to A7 - F0 stars. Very few stars are classified A7 - A9 and reliable kinematical parameters cannot be derived for each of these classes.

KINEMATICS AND SPATIAL DENSITY OF MAIN SEQUENCE SMR STARS

M. Grenon
Geneva Observatory
1290 Sauverny
Switzerland

ABSTRACT. A population of supermetal-rich stars is searched among nearby and proper-motion stars. The metallicities, Teff, Mv are obtained photometrically and lead to estimates of the age and space density of the SMR subpopulation. Its origin is deduced from the evolutionary and kinematical properties of the SMR candidates.

1. INTRODUCTION

A significant fraction of red giants appears to be metal-rich to super metal-rich with strong CN bands and sometimes also conspicuous C_2 bands. It appears important to know wether these peculiar abundances are primordial or the consequence of surface chemistry modifications resulting from the stellar evolution.

As a deep mixing occurs at the beginning of the giant branch, it is necessary to consider the red giants progenitors and to check if a population of very metal-rich stars exists among the subgiants and dwarfs which may be identified as the counterpart of the SMR giants group.

This work will describe some of the physical and kinematical properties of the objects found to be metal-rich or SMR among the Gliese's, Woolley's and Luyten NLTT stars.

2. DEFINITION

According to our photometric calibrations, the Hyades stars have a [M/H] ratio = +.15. In the following the stars having $[M/H] \in [+.20,+.29]$ will be named metal-rich or MR and these with $[M/H] \geqslant +.30$, super metal-rich or SMR.

M. Jaschek and P. C. Keenan (eds.), Cool Stars with Excesses of Heavy Elements, 147–152.

3. THE STAR SAMPLE

The Geneva calibrations, Grenon (1978,1982) have been applied to all stars from Gliese's and Woolley's catalogues brighter than mv=11.0 and measured in the Geneva photometric system. Southern NLTT stars with $\mu \geqslant .19''/y$ and $m_R \leqslant 11.0$ have also been partially measured. The declination zone (0°,-10°) is now almost completely observed and will be used as a test zone. The selection according to a minimum value of μ provides a sample defined by a minimum tangential velocity, function of the distance ($V_T \geqslant 90$ km/s at 100 pc). Between 25 pc and 100 pc mainly old disk and halo stars are selected.

For all stars later than F2 and with Teff $> 4750^oK$ if dwarfs,an accurate estimation of [M/H] (s.e + 0.05) is obtained in addition to the parameters Mv and Teff.

4. THE FREQUENCY OF MR AND SMR STARS

648 stars from Gliese's catalogue have photometrically defined [M/H] ratios. We have found 68,50 and 27 stars having respectively [M/H] $\geqslant$ +.20, +.25 and +.30. Among the nearby stars the fractions of MR and SMR stars are then:

[M/H] $\geqslant$ +.20	:	10.5 $\pm$1.7 %
[M/H] $\geqslant$ +.25	:	7.7 $\pm$1.4 %
[M/H] $\geqslant$ +.30	:	4.2 $\pm$1.0 %

These percentages are very similar when the sample is restricted to the stars with a distance confirmed to be less than 22.2 pc ($\pi \geqslant .045''$), but they have to be considered as upper limits for the solar neighbourhood as there is some evidence that part of mild metal poor stars have been missed in the Gliese's compilation, cf. Grenon (1984).

In the NLTT sample, the MR and SMR are more frequent i.e. 40 and 26 stars among 356 have [M/H] $\geqslant$ +.25 and +.30 respectively :

[M/H] $\geqslant$ +.25	:	11.2 $\pm$2.4 %
[M/H] $\geqslant$ +.30	:	7.3 $\pm$1.8 %

When the distances are restricted to the interval 25 pc to 50 pc these proportions increase to 16.0 and 9.2% , an indication that MR and SMR stars have an old disc kinematical behaviour.

5. THE MR & SMR HR DIAGRAM

The HR diagram for the nearby stars show that MR and SMR stars are mostly K dwarfs. The stars with [M/H] $\in$ (0.20, 0.24) are mainly young stars having motions similar to those of Hyades group or to other younger groups. The most metal-rich stars form an HR diagram similar to that of the metal-rich globular clusters. The hottest SMRs are either binaries or high velocity early G type subgiants which may be classified as blue stragglers. In these later cases an effect of mixing may be suspected.

The HR diagram for NLTT stars with [M/H]>+.25 is given in fig 1.; they are older than M 67 and in particular the SMRs form an homogeneous group in age with a well defined turn-off at G7IV or (B_2-V_1)= +.44. The low main sequence is densely populated.

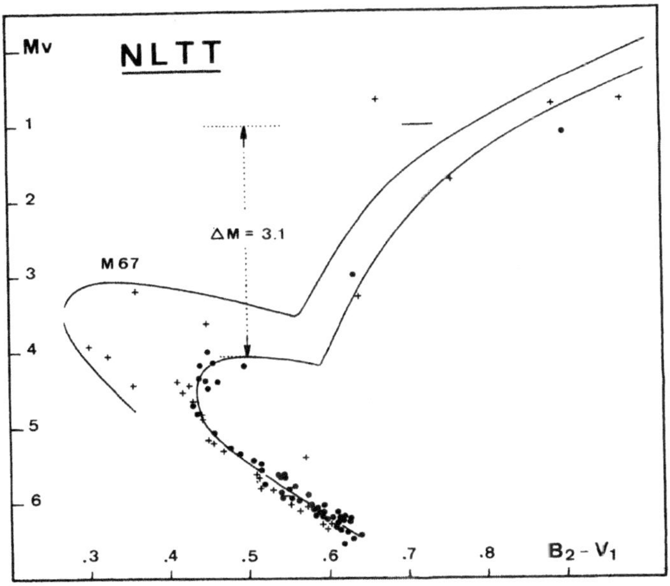

Fig. 1. The HR diagram of NLTT stars having distances exceeding 25 pc. Symbols: + : [M/H] ∈ +.25,+29
• : [M/H] $\geqslant$ + 30
Note the large ΔMv between the horizontal branch and the subgiant branch.

6. THE AGES

Metal-rich star show a wide range of evolutionary ages from young disk to very old disk. The SMR age distribution appears somewhat bimodal: part of them are associated with young population I groups, whereas the majority is definitely very old. For [M/H] = +.35 or Z = 0.04. the turn-off temperature is about 5550 °K. The top of the subgiant branch is at Mv = 4.1. The derivation of an age from the M bol/log Te diagram with evolutionary models from Ciardullo et.al. (1977) is rather unsafe as the agreement between the observed and theoretical HR diagrams is poor, partly due to the treatment of convection. For obtaining the main sequence fitting a high helium content is needed, i.e. y=0.35. This also implies a ratio of the relative increase of helium on heavy elements $\Delta Y / \Delta Z = 3.0$. The only reliable parameters for the age determination is the subgiant branch luminosity which indicates an age $a = 9.10^9$ y, a model dependent value.

7. THE SPACE VELOCITIES AND SMR ORIGIN

The MR and SMR stars show quite a variety of kinematical properties. The excentricities of their galactic orbits range from 0.00 to 0.30. Most SMRs, young according to the HR-diagram, are also kinematically young. The mean galactocentric distances of MRs and SMRs are smaller than that of the sun and are as small as 7.8 kpc if the sun lies at 10 kpc from the galactic center. The distribution of pericentric distances Rp, ranges from 5.2 kpc to 10.0 kpc and peaks at about 6.5 kpc for SMRs. A second peak seems to occur near Rp = 9.0 kpc corresponding to the young disk metal-rich population. For the SMRs the mean velocities and the velocitiy dispersions are as follows :

	<U>	<V>	<W>	σ_u	σ_v	σ_w
Gliese's stars :	+23	-28	-11	45	30	18
All stars :	+26	-41	-11	41	31	18

The SMRs appear to be a mixture of stellar populations: 1) a young disk component associated with young stellar groups in almost circular motion. 2) an old disk component characterized by $e \lesssim 0.3$, $\langle Rp \rangle \simeq 6{,}5$ kpc, $\langle Ra \rangle \lesssim 11.0$ kpc. It is noticeable that the most metal-rich stars have also, in the mean, the smallest galactocentric distances. This fact is a signature of the radial gradient of metal abundances. The high velocitiy SMRs may be identified as members of the galactic bulge observed, in the solar vicinity, near their apocenters.

On the other hand the low velocity SMRs seem to have been recently formed from interstellar matter progressively enriched in metal during the galactic disc evolution. From the maximum [M/H] distribution as a function of the mean galactocentric distance it results that the maximum metallicity presently observed in the local interstellar matter [M/H] = +.30, remains lower than that reached in the bulge about 10 billion years ago, i.e. [M/H] = +.50.

8. CONCLUSIONS

A population of SMR stars has been identified among main sequence and subgiant stars, with well defined kinematical and physical properties.

In a given volume of space near the sun, their proportion amounts to about 4%. The fraction of true SMR among giants is probably slightly less because -part of mild metal poor stars are missing in the reference sample of nearby stars, -some of the F and G subgiants may be mixed objects (binaries or blue stragglers) and finally because the density of young red giants progenitors is less than that of main sequence G and K stars, a mass function effect.

When the sample of giants is biased in favour of high velocity stars the percentage of true SMR stars may reach about 9%. The giants apparently SMR, and with V component of space velocity between -50 and -80 and with $|U| \leqslant 100$ km/sec are highly probably true super metal-rich stars.

BIBLIOGRAPHY

Ciardullo R.B., Demarque P., 1977, Trans. Yale University 33.
Grenon M., 1978, Publ. Obs. Genève Ser. B, 5.
Grenon M., 1982, I.A.U. Coll. 68, 393.
Grenon M., 1983, Van Vleck Observ. Contrib. 1, 439.

DISCUSSION

HECK - What is the main source of absolute magnitudes for these stars ?

GRENON - I used photometric values of M_V. The calibration is based on large trigonometric parallax stars in the late G and K dwarf domain, and is extended to the hotter stars and red giant domain using the galactic cluster sequence.

JASCHEK - I am sligthly puzzled that you obtained more metal rich stars (5-10%) than metal poor stars. How do you relate this to the change of metallicity with age ?

GRENON - In the solar neighborhood the number of stars with $|me/H| > 0.3$ is very similar to that of stars with $|me/H| < 0.3$. When high velocity stars are considered, the fraction of SMR stars or metal poor stars is a decreasing function of the space velocity amplitude. The apparent discrepancy in the age-metallicity relation results from the fact that at least the high velocity SMR stars are not born at a solar galactocentric distance, but originate from denser and more metal rich regions closer to the galactic center.

KEENAN - In view of the enormous discrepancies between the best analyse of the best known SMR stars, do you think that the distinction between metal-rich and SMR stars is really significant ?

GRENON - From what we learn from main sequence stars, mostly not analysed spectroscopically, the distinction between MR and SMR ramains useful as the old

SMR's form a subpopulation, related to the galactic bulge, having abundancy without counterpart among the young disc objects. The local interstellar matter does not seem now enough metal-enriched to produce similar stars.

SOME STATISTICS ON BARIUM STARS

M. Holtzer
Observatoire de Strasbourg
11, rue de l'Université
67000 Strasbourg
France

INTRODUCTION. The starting point of this statistical investigation was the "Catalog of Stellar Groups" (Jaschek and Egret, 1985). The 170 certain Ba stars and the 140 marginal Ba stars contained in this catalog were compared to a sample of 200 giants of the same spectral type, G5 to K5, taken from the "Bright Star Catalog" (Hoffleit, 1982). Keenan's criteria for defining barium stars was used. The aim of our study was to determine if any differences exist between what have been named the "certain" barium stars and the marginal barium stars. However, we must caution that the fact that some of the stars we used as Ba stars could be adjudicated to other groups such as CN, CH, or C introduces somewhat of a doubt as to their classification as barium stars.

1. SURFACE DISTRIBUTION

The samples of bright giants as well as the marginal Ba stars are homogeneously distributed, but the certain Ba stars show an irregular distribution for the Northern Galactic hemisphere. The resulting statistical bias is mainly due to the fact that most Ba stars were discovered in the Southern Hemisphere, where coverage is uniform until -26°, good until 0°, and very spotty afterwards.

2. MAGNITUDE DISTRIBUTION

Figures 1a and 1b represent the certain Ba stars and all known marginal Ba stars, respectively. Figure 2b shows the marginal Ba stars which appear in the Michigan Catalog south of declination -26°. These figures have been broken down into magnitudes smaller than or equal to 6.5 (Bright Star Catalog) and into magnitudes smaller than or equal to 9 (Michigan Catalog, Vols. I - III, 1982). Figures 1b and 1a will enable us to compare the magnitudes of these two subgroups. Figure 1b reveals the presence of a peak between magnitudes 4.5 and 5 which does not occur in Figure 1a. In Figure 2b, the peak, composed of 28 stars between magnitudes 4 and 5 has greatly diminished, implying that it is an artifact due to unequal discovery techniques. A comparison of Figures 1a and 2a shows that the total number of the certain Ba stars is about the double of those discovered south of -26°. Since,

M. Jaschek and P. C. Keenan (eds.), Cool Stars with Excesses of Heavy Elements, 153–157.

however, this zone represents only one quarter of the sky, one can easily expect that their real number is at least the double of their known number today, i.e., about 350 or 1.2 % of all K-type giants. If we superimpose Figure 1b, taken from the C.D.S. catalog, onto Figure 2b, taken from the Michigan Catalog, we see that they agree in distribution. This implies that the Michigan Catalog contains practically all of the marginal Ba stars which have been discovered in the whole sky -a highly unlikely implication.

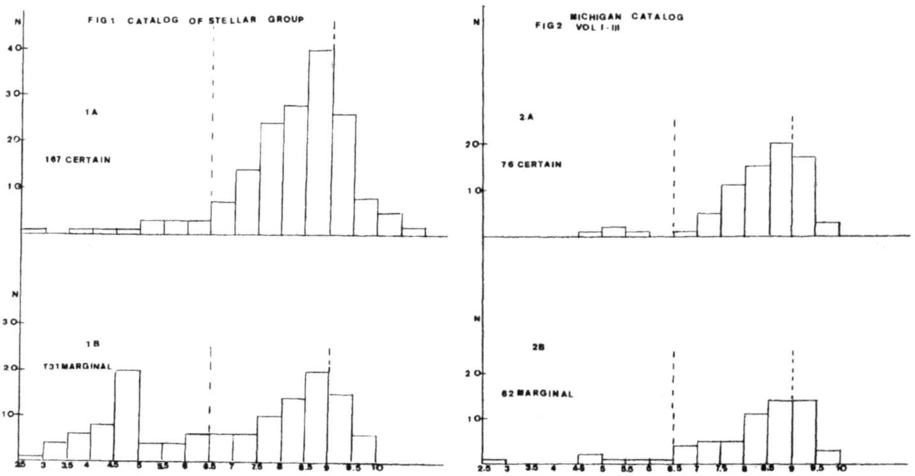

3. RADIAL VELOCITY DISTRIBUTION

We used all radial velocities available in the literature. They number 85 for the certain Ba stars and 68 for the marginal Ba stars. These velocities were used to derive the solar apex and the distribution of residual velocities. This is shown in Table I.

TABLE I

	L	b	V_0 km/s	σ km/s
Certain	61°	11°	15	20
Marginal	62°	16°	14	19
B.5 Giants G5 - K5 *	70°-50°	34°-14°	24-12	23-19

*Taken from Ochsenbein and Holtzer (this volume)

For the dispersion derivation, the residual radial velocities were corrected for a galactic rotation assuming $\bar{M} = -0^m.2$ (Jaschek et al, 1985) and $M = +0^m.3$ (Mac Connel, 1972) for the certain and the marginal Ba stars, respectively. However, the influence of the galactic rotation correction is small. The following conclusions seem justified.

3.1. A comparison of giants and the certain Ba stars shows that the galactic latitude of the apex is slightly ouside the apices of the normal giants. This may be due to the distribution anomaly of the Ba stars on the sky. The solar velocity and the dispersion are, however, in good agreement. Therefore, we may conclude that the kinematic behavior of the Ba stars closely resembles that of normal giants, and precludes any great difference in age.

3.2. The only indication that differences may exist between giants and the Ba stars is that out of 85 residual radial velocities, 4 are larger than 60 km/s (or 5 %). The frequency of the residual velocities shows a slightly asymmetric distribution with an excess of negative residual velocities, perhaps indicating a mixture of two different radial velocity distributions. This is shown in Table II.

TABLE II

RESIDUAL VELOCITIES

	$V_{RS} < 0$	$V_{RS} > 0$
Certain	57 % ± 5,3 %	43 % ± 5,3 %
Marginal	50 % ± 8,6 %	50 % ± 8,6 %
Giants G5 - K5	51 % ± 4 %	49 % ± 4 %

3.3. If we compare the certain Ba stars to the marginal Ba stars, we find little difference between them, just as we find little difference between the Ba stars and the normal giants.

4. COLORS

We started our work using the UBV photometric system, but unfortunately, only a small number of the certain Ba stars are available in this system. Figure 3 shows that the marginal Ba stars and the giants are distributed very similarly on the same sequence, but that the certain Ba stars are shifted towards the red.

The shift observed seems to be caused by a depression in the continuum of the certain Ba stars, which affects the B-filter. To evaluate the effect of such a depression on this filter, we used a spectrophotometry of Gow. It provides the mean relative spectral energy distribution of three couples of barium stars compared to standard stars. By convolving this energy distribution with the known transmission of the UBV filters one can calculate the effect of the depression on the filters. For a K0 III star ($B - V = 1^m.0$, $U - B = 0^m.85$, $V = 0$) a shift of $0^m.09$ with a slope of 40°

results in the UBV diagram. The observed shift of the two sequences in Figure 3, obtained with the least square line through the plotted certain Ba stars on the UBV diagram, is $0\overset{m}{.}10$, which agrees very well with the calculation, suggesting, therefore, that the deplacement of the certain Ba stars from the sequence for giants is due to a depression in the energy distribution curve.

We can discard interstellar extinction as an alternative explanation because a close examination of the two color plots shows that among the stars lying on the "shifted" sequence, one finds both nearby and far away stars. This eliminates interstellar reddening as a major explanation of the shift.

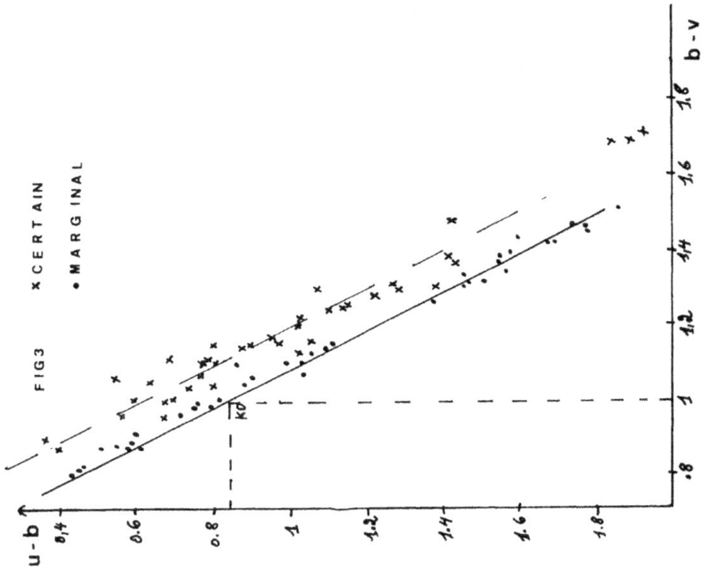

A close look at Figure 3 also suggests the possible existence of a gap in the distribution over the two color diagrams, with very few, if any, points in the color interval. A similar "gap" is found in Figure 4, which portrays the distribution of 1500 giants. The possible gap is found at U - B = 1.2 - 1.4.

5. CONCLUSION

We have shown that the statistics of the Ba star group suffers from observational and classification biases. The marginal Ba stars and the certain Ba stars seems to be homogeneous in both age and population type.

FIG 4

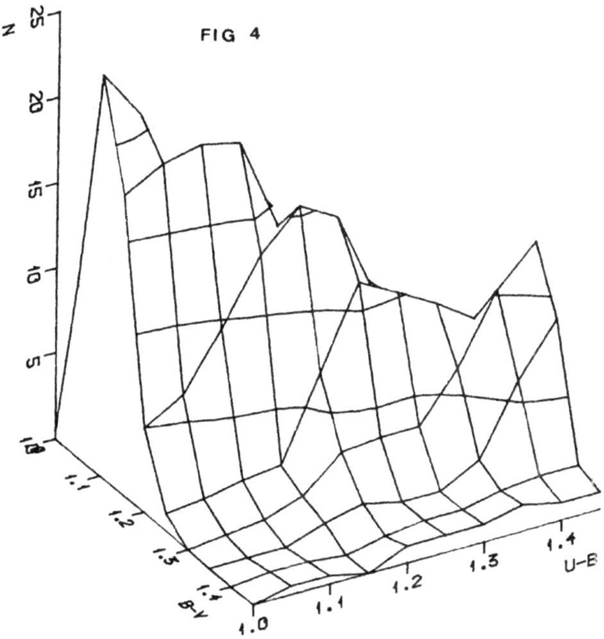

REFERENCES

Gow E. : 1976, "Spectrophotometry of Ba, CH and R type carbon stars", Ap J **81**, 993

Hoffleit D. - Jaschek C. : 1982, "The bright star catalog" - Yale University Observatory

Houk N. : 1982, "Michigan catalog", Vol. I - III

Jaschek M. - Jaschek C. - Grenier S. - Gomez A. - Heck A. : 1985, "Absolute magnitude of Ba stars" (this colloquium)

Jaschek M. - Egret D. : "Catalog of stellar groups", C.D.S. publ. (in preparation)

Mac ConnellD. - Frye G. - Upgren A. : 1972, "The absolute magnitue of Ba stars" Ap J **77**, 384

DISCUSSION

JOHNSON - You seem to regard the kinematic difference between the certain Ba stars and the marginal Ba stars as significant. Is that correct ? Does that mean that certain Ba stars are older ?

HOLTZER - The answer is "yes" to the first question and "perhaps" to the second.

CARBON STARS AND THE CENTER OF OUR GALAXY

M. F. McCARTHY S.J.
Specola Vaticana

V.M. BLANCO
CTIO

ABSTRACT. Red giant stars are studied on infrared sensitive plates taken with a grism and 4 m CTIO reflector . Each of our regions is 24' diameter and covers 0.12 sq degrees. We find M5 to M9 stars abundant and carbon and S stars almost non existent in the direction of the galactic center. In two regions (Baade's window at NGC 6522 and in Sgr I) we have counted more than 1 000 very cool M giants and no C or S stars. Supplementary photography of 10 other regions of the Nuclear Bulge region reveals one genuine C star at $l = 0^{\circ}$ $b = -5^{\circ}6$, no S type stars and very large numbers of M stars.

More than 1 000 red giant stars, none of which is a carbon star, have been marked by the authors on infrared sensitive plates obtained at CTIO with a grism and the 4 meter telescope. Two regions in the direction of the galactic center were surveyed, each of which covers 0.12 square degrees in area. One is Baade's window near NGC 6522 at $l = 1^{\circ}0$ and $b = -3^{\circ}9$. The other region, named Sgr I at $l = 1^{\circ}4$ and $b = -2^{\circ}6$.

The scarcity of carbon stars is striking and points up a fundamental difference in the red giant populations of the bulge relative to those found in the galactic periphery by Blanco (1965) and by Furenmayor (1981) and in the Magellanic Clouds by Blanco, Blanco and McCarthy (1978).

Two stars which had been noted for C star characteristics in these two regions were eliminated from our lists when subsequently obtained plates showed that overlapping had caused the blending of an M star image with one of earlier type in these most crowed regions.

Recently Blanco has obtained 10 additional grism plates in the direction of the galactic center. These are taken with the same equipment and cover fields of the same size as those mentionned above. The new fields are centered near $l = 0^{\circ}$ and at b ranging from -1° to -11°.

M. Jaschek and P. C. Keenan (eds.), Cool Stars with Excesses of Heavy Elements, 159–161.

One genuine carbon star was found by Blanco at l = 0° and b = -5°6. This star is now being studied by Frogel photometrically at JHK wavelengths. It will be very important to obtain spectra at higher dispersion to see if this may be an early R type star similar to those discussed recently by Dominy (1984).

These 12 fields look out through different concentrations of absorbing material in the direction of the nuclear bulge and, as far as absorption permits, approach and go beyond the bulge at galactocentric distances no closer than 500 parsecs. We have published (1984) our finding chart, positions, magnitudes and spectral classification for the stars in Baade's window near NGC 6522. 303 red giants of type M5 and later, including 22 stars of types M8 and M9, were marked. There are no C or S stars discovered in this field. For our study of the Sgr I field we are preparing for publication finding charts and positions plus spectral classifications. Here, 765 red giant stars of types M5 and later have been found and again there are no C or S stars. Work is only just beginnning on the new material obtained recently by Blanco. In these fields the M giants are very abundant, 1 carbon star and no S type stars have been detected. A marked contrast in the red giant populations, indicated earlier by Blanco, Hoag and McCarthy (1979) between the centers of the Magellanic Clouds and the center of our galaxy is emphasized by these new observations made in the direction of the bulge.

The relative scarcity of carbon stars in starfields known for prominent metal richness is well discussed but requires further study. Furthermore, age effects point to an intermediate age for most C stars thus far observed. Thus the absence of red giant carbon stars close to the center of the galaxy may indicate an older and well stabilized population of metal rich red giant stars which does not show major signs of large scale star formation within the past few 10^8 years. This state of tranquility is noted well outside of the very active nuclear region close to the geometric center of the galaxy. Our observations of red giant stars do not reach galactocentric distances closer than 500 parsecs.

REFERENCES

Blanco V.M. : 1965, "In galactic structure" eds. A. Blaauw and M. Schmidt (Univ. of Chicago Press, Chicago) p. 241
Blanco V.M. - Blanco B.M. - McCarthy M.F. : 1978, Nature **271**, 639
Blanco V.M. - Hoag A.A. - McCarthy M.F. : 1978, "In the HR diagram" IAU Symp. 80, eds A.G.D. Philip and D.S. Hayes (Reidel, Dordrecht) p. 33
Blanco V.M. - McCarthy M.F. - Blanco B.M. : 1984, Astron. J. **89**, 636
Dominy J.F. : 1984, Astrophys. J. Suppl. 57, 27
Furenmayor F.J. : 1981, Rev. Mexicana Astron. Astrof. **6**, 83

DISCUSSION

CATCHPOLE - Does the observed distribution of stars with magnitude reflect the fact that you still have an observational cut-off or are you seeing beyond the distance of the galactic centre and seeing a real reduction of star numbers ?

McCARTHY - Yes, we think we are seeing beyond the galactic centre.

JASCHEK - Could you make some comments on the space distribution of your stars ?

McCARTHY - We regard the maximum frequency for late M-stars to occur about I = 13.3. This can be regarded as caused by the large stellar space densities found where the line of sight passes close to the galactic center. This sharp maximum corresponds approximately at the point where the line of sight passes closest to the galactic center. This will allow us to estimate the absolute magnitudes of the giant M stars in the nuclear bulge. Our observations of apparent infrared magnitudes are average values, obtained from iris diaphragm photometer reductions of I plates and are based on a photoelectric sequence based on Cousins RI magnitude system. Differences among sub-regions of same 0.12 square deg. area on Baades window indicate varying absorption in the area and we know that the final value for the distance modulus of the galactic center is still under investigation ; with good magnitude, color and distance modulus values coming soon, the calculation of stellar densities in the direction to the galactic center should be straightforward.

LU - Numerous M-stars are found on the 4 m survey toward galactic center. Can you determine what percentage of them are M-giants ?

McCARTHY - Our dispersion (2350 A/mm) does not allow us to make luminosity classifications of our M stars. If we assume that the space density found in the solar vicinity for dwarfs M6 and later holds true along the line of sight to the limiting magnitudes reached by our grism plates, the expected number is less than 1 star per 0.12 sq. degree. So here $<$ 1 star out of the 301 M stars in Baade's window and $<$ 1 star out of 765 M stars in Sgr I would be expected to show dwarf characteristics. The number of supergiants should also be negligible especially since they are usually of types earlier than type M5. We discuss this in our papers.

THE EVOLUTIONARY CONNECTION BETWEEN S AND C STARS: EVIDENCE FROM STAR CLUSTERS AND THE MAGELLANIC CLOUDS

T. Lloyd Evans
South African Astronomical Observatory
P.O. Box 9 Observatory 7935 Cape
South Africa

ABSTRACT. Observations of intermediate-age clusters in the Magellanic Clouds indicate an evolutionary sequence, M→S→C, for stars on the asymptotic giant branch. Trends with age of the clusters are generally in agreement with theoretical prediction but the lifetime of the S star stage appears to be longer than expected. CS stars have been found among high luminosity field variables and can be accounted for by the envelope-burning process. The presence of faint S stars in ω Cen, where they may be interpreted as the result of a high primordial heavy-element abundance, shows that S stars are not necessarily all self-enriched in s-process material, while the low luminosity J stars have probably evolved up the giant branch as carbon stars without passing through an S star stage.

1. THE AGB IN MAGELLANIC CLOUD CLUSTERS.

The stars on the upper AGB in many Magellanic Cloud star clusters of intermediate age have been observed in order to construct a picture of evolution from M via S to carbon stars (Bessell, Wood & Lloyd Evans 1983; Lloyd Evans 1983a, 1984a). The observational definition of the M→S transition depends on the ability to detect the 6473 Å band of ZrO and hence on the quality of the spectra, but most stars can be classified as O-rich or C-rich by infrared colours alone and the few marginal cases are easily tackled spectroscopically. The spectral type at the M→S transition depends on metal content (Bessell, Wood & Lloyd Evans 1983) while both transition points are brighter in younger clusters (Lloyd Evans 1983a) and probably also in more metal rich clusters (Lloyd Evans 1984a). This is in accordance with theoretical predictions (Iben & Truran 1978; Renzini & Voli 1981; Wood 1981) but the S stars occupy a range of several tenths of a magnitude in M_{bol} and are more numerous than carbon stars in the younger clusters, contrary to the prediction that neutron-capture nucleosynthesis would follow the initial dredge-up of carbon to the surface and that the changeover from an O-rich to a C-rich atmosphere would be very rapid (Iben & Renzini 1982a, b). Stars intermediate between the early S

M. Jaschek and P. C. Keenan (eds.), Cool Stars with Excesses of Heavy Elements, 163–166.

stars and the mostly strong banded carbon stars are rare: if Keenan & Boeshaar's (1980) calibration of various spectroscopic groups as a function of C/O is correct, this supports the idea of a rapid changeover from O-rich to C-rich but also implies a long period during which enrichment of the envelope with heavy elements occurs.

The galactic globular cluster NGC 6723 contains two MS stars with $M_{bol} \sim -3.3$, nearly a magnitude fainter than the faintest Magellanic Cloud S star (Lloyd Evans 1984b). Detection of the ZrO band is favoured by the brighter apparent magnitude of these stars but it seems likely that the M→S transition is fainter in this cluster which is probably much older than the Magellanic Cloud clusters with S stars. It would be interesting to know why other galactic globular clusters do not contain S stars.

Early S stars are also found among the luminous variables in the fields of both Magellanic Clouds (Wood, Bessell & Fox 1983), with luminosities mostly in the range $M_{bol} = -6$ to -7. Two CS stars have $M_{bol} \sim -6$ and probably result from the envelope-burning process which is expected to slow the carbon-enrichment process in more massive stars (Iben 1975; Renzini & Voli 1981). The surface distribution of the luminous red variables in the SMC indicates masses $\lesssim 6\ m_{\odot}$ (Lloyd Evans 1971, 1984c).

BH Cru, which was classified as an SC star with weak ZrO and no C_2 in 1965-73 (Catchpole & Feast 1971; Keenan & Boeshaar 1980), had moderately strong C_2 and probably no ZrO in 1980 while C_2 was strong and ZrO absent in 1984 (Lloyd Evans, in preparation). The spectra cover a substantial part of the light curve during both time intervals, so it is unlikely that the spectral changes represent the ordinary cyclic behaviour of the star. Explanations in terms of the development of an extended atmosphere or the rotation of a spotted star need to be excluded before it can be said that we have seen stellar evolution in action.

2. DO S AND C STARS ALL ARISE FROM THE SAME PROCESS?

The Magellanic Cloud clusters show that S and C stars represent a normal stage in the evolution of single stars and the luminous variables as well as the fainter S stars in NGC 6723 can probably be accounted for in the same way. However faint ^{13}C-rich carbon stars (J stars) such as that in NGC 419 (Bessell, Wood & Lloyd Evans 1983) and a second example in NGC 2213 (Lloyd Evans, in preparation) as well as those previously found in the field of the LMC (Richer 1981) are not readily explained by the envelope-burning model which accounts for the luminous J stars among LMC field carbon stars (Richer, Olander & Westerlund 1979). A re-examination of published data shows that the early and late R stars form a physical group characterised by a low $^{12}C/^{13}C$ ratio and it is probable that the low luminosity J stars of the Magellanic Clouds are the brightest and coolest R stars. They may have evolved up the giant branch from the red giant clump without ever passing through an S star stage, as early R stars in the Galaxy do not show enhanced s-process elements (Dominy 1984). An unusual development of the He flash, which does not normally affect the envelope composition in a conspicuous manner, may be responsible.

It is also possible in principle for stars of abnormal composition to reflect the composition of their primordial material and not the nuclear reactions which take place in their own interiors. The galactic globular cluster ω Centauri has a non-uniform composition, with strong BaII in many stars (Dickens & Bell 1976), and the MS and early S stars with M_{bol} in the range -2.4 to -3.6 find a natural explanation as the coolest representatives of the most metal rich population in the cluster (Lloyd Evans 1983b). The heavy elements which are overabundant in these stars were presumably created in an earlier stellar generation. The only possible difference found so far between these and the S stars in the Magellanic Clouds and galactic field is that the ω Cen stars do not show a violet deficiency.

REFERENCES

Bessell, M.S., Wood, P.R. & Lloyd Evans, T., 1983. Mon. Not. R. astr. Soc., **202**, 59.
Catchpole, R.M. & Feast, M.W., 1971. Mon. Not. R. astr. Soc., **154**, 197.
Dickens, R.J. & Bell, R.A., 1976. Astrophys. J., **207**, 506.
Dominy, J.F., 1984. Astrophys. J. Suppl., **55**, 27.
Iben, I., 1975. Astrophys. J., **196**, 525.
Iben, I. & Renzini, A., 1982a. Astrophys. J., **259**, L79.
Iben, I. & Renzini, A., 1982b. Astrophys. J., **263**, L23.
Iben, I. & Truran, J.W., 1978. Astrophys. J., **220**, 980.
Keenan, P.C. & Boeshaar, P.C., 1980. Astrophys. J. Suppl., **43**, 379.
Lloyd Evans, T., 1971. Observatory, **91**, 118.
Lloyd Evans, T., 1983a. Mon. Not. R. astr. Soc., **204**, 985.
Lloyd Evans, T., 1983b. Mon. Not. R. astr. Soc., **204**, 975.
Lloyd Evans, T., 1984a. Mon. Not. R. astr. Soc., **208**, 447.
Lloyd Evans, T., 1984b. Mon. Not. R. astr. Soc., in press.
Lloyd Evans, T., 1984c. Submitted.
Renzini, A. & Voli, M., 1981. Astr. Astrophys., **94**, 175.
Richer, H.B., 1981. Astrophys. J., **243**, 744.
Richer, H.B., Olander, N. & Westerlund, B.E., 1979. Astrophys. J., **230**, 724.
Wood, P.R., 1981. Physical Processes in Red Giants, p 135, eds Iben, I. & Renzini, A., Reidel, Dordrecht, Holland.
Wood, P.R., Bessell, M.S. & Fox, M.W., 1983. Astrophys. J., **272**, 99.

DISCUSSION

WING - Now that a substantial number of S stars are known in the Magellanic Clouds, would you care to speculate on why no pure S stars have been found ? They should be easier to recognize than the ones you find. Perhaps another way of asking the same question is, what is a pure S star, in the evolutionary sense ?

LLOYD EVANS - Possibly a pure S star is confined to a narrow branch of C/O and is rather rare, as are the SC stars. Pure S stars, SC stars and weak banded C stars are found among luminous Magellanic Cloud stars, but not in the intermediate age clusters, where there seems to be a rapid transition from mild S stars to C stars with fairly strong bands.

WOOD - There are now two pure S stars known in the Magellanic Clouds (Blanco et al, 1981, PASP **93**, 532) ; (Wood et al, 1984, Ap. J., submitted).

LLOYD EVANS - That makes two S stars, one SC star and two Cs stars in the Magellanic Clouds so far. The relative numbers may be quite normal in view of the small sample.

CATCHPOLE - Did I understand that you suggested the Cloud SC stars are actually occurring as a result of C burning in C stars, so the sequence is really S, C, SC ?

LLOYD EVANS - No. The SC stars are intermediate between S and C stars, but represent a narrow range in C/O, so are only seen if a star changes slowly from S to C. This may occur in a high-mass star where envelope burning of C^{12} to C^{13} and N^{14} occurs to slow the process, and even prevent a star from ever becoming a carbon star, but not in a low-mass metal-poor cluster star, where the S→C change is very rapid.

KEENAN - Do you think that the extraordinary changes in the spectrum of BH Cru have any evolutionary significance ?

LLOYD EVANS - It would be nice if they be ! However we would be lucky to see this and if, as claimed, TT Cen has shown similar change, it may be rather a normal process in such stars. It seems unlikely to occur every pulsation cycle. All early spectra (1965 - 73) of BH Cru show no C_2, later spectra (1980 - 84) do, but it could be due to rotation of a spotted star or to formation of an extended envelope. If a star changed back in CS→SC, not SC→CS, we would probably not want to attribute it to evolution.

KINEMATICS OF CARBON STARS IN THE LARGE MAGELLANIC CLOUD

Harvey B. Richer, Michael W. Ovenden and Gary D. Joslin
Department of Geophysics and Astronomy
University of British Columbia, Vancouver, Canada

B.E. Westerlund and N. Olander
Astronomical Observatory, Uppsala, Sweden

ABSTRACT. Published radial velocities of HII regions, luminous early- and late-type stars, and planetary nebulae, are consistent with a rotation law for the Large Magellanic Cloud $V \propto R$ for $R < 5.5°$. New radial velocities for some of the carbon stars in the WORC Catalogue extend the determination of rotation velocities to $R \simeq 7°$. These observations are consistent with $V \propto R$ for $R < 5.5°$, with a rapid decrease of V for greater distances. If these results are interpreted as indicating solid-body rotation of a finite disc, and if a true distance modulus of 18.6 is assumed for the LMC, the mass of the LMC is found to be $\sim 2.6 \times 10^{10}$ solar masses.

We have analysed radial velocities of a number of objects in the Large Magellanic Cloud included in the catalogue of Feitzinger and Weiss (1979). The radial velocities of planetary nebulae, HII regions and luminous stars are consistent with solid-body rotation of a disc. New radial velocities of some of the carbon stars in the catalogue of Westerlund, Olander, Richer and Crabtree (1978, hereinafter WORC) are also consistent with such an interpretation, and are far enough from the dynamical centre of the LMC to show a velocity cut-off at the edge of the disc.

In the analysis, it is assumed that the LMC is in the form of a thin disc, and that the radial velocity (ρ) of a star is the sum of the line-of-sight systemic velocity component (ρ_0) and the line-of-sight component of a velocity (V) directed at right angles to the radius vector (R) to the star from the dynamical centre of the LMC. Since there will be peculiar velocities as well as errors of observation in finding V as a function of R, we apply to each observed radial velocity the appropriate geometrical statistical weight. For each analysis, the weighted observations were collected into smoothed normal points, which were then fitted to a straight line $V = a + bR$, R being measured in equivalent degrees subtended at the observer.

To convert from an observed position (α, δ) and observed radial velocity (ρ) to distance (R) and cross-velocity (V) requires a knowledge of the inclination to the line-of-sight (i) of the normal to the plane of the LMC, the position angle (ϕ_c) of the line of nodes, the

M. Jaschek and P. C. Keenan (eds.), Cool Stars with Excesses of Heavy Elements, 167–170.

systemic radial velocity (ρ_0) and the coordinates (α_0, δ_0) of the dynamical centre of the LMC. From the apparent shape of a well-defined ring of HII regions in the LMC, $i=27°$, $\phi_0=171°$ (Feast, 1964). The parameters ρ_0, α_0 and δ_0 were determined to be +273.8 km/s, $5^h19.9^m$ (1950) and $-68°37.2'$ (1950) respectively; this combination of parameters giving the best fit of the HII observations to the model.

The catalogue values of radial velocity for the HII regions and the planetary nebulae separately gave good linear correlations. The results are given in Table I, where the goodness of fit of the linear regression is indicated by the coefficient of determination $r^2 = [\Sigma xy - \Sigma x.\Sigma y/n]/[\Sigma x^2-(\Sigma x)^2/n][\Sigma y^2-(\Sigma y)^2/n]$. Values close to 1 indicate a good fit.

TABLE I

OBJECT	a	b	r^2
HII Regions	0.0	+30.12	0.998
Planetary Nebulae	+1.78	+20.52	0.962
Luminous Stars (selected)	(-74.47)	+41.22	0.944

In the case of the luminous stars, the values of V deduced from the catalogue radial velocities did not show a good linear correlation. This was interpreted as indicating that the stars were not confined to a thin disc. For a thick disc, with cylindrical solid-body rotation, in any strip perpendicular to the line of nodes, the least value of the radial velocity would correspond to a star in the central plane of the disc. In practice, with the luminous stars, the five smallest values of ρ, in each 1° strip, were taken as representative of the central plane. The linear fit is now good (see Table I), but the value of 'a' is significantly negative. In the presence of peculiar velocities for individual stars within the range $\pm$K km/sec, the selection technique will yield a value for each strip which is $-K \sin i$. The observed negative value of 'a' is thus consistent with peculiar velocities for the stars plus errors of observation within the range $\pm$33 km/sec.

In Figure 1 the normal points for the luminous stars have been adjusted to a=0. The differences in the slopes of the regression lines are interpreted as indicating that for the different classes of objects, the average osculating orbits have different eccentricities. The greatest slope would correspond to the smallest eccentricity.

The stars from WORC cover (rather non-uniformly) the entire area covered by this survey. Most of the spectra were secured with the Carnegie Image Tube Spectrograph attached to the CTIO 4-meter telescope. These were generally at a dispersion of 90 Å/mm, with a few at 50 or 25 Å/mm. Some spectra were obtained at 126 Å/mm with the ESO 3.6-meter telescope and the Boller and Chivens Spectrograph. Nine spectra were obtained with the 1.5-meter telescope at CTIO with a GEC CCD detector.

The CTIO 4-meter and ESO spectra were reduced in standard fashion. Standard stars were used to define a set of stellar wavelengths, which were then used to determine the radial velocities of the LMC stars. The internal standard deviation for the 90 Å/mm spectra was typically 25 km/sec from about a dozen lines. The 1.5-meter CCD spectra were reduced using a simple cross-correlation technique which produced velocities

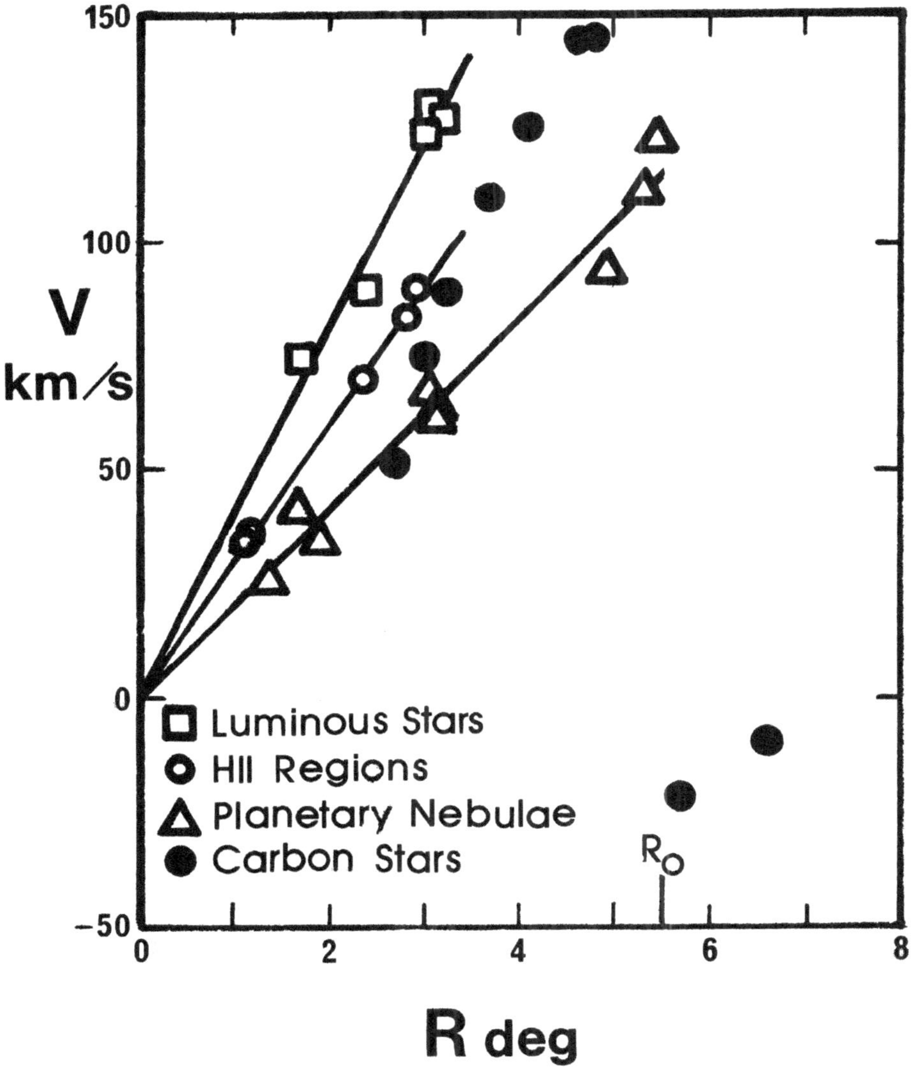

Figure 1. The cross velocity V plotted against radial distance R for different objects in the LMC. Each plotted point represents a smoothed normal point. Regression lines are indicated for all objects except WORC carbon stars. The normal points for the luminous stars have been adjusted to zero intercept.

accurate to about one tenth of a pixel (∿ 13 km/sec). The total number of radial velocities of WORC stars used in this analysis was 52.

The normal points for the WORC stars have been plotted on to Figure 1. For R < 5°, the normal linear fit is reasonably good, but with a significant negative value of "a". This may be because the carbon stars are also not confined to a plane, but the number of available radial velocities is too small for the selection technique that was used for the luminous stars to be used for the WORC stars. No regression line is plotted for the WORC stars. The importance of the carbon star radial velocities is that for R > 5.5°, there is a very rapid drop in the deduced orbital velocity V. This is consistent (see Hunter, Ball and Gottesman, 1984) with a sharp cut-off to the disc of the LMC at a projected radius of about 5.5°.

This cut-off enables an estimate to be made of the mass of the LMC. The mass of a thin disc of radius R, rotating as a solid body with angular velocity Ω, is $M = 4\ (R_o)^3\ \Omega^2/3\pi$ G. Hence the mass of the LMC, in units of the mass of the Sun, is given by $M=1.72 \times 10^3\ d(R_0)^3$ (dV/dR) where d is the distance of the LMC in kpc, R is the radius of the disc in degrees, and (dV/dR) is measured in km/sec/deg. We assume that the steepest slope (i.e. that for the luminous stars) corresponds to circular velocities ; thus we take dV/dR = 41.2 km/sec/deg. We take the distance of the LMC to be 52.5 kpc, and the cut-off radius (from the WORC radial velocities) to be 5.5° ; hence $M = 2.6 \times 10^{10}$ solar masses.

REFERENCES

Feast M.W. : 1964, M.N.R.A.S. 127, 195
Feitzinger J.V. - Weiss G. : 1979, Astron. Astrophys. Suppl. 37, 575
Hunter J.H., Ball R. and Gottesman S.T. : 1984, M.N.R.A.S. 208, 1
Joslin G.D. : 1983, M. Sc. Thesis, University of B.C.
Westerlund B.E., Olander N., Richer H.B. and Crabtree D.R. : 1978, Astron. Astrophys. Suppl. 31, 61

DISCUSSION

McCARTHY - Have you noted any spectral peculiarities observed in the outer regions of the LMC ?

RICHER - No. We have not noted anything to distinguish these from C stars closer to the disc.

LLOYD EVANS - Does the small dispersion of absolute magnitude of LMC carbon stars relate to a complete sample or only to those in the Westerlund survey ?

RICHER - The small dispersion ($\sigma_I = \pm\ 0^m.4$) which I mentioned for the LMC carbon stars is based on the Blanco, McCarthy, Blanco surveys which should be complete for the AGB carbon stars.

CARBON AND M STARS IN NGC 205

Harvey B. Richer† and Dennis R. Crabtree†
Department of Geophysics and Astronomy
University of British Columbia
Vancouver, Canada

and

Christopher J. Pritchet†
Department of Physics
University of Victoria
Victoria, Canada

ABSTRACT. Broadband VRI and narrowband CCD frames were obtained at CFHT of the central region of NGC 205, the dwarf elliptical companion of the Andromeda Galaxy. The narrowband images isolate TiO and CN and colors formed from the photometry of these images allow carbon and M stars to be unambiguously separated. Among the stars with reliable photometry, 7 carbon stars were found together with 21 late M stars with spectral types as cool as ~M5. Up to 4 of these latter objects may lie in the foreground. If the relative number of carbon and M giants is solely correlated with metallicity, then the metal abundance of NGC 205 lies in the range of that of the Magellanic Clouds. The presence of carbon stars in NGC 205 implies that star formation has been an ongoing process in this dwarf elliptical galaxy.

INTRODUCTION

In searching for late-type stars in the non-dwarf spheroidals of the Local Group, the limit to detectability of the stars is not their faintness, but the crowding. From this it is clear that a direct imaging search technique will be more efficient than a grism search. Broadband colors are not sufficient in themselves to clearly separate different spectral classes (e.g. carbon and M stars) of late-type stars. Hence we have established a three filter narrowband system which closely mimics bandpasses in Wing's (1971) eight-color system. The three filters are centred at 7100, 7800, and 8100 Å and each has $\Delta\lambda \simeq 100$ Å.

†Visiting astronomers at the Canada-France-Hawaii Telescope, operated by the National Research Council of Canada, the Centre National de Recherche Scientifique of France, and the University of Hawaii.

M. Jaschek and P. C. Keenan (eds.), Cool Stars with Excesses of Heavy Elements, 171–174.

The 7100 filter measures the $\gamma(0,0)$ band of TiO, the 7800 filter is a good continuum filter for carbon stars but suffers from significant TiO absorption in M stars, while the 8100 filter measures CN ($\Delta\nu$ = +2). Figure 1 can be used to envision how the filter system operates. The spectra plotted here are from Bessell, Wood and Lloyd Evans (1983) and the bandpasses of the three filters have been superimposed. In a two color diagram [(71-78) versus (81-78)] carbon stars will have large (81-78) indices (CN strong) with modest (71-78), while late M stars will have large (71-78) and small (81-78) compared to the carbon stars. This

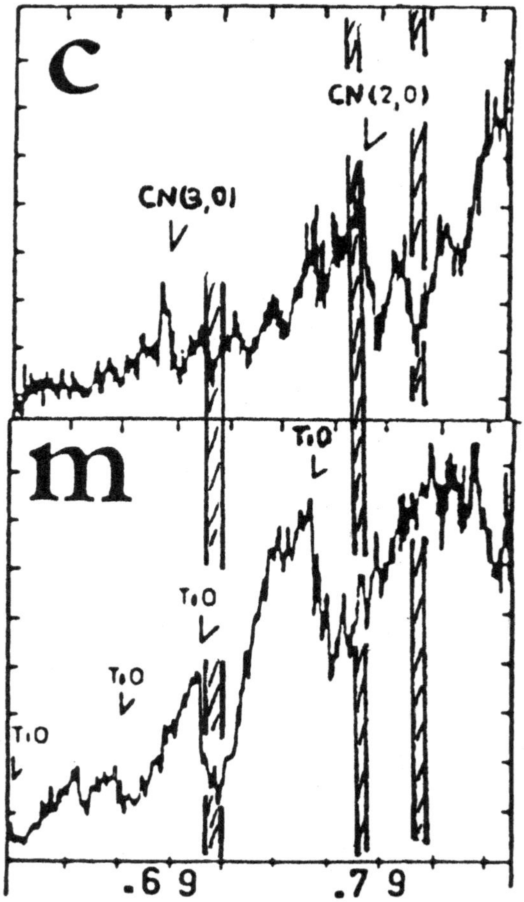

Figure 1. Spectrum of a typical carbon and M star. The bandpasses of the narrowband filters have been superimposed.

filter system, together with a broadband color (e.g. (V-I)) can separate out carbon, M, and earlier type stars. We have used these filters in front of the CFHT CCD system in a survey of several Local Group Galaxies (M31, NGC 6822, Fornax, NGC 205) for late-type giants.

SURVEY RESULTS FOR NGC 205

A 3.'5 x 2.'5 region centred on NGC 205 was imaged through the narrow-band filters described above. Broadband VRI images were also obtained. In Figure 2 we present a plot of (81-78) versus V-I for 112 mostly red stars in the central part of NGC 205. These represent all red stars whose photometry was judged to be uncontaminated, together with a sample of luminous blue stars (Baade 1951, Hodge 1973). Stars with (V-I) < 1.0 which corresponds to spectral types earlier than K have basically featureless spectra and define a horizontal band in this diagram. Seven NGC 205 stars lie more than 5σ above the mean value of these early type stars and hence have strong CN absorption in their spectra. Stars lying significantly below the horizontal band suffer from TiO absorption in the 7800 Å filter and can be identified as M stars. For comparison, in

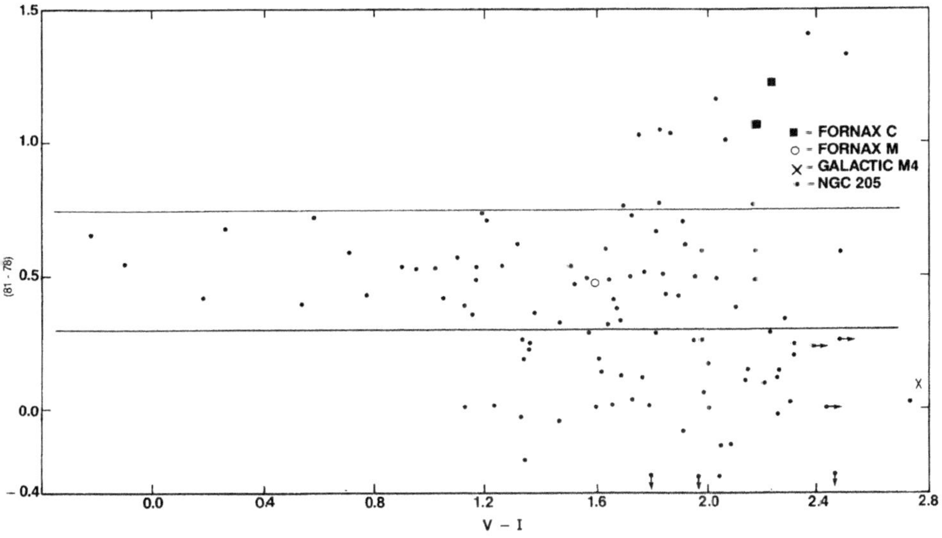

Figure 2. (81-78) versus (V-I) for NGC 205 stars with reliable photometry. The horizontal band is defined by the early type stars which have basically featureless spectra. Fornax C and M stars as well as a Galactic M4 giant are plotted for comparison.

Figure 2 we have included 2 known carbon stars in Fornax, one Fornax early M star (Frogel et al. 1982), and a Galactic M4 giant.

If the ratio of carbon to M stars (of spectral type M2 or later) is correlated solely with metallicity (Blanco, McCarthy and Blanco 1978, Blanco and McCarthy 1983) then the numbers of C and M stars in NGC 205 lead directly to an estimate of [Fe/H] for this galaxy. By comparing the NGC 205 C/M ratio with the ratio and metallicities of the Galactic Center (Blanco, Blanco and McCarthy 1978), the LMC (Blanco and McCarthy 1983), the SMC (Blanco and McCarthy 1983), Fornax (Frogel et al. 1982) and Carina (Mould et al. 1982, Aaronson et al. 1983) we can derive [Fe/H] $\simeq$ -0.8 for NGC 205.

NGC 205 contains luminous blue stars in its central region (Baade 1951, Hodge 1973). This implies that recent star formation has taken place in this galaxy. The present discovery of carbon stars, which are relatively low mass (1-3 $M_\odot$) AGB stars, implies that star formation has been an ongoing process in this dwarf elliptical companion of the Andromeda Galaxy.

Further details concerning these observations can be found in Richer, Crabtree and Pritchet (1984).

REFERENCES

Aaronson, M., Olszewski, E.W., and Hodge, P.W. 1983, *Ap. J.*, **267**, 271.

Baade, W. 1951, *Pub. Michigan Obs.*, **10**, 7.

Bessell, M.S., Wood, P.R., and Lloyd Evans, T. 1983, *M.N.R.A.S.*, **202**, 59.

Blanco, B.M., Blanco, V.M., and McCarthy, M.F. 1978, *Nature*, **271**, 638.

Blanco, V.M., and McCarthy, M.F. 1983, *A. J.*, **88**, 1442.

Frogel, J.A., Blanco, V.M., McCarthy, M.F., and Cohen, J.G. 1982, *Ap. J.*, **182**, 671.

Mould, J.R., Cannon, R.D., Aaronson, M., and Frogel, J.A. 1982, *Ap. J.*, **254**, 500.

Richer, H.B., Crabtree, D.R., and Pritchet, C.J. 1984, *Ap. J.*, in press (Nov. 15).

Wing, R.F. 1971, in *Proceedings of the Conference on Late-Type Stars, KPNO* Contrib. No. 554, eds. G.W. Lockwood and H.M. Dycks, p. 145.

CARBON STARS IN THE CARINA DWARF GALAXY

R.D. Cannon
Royal Observatory, Edinburgh

This note summarises recently obtained data on carbon stars in the Carina dwarf spheroidal galaxy, and the implications for the origin and evolution both of carbon stars and of dwarf galaxies.

1. THE DISCOVERY OF CARBON STARS IN CARINA

Not long after the discovery of the Carina dwarf galaxy, spectra were obtained for two potential red giant members selected from a preliminary CM diagram. Totally unexpectedly, both stars were found to be carbon stars (Cannon, Niss & Nørgaard-Nielsen, 1980). At almost the same time, carbon stars were discovered in the Sculptor and Fornax dwarfs, and soon carbon stars had been found in all seven known dwarf spheroidal satellites of our Galaxy (Aaronson, Olszewski & Hodge, 1983). A systematic search of the Carina dwarf was carried out on UK 1.2m Schmidt Telescope low-dispersion objective prism plates, and half-a-dozen further candidates were identified. Infrared photometry (Mould, Cannon, Aaronson & Frogel, 1982) and optical spectra (Lynden-Bell, Cannon & Godwin, 1983) have now been obtained for all the candidates, and a total of six carbon stars have been confirmed in the Carina dwarf.

The carbon stars occupy two distinct positions in the CM diagram of the Carina dwarf; three of the four carbon stars for which optical photometry exists lie at the extreme red tip of the giant branch with B-V ~ 2.0 while the fourth (C2) is much bluer and somewhat less luminous. In an M_{bol} - log T_e diagram derived from infrared photometry (Mould et al., 1982) five of the six carbon stars lie on the upper asymptotic giant branch, well above the maximum luminosity attained by normal hydrogen burning giants, while C2 again appears to be fainter and hotter. It appears that C2 may be analogous to the CH stars found in the globular cluster ω Centauri (Dickens, 1972), but the brighter Carina carbon stars have no counterparts in galactic globular clusters.

M. Jaschek and P. C. Keenan (eds.), Cool Stars with Excesses of Heavy Elements, 175–179.

Although the sample is small, the frequency of carbon stars is clearly very much higher in Carina than in galactic globular clusters. Inspection of a recent substantially complete CM diagram, based on COSMOS measures of AAT plates (Godwin, private communication) which must contain well over half of all the red giants that exist in Carina, shows that the system is less populous than many globulars. Thus although the upper giant branch and horizontal branch of Carina look similar to those of many globular clusters, and although the metallicity seems to be low (as judged from $(B-V)_{og} \leq 0.8$), there must be some other fundamental difference. Indeed, the low metallicity ($[M/H] < -2$) combined with the red horizontal branch already makes Carina somewhat peculiar in the same sense as the prototype 'second parameter' globular cluster NGC 7006 (Sandage & Wildey, 1967). If globular clusters contained carbon stars as frequently as does Carina, there should have been dozens of bright carbon stars in the galactic globular cluster system.

2. STELLAR EVOLUTION AND THE AGE OF CARINA

The explanation for the occurrence of carbon stars seems to be that the Carina dwarf is much younger than the globular clusters, or at any rate underwent a very major episode of star formation much more recently. The evidence for this comes from the recent CCD CM diagram by Mould & Aaronson (1983) which reaches to $V \sim 24$. This shows a large number of relatively blue stars only some 2-3 magnitudes below the horizontal branch, lying approximately on isochrones corresponding to ages of 6-9 billion years, i.e. about half the age of globular clusters. Thus the evolving red giants (or at least a large fraction of them) probably have masses of around 1.0 $M_{\odot}$ rather than 0.7-0.8 $M_{\odot}$ and it seems that low metallicity 1.0 $M_{\odot}$ stars are more likely to become carbon stars (i.e. to attain surface abundances with $C > O$) than stars which have either higher metallicity or lower mass. (A comprehensive review of the possible evolutionary processes is given by Iben & Renzini, 1983.) Whether all of the Carina stars are younger than globular cluster stars remains to be determined, and it may also be that many of these blue stars are in fact 'blue stragglers' rather than conventional main sequence stars. Such blue stragglers can be very common in old open clusters of ages 5-10 billion years (NGC 188 seems to contain almost as many blue stragglers as red giants), but their evolutionary status is still not understood. These questions can only be resolved by obtaining still fainter photometry, of high precision, for a large sample of Carina stars.

Even if Carina is relatively young, the same does not seem to be true for all the dwarf spheroidals. Although Fornax contains many carbon stars, their relative frequency is only about a tenth as high as in Carina. Several of the dwarf spheroidals contain RR Lyrae stars, while the Ursa Minor dwarf is unique in possessing a strong blue horizontal branch (van Agt, 1967); both of these properties are still regarded as indicative of great age, and would be difficult (although not impossible) to explain in the current stellar evolutionary theory for ~ 1 $M_\odot$ stars; the requirement would essentially be for a very high rate of mass loss, so that stars could reach the hot end of the horizontal branch with about half their initial mass. It thus appears that there is a large age spread amongst the seven dwarf spheroidals. Such an age spread may be difficult to reconcile with the elegant picture of Lynden-Bell (1982), in which he suggests, on the basis of their far from random distribution on the sky, that all of the dwarf spheroidal galaxies are members of one or other of two 'Magellanic Streams'.

The dwarf galaxies also provide an ideal opportunity to understand the origin and evolutionary status of the carbon stars themselves since they give large samples of stars all at the same distance. Thus not only can the carbon star luminosities be well determined, and their relationship to other types of star demonstrated in the CM diagram, their relative frequency of occurrence and hence the lifetimes of the carbon star phases of evolution can be estimated. This property has yet to be exploited fully, but one important aspect of this type of study is the determination of $^{12}C/^{13}C$ isotopic ratios. Crudely, if the excess carbon was created by CNO-cycle nuclear processing then the ^{13}C abundance will be relatively high, whereas the triple-alpha reaction produces ^{12}C. There are complicated possible mixing and evolutionary sequences which could affect the $^{12}C/^{13}C$ ratio, as proposed for example by Wood, Bessell & Fox (1983) with reference to the Magellanic Clouds. No quantitative determinations have yet been made for the Carina carbon stars, but there are interesting qualitative differences between the 4737 Å Swan band as seen in spectra obtained recently with the IPCS on the AAT. For example, star C1 shows a stronger $^{12}C^{13}C$ bandhead, although the overall strength of the bands looks greater in C5.

3. DYNAMICAL IMPLICATIONS

The existence of carbon stars in the dwarf spheroidals has accidentally made it much easier to determine the radial velocities of these distant systems, since the strong Swan bands can be observed even in very faint stars. The precision attainable is still a matter for debate, but it is clearly possible to attain accuracies of order 10 km s^{-1}. This is significant since earlier velocity determinations for the dwarfs were often uncertain by ~ 100 km s^{-1}.

The velocities of the dwarfs are very important for setting at least a lower limit to the mass of our Galaxy, since it seems that the dwarfs must be gravitationally bound satellites of the Galaxy. This follows from the fact that the UK Schmidt Telescope Southern Sky Survey on IIIa-J emulsion revealed only one new dwarf spheroidal (Carina itself) in the southern third of the sky, despite penetrating about 1.5 magnitudes fainter and hence surveying a four times larger volume of space than the earlier Palomar survey of the northern and equatorial two thirds of the sky. The latter survey had led to the discovery of four dwarf spheroidal systems, so that if they were uniformly distributed throughout the Local Group the UKST survey might have found more than a dozen new dwarfs. In the case of Carina, a recent radial velocity determination (Lynden-Bell et al. 1983) from all six carbon stars gives a mean galactocentric velocity of only about +24 km s^{-1}. An earlier velocity estimate had indicated a very high velocity which seemed to imply a very high mass for the Galaxy. There has been a similar downwards revision of the Sculptor dwarf's velocity. The total mass of the Galaxy now need be no more than a fairly 'conventional' $2.6 \pm 0.8 \times 10^{11}$ $M_{\odot}$, out to a radius of 100 kpc.

Another intriguing possibility is to use the carbon stars within each dwarf to determine the masses of the dwarf spheroidals themselves, from the internal dispersion of the carbon star velocities. A first attempt to do this for the Draco dwarf (Aaronson, 1983) suggested that the velocity dispersion was much higher than expected and that Draco might be over-massive by an order of magnitude, although there are several possible effects such as mass motions within the carbon star atmospheres and binary stars which may mean that they cannot be used to determine radial velocities accurate to 1 km s^{-1}. The velocity dispersion in Draco remains ambiguous (Aaronson & Cook, 1983), but Cook, Schechter & Aaronson (1983) now report that Carina too may have a significant mass excess.

REFERENCES

Aaronson, M.: 1983, Astrophys.J. 266, L11.
Aaronson, M. & Cook, K.: 1983, Bull.Amer.astr.Soc. 15, 907.
Aaronson, M., Olszewski, E.W. & Hodge, P.W.: 1983, Astrophys.J. 267, 271.
Cannon, R.D., Niss, B. & Nørgaard-Nielsen, H.U.: 1980, Mon.Not.R.astr.Soc. 196, 1P.
Cook, K., Schechter, P. & Aaronson, M.: 1983, Bull.Amer.astr.Soc. 15, 907.
Dickens, R.J.: 1972, Mon.Not.R.astr.Soc. 159, 7P.
Iben, I. & Renzini, A.: 1983, Ann.Rev.Astr.Astrophys. 21, 271.
Lynden-Bell, D.: 1982, Observatory, 102, 202.
Lynden-Bell, D., Cannon, R.D. & Godwin, P.J.: 1983, Mon.Not.R.astr.Soc. 204, 87P.
Mould, J.R. & Aaronson, M.: 1983, Astrophys.J. 273, 530.
Mould, J.R., Cannon, R.D. Aaronson, M. & Frogel, J.A.: 1982, Astrophys.J. 254, 500.
Sandage, A.R. & Wildey, R.: 1967, Astrophys.J. 150, 469.
van Agt, S.: 1967, Bull.astr.Soc.Netherlands 19, 275.
Wood, P.R., Bessell, M.S. & Fox, M.W.: 1983, Astrophys.J. 272, 99.

DISCUSSION

JØRGENSEN - With what accuracy do you measure the radial velocity of the C stars observed, and from which spectroscopic features ?

CANNON - I obtained an accuracy of only about 20 km s^{-1} for each star, although I believe careful analysis of the AAP spectra could give 10 km s^{-1} accuracy. I determined the velocities in several ways, using the Swan carbon band head and fitting them to galactic standard stars, both by visual fitting of large scale tracings and by using computer cross correlation methods. I also cross correlated sections of the spectra excluding the Swan band heads.

WING - Since the brightest and reddest stars of some galaxies are C stars while the nuclear region of our own galaxy is dominated by M stars, would it not be useful to measure the integrated of more distant systems in a CN/TiO index, such as Richer's 78 - 81 index ? Or does too much of the light come from ordinary giant's?

CANNON - I agree this would be useful, if a large enough proportion of the light comes from carbon stars. I believe it would certainly work with some of the MC stars clusters, but we have not yet dare the calculation for a Carina-like population.

McCARTHY - In your picture of the Carina dwarf galaxy can you comment on any of the stars between the brighter very red (B - V > 2) and the fainter less red C stars ?

CANNON - No. We have no spectra for those.

CARBON STARS AND S STARS NEAR OPEN CLUSTERS
--- *a statistical approach.*

U. G. Jørgensen
Astronomiska Observatoriet
Box 515
S - 751 20 Uppsala
Sweden

ABSTRACT. The positions of 3218 C stars, 741 S stars and 1140 open clusters have been compared. 59 C stars and 15 S stars have been found to lie in the line of sight to open clusters. This is no more than could be expected from a random distribution of the stars.

INTRODUCTION.

Evolutionary implications and the well known discrepancy between observed and theoretically obtained absolute magnitudes of carbon stars in the Magelanic Clouds have stimulated interest in identifying possible high mass carbon stars in the Galaxy. For this reason a number of publications deal with the observation of carbon stars close to the line of sight of open clusters. Generally radial velocity and proper motion are not known for the stars in question, and often these quantities are not known for the corresponding cluster either. The probability of finding a given star within the observed distance to the line of sight to the cluster center is unity when the star is observed *a priori*, and these observation therefore often contain only little information.

METHODS AND RESULTS.

In this work the positions of the 3218 carbon stars from A General Catalogue of Cool Carbon Stars (C.B.Stephenson, 1973) have been compared with the positions of the 1140 open clusters in A Catalogue of Open Cluster Data (G.Lyngå, 1983). A star has been considered "in the line of sight to a cluster" if it lies within an angular distance Θ from the cluster center given by

$$\Theta = 2.5 + 0.7 \cdot d - 6.4 \cdot 10^{-4} \cdot (d)^2 \qquad (1)$$

where d is the angular diameter of the cluster, and where Θ and d are to be measured in arc min.

In the same manner the positions of the 741 S stars in A General

M. Jaschek and P. C. Keenan (eds.), Cool Stars with Excesses of Heavy Elements, 181–184.

Catalogue of S Stars (C.B.Stephenson, 1976) have been compared with the cluster positions, and 59 carbon stars and 15 S stars have been found in the line of sight to an open cluster.

To study the statistical significance of this result, the galactic longitude and latitude distribution of carbon stars, S stars and open clusters from the catalogues have been plotted. The lattitude distribution is most narow for the open clusters, intermediate for the carbon stars and widest for the S stars. On the contrary the longitude distribution, as shown in Fig. 1, is quite similar for C stars, S stars and open clusters. It is remarkable that this distribution is clearly different from the distribution of late M stars found by V.M. Blanco (1965 - see also the discussion by Yorka and Wing, 1979), and it is concluded that whereas the ratio of M stars to open clusters is different in different part of the Galaxy, the ratios of C stars and of S stars to open clusters are more constant in different directions of galactic longitude.

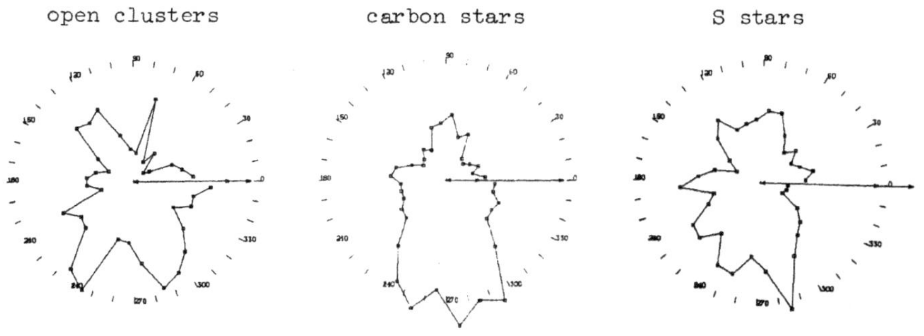

Fig. 1. The galactic longitude distribution of 1140 open clusters, 3218 carbon stars and 741 S stars.

In this study the stars have not been divided into groups (R, N, Miras, etc.), as it has been found most important to include as many stars as possible, and a great number of the stars are not classified more precisely in the catalogues than as being for instance a carbon star. Considering the catalogues as beeing not to heavily biassed, the longitude distributions of clusters, carbon stars and S stars, as shown in Fig. 1, are reasonable similar to allow the probability of a given number, N, of stars falling in the line of sight to a cluster to be expressed by a binomial distribution for each latitude interval. For a latitude interval of 1° the probability of a given star falling in the line of sight to an open cluster is then simply

$$p = \frac{\Pi}{4} \left(\frac{\Theta}{60}\right)^2 / 360 \qquad (2)$$

with Θ given from (1). The results are shown in Table 1, where also the numbers, N, actually found by comparing positions in the catalogues, are

Table 1. The probability from (2) is given for each galactic latitude interval of width 1^0, together with the number, N, of stars found in the line of sight to a cluster, the binomial expectation value, $\mu = M \cdot p$ (where M is the total number of C and S stars, respectively, in the interval), the variance $\sigma = \sqrt{Mpq}$ and the probabitity, $P(X \geq N)$, of coincidentially finding N or more stars in the line of sight to a cluster.

GALACTIC LAT	P	CARBON STARS				S STARS			
		N	MY=M*P	SIGMA	P(X>=N)	N	MY=M*P	SIGMA	P(X>=N)
-15 - 14	0.00	0	0.03	0.17	1.00	0	0.01	0.08	1.00
-14 - 13	0.00	0	0.03	0.17	1.00	0	0.01	0.11	1.00
-13 - 12	0.01	0	0.29	0.54	1.00	0	0.07	0.26	1.00
-12 - 11	0.00	0	0.13	0.36	1.00	0	0.02	0.15	1.00
-11 - 10	0.05	1	2.05	1.39	0.88	0	0.97	0.96	1.00
-10 - -9	0.01	2	0.51	0.71	0.09	0	0.14	0.37	1.00
-9 - -8	0.00	2	0.17	0.41	0.01	3	0.06	0.24	0.00
-8 - -7	0.14	1	6.48	2.36	1.00	0	2.11	1.35	1.00
-7 - -6	0.01	3	0.77	0.87	0.04	0	0.38	0.61	1.00
-6 - -5	0.01	2	1.20	1.09	0.34	1	0.36	0.60	0.31
-5 - -4	0.03	2	4.98	2.20	0.96	1	0.93	0.95	0.61
-4 - -3	0.02	9	5.27	2.27	0.08	0	0.55	0.73	1.00
-3 - -2	0.02	6	6.11	2.44	0.57	0	0.72	0.84	1.00
-2 - -1	0.04	4	10.83	3.22	1.00	1	1.69	1.27	0.82
-1 - 0	0.04	6	8.65	2.89	0.87	1	1.52	1.21	0.79
0 - 1	0.04	10	10.11	3.11	0.56	2	1.38	1.15	0.40
1 - 2	0.05	3	11.16	3.25	1.00	1	2.42	1.51	0.92
2 - 3	0.03	3	4.75	2.15	0.86	1	0.88	0.73	0.59
3 - 4	0.03	1	5.35	2.27	1.00	1	1.25	1.10	0.72
4 - 5	0.01	2	0.80	0.89	0.19	1	0.21	0.46	0.19
5 - 6	0.01	0	0.74	0.86	1.00	0	0.21	0.45	1.00
6 - 7	0.00	0	0.13	0.36	1.00	1	0.10	0.32	0.10
7 - 8	0.00	0	0.06	0.24	1.00	0	0.02	0.14	1.00
8 - 9	0.00	0	0.05	0.22	1.00	0	0.01	0.10	1.00
9 - 10	0.00	0	0.08	0.28	1.00	0	0.04	0.20	1.00
10 - 11	0.00	0	0.04	0.19	1.00	0	0.01	0.07	1.00
11 - 12	0.00	1	0.04	0.21	0.04	0	0.01	0.12	1.00
12 - 13	0.05	0	0.41	0.62	1.00	0	0.27	0.51	1.00
13 - 14	0.02	0	0.26	0.50	1.00	0	0.10	0.31	1.00
14 - 15	0.00	0	0.05	0.22	1.00	1	0.01	0.10	0.01

tabulated together with the expected number, μ, from the binomial distribution, the variance, σ, and the probability, $P(X \geq N)$, of finding N or more stars in the line of sight to a cluster if the stars are randomly distributed.

DISCUSSION.

There are a few latitude intervals where the number of stars in cluster areas are actually greater than what should be expected from a random distribution (f.ex. C stars in the interval $b = -4^0$ to -3^0), but there are also intervals where the stars succeed in avoiding" the clusters (f. ex. C stars in the interval $b = -2^0$ to -1^0).

Column 3 and 4 give the number of C stars near clusters from comparing the catalogue positions and from assuming a random distribution, respectively. Column 7 and 8 gives the same numbers for the S stars. Summarizing in these columns give a total of 59 C stars and 15 S stars with a position in the line of sight to a cluster, compared to 81.5 and 16.5, respectively, found from the statistical assumption.

Varying, as a final statistical check, the cluster positions artificially by rotating the cluster frame a small number of degrees of cause gives the same expectation value, but also, it is found, approximately the same number of positional coincidences. Adding f.ex. 20^0 to all l values of the clusters gave 49 coincidences between cluster and C star positions and 13 with S star positions.

CONCLUSIONS.

This study has more than tripled the number of C and S stars those positions are known to be in the line of sight to an open cluster (compared to what has previously been given in the litterature. For the latest review see Scalo and Miller, 1979). Even with this higher number it has been shown that there are no statistical reason to belive that any of the C or S stars, as has occasionally been indicated in the litterature, can be regarded as a cluster member only because it lies in the line of sight to a cluster. To make a cluster membership of a C or S star probable, more observations like radial velocity, prober motion or absolute magnitude estimate, are needed in each individual case.

ACKNOWLEDGEMENTS. It is a pleasure to thank Dr. B. Reipurth for important initial suggestions on this project. Thanks are also due to Dr. G. Lyngå for supplying me with the latest eddition of the cluster catalogue, and to Dr. H. Richer and Dr. R.F. Wing for valuable comments.

REFERENCES.

Blanco,V.M. 1965, in *Galactic Structure,* ed. A.Blaauw and M.Schmidt, Univ. Chicago Press, Chicago.
Lyngå,G. 1983, *A Catalogue of Open Cluster Data,* Lund Observatory.
Scalo,J.M., Miller,G.E. 1979, *Ap.J.*,**233**,596.
Stephenson,C.B. 1973, *Publ. Warner Swasey Obs.*vol.**1**,no.4
Stephenson,C.B. 1976, *Publ. Warner Swasey Obs.*vol.**2**,no.2.
Yorka,S.B., Wing,R.F. 1979, *Astron.J.* ,**84**,1010.

THE ABSOLUTE MAGNITUDE OF BARIUM STARS

C. Jaschek* - M. Jaschek* - S. Grenier+ - A. Gomez+ - A. Heck**
* C.D.S. - Observatoire de Strasbourg - France
+ Observatoire de Meudon, LA 337 - Meudon - France
** Observatoire de Strasbourg - France

SUMMARY. We have derived the mean absolute magnitude of a sample of "certain" and of "marginal" Ba stars. The certain Ba stars behave as normal giants, both in luminosity and kinematical properties.

Previous work on the absolute magnitude calibration of the barium stars indicated a rather wide range in luminosity. The values found in the literature are collected in Table I.

TABLE I

Method	Authors	M_V of Barium Stars			
		N	Certain	N	Marginal
Trigonometric parallaxes	Warner (1969) Eggen (1972)	8	$-3.7 < M < +2.2$		
Hα widths	Kemper (1975)	33	$-2.8 < M < +5$	9	$-2.3 < M < +2.9$
K-emission width	Williams (1975)	6	$-3.1 < M < +2.2$		
	Keenan and Wilson (1977)	2	0.3 and 1.3	10	$-0.2 < M < +1.6$
Statistical parallaxes	. Mac Connell, Frye, Upgren (1972) (1)	83	− 0.4	51	+ 1.0
	. Upgren, Lü, Mac Connell(1977) (2)	71	− 0.5	42	+ 0.3
	. Catchpole, Robertson, Warren (1977) (3)	40	+ 0.6 Ba strong		
		25	− 2.5 Ba weak		
DDO photometry	Lü, Upgren (1979)	82	Between classes III and IB		

N = Number of stars
* (1) using proper motions from SAO ; (2) Using proper motions from Yale ;
(3) using radial velocities

M. Jaschek and P. C. Keenan (eds.), Cool Stars with Excesses of Heavy Elements, 185–188.

In view of these discrepancies we decided to undertake a new determination of the mean absolute magnitude of the stars of this peculiarity group.

1. MATERIAL

We have used the "Catalog of stellar groups" by Jaschek and Egret (1984) to select a sample of "certain" and of "marginal" Ba stars. The mean absolute magnitude was derived from an algorithm based on the principle of maximum likelihood developed by Heck (1975, 1978). To apply this method one needs for all stars positions (not critical), apparent magnitudes, proper motion components and their errors and radial velocities and their errors. These data were taken from the following sources :

- positions : from the Catalog of stellar identifications od the Centre de Données Stellaires (C.D.S.)
- apparent magnitude : from Nicolet (1978) or Ochsenbein (1973)
- proper motion components : from Menessier et al (1978)
- radial velocities : Wilson (1953), Abt and Biggs (1972), Barbier (1977), Evans (1978), Catchpole et al (1977), Culver and Ianna (1977, 1980), Mac Clure (1983).

Because some of these data do not exist for all stars of our sample, the number of stars in the final sample is rather small when compared to the original one, namely 31 in the group of "certain" and 47 in the group of "marginal" Ba stars.

2. RESULTS

The resulting absolute magnitudes are quoted in Table II. and represented in fig. 1 in a HR diagram. In the latter the mean absolute magnitudes for normal G and K giants are from Egret et al (1982) who used the same statistical procedure as we did.

TABLE II.

SAMPLE	N	M_V	σ_M
"CERTAIN"	31	0.0 ± 0.2	1.2 ± 0.2
"MARGINAL"	47	-1.3 ± 0.3	0.7 ± 0.3

N : number of stars

Our results show that the "certain" Ba stars have luminosities of normal G and K giants, whereas the "marginal" Ba stars are more luminous. The result is in disagreement with the previous determinations collected in Table I. However not too much weight should be attributed to it since the group of "marginal" Ba stars is rather heterogeneous and is likely to contain an admixture of the higher luminosity stars, in which Ba is enhanced.

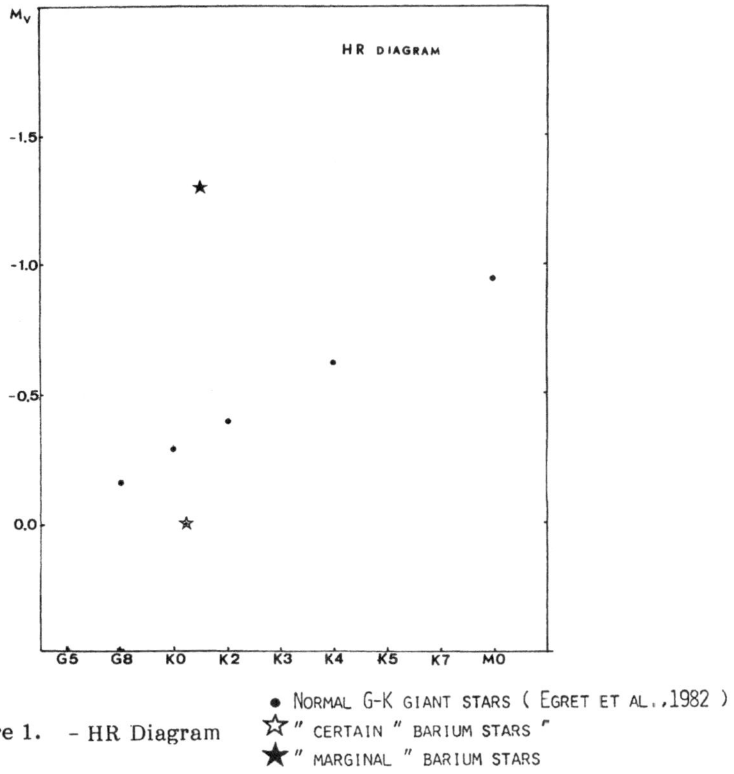

Figure 1. - HR Diagram

The statistical method we used provides as a by-product some kinematical parameters of the sample. These are given in Table III, together with those corresponding to normal giants (Egret et al, 1982). The obvious result is that the Ba star of our samples are typical population I objects.

TABLE III

SAMPLE	N	U	V	W	σ_U	σ_V	σ_W
"CERTAIN"	31	1 ± 4	-9 ± 4	-8 ± 3	24 ± 4	19 ± 3	13 ± 4
"MARGINAL"	47	-1 ± 3	-11 ± 3	-6 ± 2	24 ± 4	16 ± 3	9 ± 2
G5 -K7 giant stars*	89	-4	-19	-8	25	23	16

*Egret et al (1982)
N : number of stars

3. CONCLUSION

Our main conclusion is that certain Ba stars behave as normal giants, as well in luminosity as in kinematical properties.

4. REFERENCES

Catchpole R.M., Robertson B.S.C., Warrent P.R. : 1977, Monthly Notices Roy. Astron. Soc. **181**, 391
Eggen O.J. : 1972, Monthly Notices Roy. Astron. Soc. **159**, 403
Egret D., Keenan P.C., Heck A. : 1982, Astron. Astrophys. **106**, 115
Heck A. : 1975, Thesis, Liège
Heck A. : 1978, Vistas in Astron. **22**, 221
Jaschek M. , Egret D. : 1984, Publ. Speciale C.D.S. N° 5 (in preparation)
Keenan P.C., Wilson O.C. : 1977, Astrophys. J. **214**, 399
Kemper E. : 1975, Publ. Astron. Soc. Pacific **87**, 537
Lü P.K., Upgren A.R. : 1979, Astron. J. **84**, 101
Mac Connell D.J., Frye R.L., Upgren A.R. : 1972, Astron. J. **77**, 384
Upgren A.R., Lü P.K., Mac Connell D.J. : 1977, Publ. Astron. Soc. Pacific **89**, 552
Warner B. : 1969, Monthly Notices Roy. Astron. Soc. **129**, 263
Williams P.M. : 1975, Monthly Not. Roy. Astron. Soc. **170**, 342

DISCUSSION

LU – Upgren and I currently study the statistical parallaxes using proper motion to see if weak Ba stars have different absolute magnitude than strong Ba stars. We are using Yale proper motions.

IV. - CHEMICAL COMPOSITIONS OF THE SEVERAL VARIETIES OF PECULIAR STARS. COMPARISONS

THE CHEMICAL COMPOSITION OF COOL STARS: I - THE BARIUM STARS

David L. Lambert
Department of Astronomy
The University of Texas
Austin, TX 78712 USA

ABSTRACT. The chemical composition of Barium stars and their putative relatives is reviewed. Main sequence stars showing heavy element excesses, subgiant CH stars, classical and mild Barium giants, Sr-Ba supergiants, and early R stars are reviewed. A cursory interpretation of the abundance anomalies in terms of the key processes of nucleosynthesis is presented. Mass transfer across a binary system is identified as a plausible explanation for composition of Barium stars.

1. INTRODUCTION

Quantitative spectroscopy of stellar atmospheres is presently enjoying a thorough renaissance. New telescopes, spectrometers, and detectors encourage spectroscopists to pursue challenging new problems demanding one or more of the following: high resolution spectra, high signal-to-noise spectra, spectra in regions outside the narrow window accessible with photographic plates, spectra of statistically significant samples of faint stars, and thorough phase coverage of variable stars.

Happily, the technological revolution whose fruits are being implanted at our observatories has been exploited too by the physicists and chemists who provide astronomers with vital atomic and molecular data. Quantum chemists may now predict accurately relevant properties of small atoms and molecules. Reports of experiments yielding accurate oscillator strengths (and other key quantities) appear regularly in the literature. Such continuing advances facilitate the extraction of chemical composition from a stellar spectrum. Advances in experimental and theoretical nuclear physics in areas pertinent to stellar nucleosynthesis are continuing thanks largely to the efforts of physicists apprised of the frontiers in nuclear astrophysics. Such advances are essential if one is to interpret fully a star's chemical composition.

Since a stellar spectrum contains signatures of both the chemical composition and the atmospheric structure, some knowledge of the latter is an essential prerequisite for an abundance analysis (and

M. Jaschek and P. C. Keenan (eds.), Cool Stars with Excesses of Heavy Elements, 191–223.

vice-versa). Through careful design limited but vital information - for example, the abundance ratio of chemically similar but nucleosynthetically disparate elements - is extractable from a spectrum without thorough knowledge of the atmospheric structure. A complete abundance analysis necessarily calls for an iterative exchange with either empirical or theoretical model atmospheres. Stellar spectroscopists participating in the current renaissance are indebted to colleagues who construct model stellar atmospheres. An extensive discussion of the uses (and abuses) of model atmospheres in the abundance analysis of cool stars is not presented here. The increasing availability of large computers, the development of new techniques in numerical mathematics, and the expanding body of relevant atomic and molecular data are the impersonal factors behind the continuing development of model atmospheres.

Other stimuli for the current renaissance are provided by theoretical astrophysicists who study the evolution of stars. Many of the contemporary thrusts in quantitative stellar spectroscopy are motivated by the prediction that cool evolved stars dredge up nuclear processed material. A spectroscopist's motto might be that of a successful general -- "Don't assume anything - Look!".

In this review, I attempt to summarize our present knowledge of Barium stars and their possible relatives. Emphasis throughout is on the reported abundances. Remarks on methodology and on the interpretation of the data in the light of the predictions of stellar evolution and nucleosynthesis are limited owing to lack of space. I hope the selection and presentation of topics illustrate the several key stimuli of the renaissance.

This review is restricted to stars with overabundances of heavy elements; i.e. [heavy el./Fe] $\geq$ 0.0 for [Fe/H] $\gtrsim$ - 1.0 where [A] = log A_* - log $A_\odot$ and 'heavy el.' refers to elements beyond the iron peak. The boundary condition [heavy el./Fe] $\geq$ 0.0 is based upon analyses of disk dwarf stars which show that [heavy el./Fe] $\simeq$ 0.0 for both the s- and r- process nuclei. I shall refer to the subjects of my review as 'peculiar stars'; of course, this is a misnomer when their evolutionary origin is understood. I make no attempt to discuss rare peculiar stars such as FG Sge. I justify the exclusion of Pop. II stars in the field and in globular clusters on the grounds that published information on their composition is limited and, in general, they have not been probed by high-resolution spectroscopy.

2. THE CHEMICAL COMPOSITION OF PECULIAR MAIN SEQUENCE STARS

2.1. Introduction

A location on or near the main sequence places tight constraints on theoretical origins for peculiar stars. Four scenarios have been sketched on the literature:

-α) The interstellar medium and, in particular, intra-cluster gas may be inhomogeneous. Massive stars through their fierce winds, and a terminal supernova explosion may contaminate their parental and other local clouds. Red giants through ejection of their envelope as planetary nebulae may contaminate the parental cloud with material enriched in carbon and the s-process nuclei. Stars formed from the contaminated regions of the interstellar cloud will be marked as peculiar from birth.

-β) Mass-transfer across a binary system from the initially more massive star to its unevolved lower mass companion must create a peculiar star when the mass-losing star is permitted to evolve to the asymptotic giant branch (AGB) and to dredge-up s-process nuclei before the mass transfer is initiated.

-γ) A violent He-core flash in a low mass star may induce such extensive mixing that a sufficiently high H content is restored in the core to return the star to the main sequence. Theoretical calculations do not support this suggestion from an observer (Bond 1974).

-δ) The anomalous composition may be produced in situ by non-nuclear process; e.g. microscopic diffusion in the outer envelope may account for the striking anomalies displayed by the several species of chemically peculiar (CP) stars. Other possibilities include general and selective accretion from the interstellar medium onto the stellar surface.

-ε) Surface nuclear reactions have been proposed to account for CP stars.

These scenarios must be tested and refined using data on the chemical composition and other properties of peculiar main sequence stars. I discuss here the chemical composition of two groups of stars: normal main sequence stars, and the chemically peculiar stars. A third group - the subgiant CH stars - are reviewed in Section 3.2.

2.2. Normal main sequence stars

The widespread impression that peculiar stars (i.e. [heavy el/Fe] > 0.0) are rare among the Galactic disk's main sequence population results apparently from the implicit assumption that such peculiar stars, if they existed, should have been uncovered in objective prism and similar surveys. Recent abundance analyses of two bright stars - the K5 dwarf ε Ind (Kollatschny 1980) and the G8 dwarf ξ Boo A (Boo A (Boyarchuk and Eglitis 1984) - indicate that an overabundance of heavy elements may be a fairly common phenomenon. With the emergence of mass-transfer across a binary as a plausible explanation for classical Barium stars, the existence of main sequence peculiar stars has assumed a special significance.

A differential curve of growth analysis of ξ Boo A relative to β Aql (Sp. type G8IV) yielded the abundances summarized in Table I. For 12 elements from Na through Ni, the mean [el/Fe] = +0.04 ± 0.11 with the extremes provided by Ti (-0.14) and Co (+0.22). The rare-earths appear to be overabundant. Note that Eu with a weak s-process component is as overabundant as the s-process dominated quartet Ba, La, Ce, and Nd. The lighter s-process elements Y and Zr are not overabundant. These results should not be overinterpreted for the authors recognize that the uncertainties and the heavy-element excesses are comparable. ξ Boo A has a remarkably high Li abundance for its spectral type (Herbig 1965): log ε(Li) = 2.6 where I convert Herbig's [Li/Ca] value to a Li abundance using a mean factor obtained from stars analyzed for [Li/Ca] by Herbig and for log ε(Li) by Duncan (1981). This abundance does not exceed the cosmic (i.e. ξ Boo's presumed initial) abundance but ξ Boo A does appear to conflict with the claim that Li depletion is more severe in the later spectral main sequence stars. It is in this sense that Boyarchuk and Egltis refer to ξ Boo A is "Li-rich". ξ Boo A is not a known spectroscopic binary. It is a visual binary with a K4V companion.

TABLE I

The Composition of ξ Boo A[a]

	Cu	Zn	Y	Zr	Ba	La	Ce	Nd	Eu
[el/Fe]	+0.07	+0.06	-0.12	+0.06	+0.20	+0.32	+0.23	+0.31	+0.23

a) from Boyarchuk and Eglitis (1984, Table 4) with βAql as the reference source.

Epsilon Indi, a bright southern K5 dwarf, was chosen by Kollatschny (1980) as "a reference star for cool dwarfs". Surprising abundance anomalies, however, were uncovered and led Kollatschny to draw parallels between ε Ind and the classical Barium stars. Now that the significance of main sequence Barium stars is appreciated, a commentary on Kollatschny's analysis falls within the scope of this review.

Photographic spectra from 3950 to 6900Å were available; their calibration was tested by comparing lunar spectra with high quality solar spectra taken under similar conditions. Stellar equivalent widths (W_λ's) with an empirical stellar atmosphere constructed to fit the wings of 4 strong Fe I lines provided log g$f\varepsilon_*$ where ε_* is the abundance $\varepsilon_* = n(el)/n(H)$. Solar W_λ's as measured off a standard spectral atlas together with an empirical solar atmosphere provided log g$f\varepsilon_\odot$. A direct comparison gave the abundance difference $[\varepsilon] = \log \varepsilon_* - \log \varepsilon_\odot$ without requiring the gf-value. Within the stated error limits, elements up through the Fe-peak and including Cu and Zn have a normal abundance, i.e. [ε] = 0.0; carbon is an apparent exception (but see below) and Sc and V may be slightly overabundant with [ε] = +0.27 ± 0.15, and +0.37 ± 0.20 (Kollatschny's Table 7). Ionized lines consistently gave a higher abundance than neutral lines of the same element: Δ(ion) = [ε(II)] - [ε(I)] = 0.12 (Sc), 0.29 (Ti), 0.22 (V), 0.31 (Cr) and 0.19 (Fe). This systematic difference appears to originate with the stellar analysis because my recalculation of the Ti abundances from Kollatschny's log gfε values and high quality experimental gf-values gave $\Delta(\text{ion})_\odot$ = -0.07 but $\Delta(\text{ion})_*$ = -0.23. I surmise that Kollatschny's Δ(ion) values result from either an inappropriate atmospheric structure, the choice of the parameters T_{eff} and g, or the presence of departures from LTE.

The remarkable result of Kollatschny's analysis was an overabundance of the heavy elements ranging from [ε] = +1.1 for Dy (1 ionized line) to +0.1 for Mo (2 neutral lines). For this group too, neutral and ionized lines gave different abundances: Δ(ion) = 0.07(Sr), 0.36(Y), and 0.28(Zr). However, inspection shows that mean abundances are in some cases constructed from sets of discordant lines; e.g. three Y I lines give [ε] = 0.01, 0.18, and 0.59, and [ε] for three La II lines ranges from 0.30 to 1.19. For neutral lines, the solar reference lines are weak and the log $gf\varepsilon_\odot$ are necessarily sensitive to unidentified blends; e.g. the strongest of five Zr I lines has just $W_\lambda(\odot)$ = 5.5 mÅ.

TABLE II

Heavy Elements in ε Indi

Source	[el/Fe]								
	Sr	Y	Zr	Ba	La	Ce	Nd	Eu	Yb
ε Indi	0.2	0.1	0.4	0.1	1.0	0.1	0.9	1.3	1.0
HR 774	1.0	1.2	1.1	1.3	1.4	1.3	1.5	0.6	-

Thanks to recent publications providing accurate measurements of gf-values for many of the lines on Kollatschny's list, stellar abundances may be obtained directly from his log $gf\varepsilon_*$-values without recourse to the solar spectrum. In Table II, I give my estimates of [el/Fe]. Neutral lines are ratioed to the Fe I lines and ionized lines to the Fe II lines. References to the adopted gf-values and results for individual lines are suppressed owing to lack of space. The entry for Ce is taken directly from Kollatschny's Table 7 (accurate Ce II gf-values are unavailable) on the grounds that the four solar Ce II lines are moderately strong (W_λ = 15 to 55 mÅ) and yield consistent values of [ε]. The Eu entry based on the 4129Å Eu II line presents a special problem. Biémont et al. (1982) find that this line gives log $\varepsilon_\odot$ = 0.48 (close to their final recommendation) but Kollatschny's log $gf\varepsilon_\odot$ implies log $\varepsilon_\odot$ = 1.19. Until this discrepancy is resolved, I consider the Eu abundance in Table II to be uncertain. However, unless the stellar W_λ is in error, a Eu overabundance seems probable. I do not include Mo I, Ru I, and Sm II in Table 2 for lack of accurate gf-values. I reject the Nb I line because there is a large discrepancy between the solar and meteoritic Nb abundances (Kwiatkowski et al. 1982; Anders and Ebihara 1982) which suggests that solar (and, hence, stellar) Nb I lines may be blended. I exclude the sole Dy II line because it appears to be blended with a molecular line (Moore et al. 1966).

Although Table II differs in detail from Kollatschny's Table 7, his conclusion that abundance anomalies occur among the heavy elements is confirmed. Note that Table II is not a fully independent confirmation because I have used his stellar W_λ's and log gfϵ_* values. Table II includes entries for the classical Barium star HR 774 (Tomkin and Lambert 1983). There are significant differences between the compositions of ε Ind and HR 774, a typical Barium star. Elements up to Ba (including Cu and Zn) show no significant overabundance in ε Ind; Kollatschny's results for Nb, Mo and Ru are compatible with this claim. With the possible exception of Ce, large overabundances are apparent for elements beyond Ba; Sm as analyzed by Kollatschny gives [Sm/Fe] = +0.5. on present evidence, ε Ind cannot be identified as a main sequence progenitor of a Barium star. No attempt is made here to tie ε Indi's composition to the nucleosynthesis of the heavy elements. An immediate task is to obtain and analyze higher quality spectra; note that all of Kollatschny's lines for La to Yb are in crowded regions ($\lambda < 4800$Å). Atomic and molecular lines of C, N, and O should also be observed; Kollatschny found [ε] = 1.5 ± 1.0 for C from a single C I line.

Perhaps, the primary lesson offered by ξ Boo A and ε Ind is that quantitative analysis of a large sample of F, G, K dwarfs is overdue. One aim of the survey would be to locate the progenitors of the Barium stars and, hence, close examination of all dwarfs with white dwarf companions is an obvious priority.

2.3 The Chemically Peculiar Stars

Certain groups of the CP stars on the upper main sequence show large overabundances of heavy elements; e.g. the Ap (Cr-Sr-Eu), Am and δ Del stars. These peculiar stars fall within the compass of this review provided that they can be shown to retain the overabundances as they evolve to the giant branch. If microscopic diffusion operating in the atmosphere (Ap) or the sub-atmospheric radiative zone (Am) creates the abundance anomalies, the giant's outer deep convective envelope will erase the anomalies and the giant will have a normal composition. Some alternative but now generally discarded explanations of the CP stars' anomalous compositions might allow the red giant to retain an anomalous composition. I recall van den Heuvel's (1968a, b) proposal that Am stars are the recipients of nuclear-processed material from an initially more massive companion. Mass transfer to the secondary (i.e. the Am star) might provide sufficient processed material that abundance anomalies are not erased by the giant's convective envelope. Today, van den Heuvel's proposal would be linked with the mass-transfer hypothesis for production of a Barium star. Unfortunately, the composition of Am and related stars and other evidence seems incompatible with the mass-transfer hypothesis.

My remarks, which are restricted to the chemical composition, are drawn from abundance analyses by Smith (1971) for Am stars and Kurtz (1976) for δ Del stars; Kurtz shows that δ Del stars are evolved Am stars. Both authors used model atmospheres and included standard stars so that an abundance difference $[\varepsilon] = \log \varepsilon_* - \log \varepsilon_{std}$ may be computed which is independent of the adopted absolute scale for the gf-values. In Figure 1, I show mean values of [el/Fe] for the Am and δ Del samples together with the results (scaled down by 0.6 dex) for the classical Barium star HR 774 (Tomkin and Lambert 1983). A presentation of the abundance ratio [el/Fe] masks the fact that the CP stars are overabundant in Fe; e.g. [ε] = +0.5 for the Am sample.

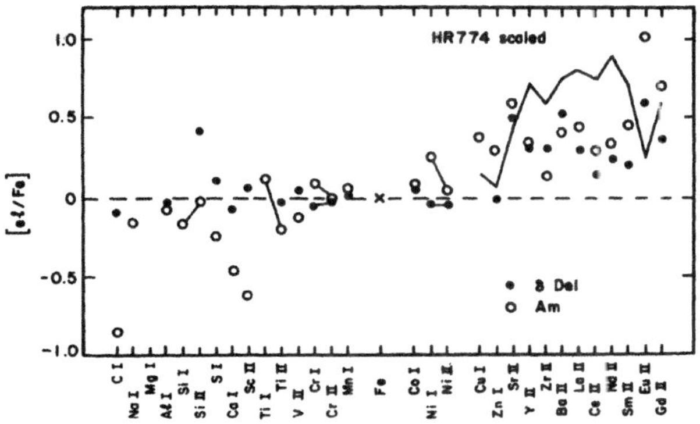

Fig. 1. - Abundances in Am and δ Del stars. The solid line from Cu to Gd connects abundances (scaled down by 0.6 dex) derived for HR 774.

The composition of Am stars is unlike that of cool peculiar stars. In particular, one notes the low C, Ca, Sc and the high Eu abundances (relative to Fe) and the marked Fe overabundance (relative to H) of the Am stars as quite uncharacteristic of cool peculiar stars.

An insistent advocate for an evolutionary connection between a class of CP stars and peculiar giants might identify the δ Del stars as progenitors of the Barium stars; a point first noted and dismissed by Kurtz. Certainly, these stars show smaller Fe overabundances ([Fe] = +0.3 in the mean), and [el/Fe] for the iron-peak and lighter elements is exceptionally normal; the Si abundance is based on just one or two Si II lines. Three difficulties confront the advocate. First, the overabundance of heavy elements in δ Del stars is below the level seen in the Barium stars; an examination of high S/N spectra is now needed to test element-by-element the abundance patterns of δ Del and Barium stars. Second, a high Eu abundance in δ Del stars sets them apart from Barium stars and all products of the s-process; however, Kurtz remarked that his abundances may be in error because he neglected the hyperfine splitting (hfs) of the Eu II lines. (Smith did consider the hfs and found Eu high in Am stars.) Third, the C abundance in δ Del stars is below the observed value for Barium stars;

new results (Roby 1984) supercede Kurtz's values used in Fig. 1 and now [C] ~ 0.4 or [C/Fe] ~ -0.5 for δ Del. Even if the abundances were to suggest a relation between Am or δ Del stars and Barium stars, a real difficulty remains: the CP star is often a member of a close binary and, in some cases, the companion is also a CP star (e.g. δ Del, Kurtz 1976). Production of a Barium star via mass-transfer requires the companion to be a stripped AGB star (e.g., a white dwarf) and a near-normal main sequence star is not readily accommodated by the hypothesis.

Inspection of Adelman's (1973) analysis of Ap (Cr-Sr-Eu) stars shows that the composition bears little resemblance to that of Barium and other peculiar giants.

In summary, the CP stars are unlikely to be progenitors of the principal classes of peculiar cool giants. Abundance anomalies displayed by CP stars appear to be erased (or diluted) as the stars leave the main sequence.

3. THE CHEMICAL COMPOSITION OF LOW LUMINOSITY GIANTS

3.1. Introduction

A single star is not expected to dredge up freshly synthesized heavy elements into the atmosphere until it ascends the AGB and experiences thermal pulses arising from the close proximity of the H and He burning shells. Several magnitudes below such luminous stars in the HR diagram lie several families of giants with abundance anomalies e.g. the classical Barium stars, the CH and CH-like giants, the warm carbon or R-type stars, and the weak G-band giants. The CH subgiants extend the domain of peculiar stars down to the main sequence. Clearly, these giants and subgiants do not fit the theoretical scenario restricting development of an overabundance of heavy elements to the AGB. This long-standing problem has attracted two proposed solutions - (i) a violent He-core flash in a low mass star; (ii) mass transfer from an AGB star to a less evolved companion which is converted to a peculiar star.

The He-core flash is a natural suspect in observers' eyes - see, for example, discussions of Barium stars running from Warner (1965) through to Tomkin and Lambert (1983) which attempt to link production and dredge-up of s-process nuclei with the flash. A more extreme requirement on the core flash was imposed by Bond (1974; see also Luck and Bond 1982) in a search for an explanation of the CH subgiants, stars lying well below the minimum luminosity for a He-core burning ('clump') giant. Bond supposed that large-scale mixing driven by the core flash restored the H abundance in the core such that the star was returned to the main sequence.

Theoretical studies of the core flash provide little encouragement for the observers' enthusiasm. Iben and Renzini (1984) review both the quasistatic and hydrodynamic calculations. In the recent quasistatic calculations (e.g. Mengel and Sweigart 1981;

Despain 1982), the initial and large flash is followed by a sequence of thin shell flashes but the thin convective zones in these shells never make contact with the base of the outer convective envelope; the surface composition is unaltered. Iben and Renzini speculate that, when more accurate opacities for carbon are incorporated into the calculations, contact with the convective envelope may be established under selected condition; i.e. rapid rotation of the core. Special conditions are a prerequisite because the peculiar giants are just a small fraction of those red giants which must have experienced a He core flash. A rapidly rotating core was earlier shown by Mengel and Gross (1976) to cause the flash to begin more off-center, a condition which facilitates mixing with the convective envelope (Paczyński and Tremaine 1977). Hydrodynamic calculations (Deupree and Cole 1983; Deupree 1984 - see these papers for references to earlier work) show the core to experience severe convective mixing and extensive nucleosynthesis and the flash to result in the ejection of a large part of the envelope. Iben and Renzini call these predictions "interesting, but hard to believe". Nevertheless, some enthusiastic observers will surely anticipate that the hydrodynamical (or quasistatic) calculations will soon predict that products of nucleosynthesis are swept into a giant's convective envelope.

A welcome alternative to the He core flash emerged with the discovery that all Barium and CH stars are members of binary systems (McClure et al. 1980; McClure 1983, 1985). This discovery encourages the idea that low luminosity peculiar red giants may be created by mass transfer across a binary system: the initially more-massive star evolves to the AGB, experiences the third dredge-up and shortly afterwards transfers part of its C and s-process enriched envelope to its companion, which is transformed to a low-luminosity peculiar star. In this scenario, the theoretical and observational evidence that a dredge-up of carbon and heavy elements occurs on the AGB and the exclusive existence of Barium stars in binary systems are combined to provide the "natural" explanation for low luminosity peculiar stars.

In the following sections, I review the chemical composition of the following kinds of stars:

--the CH subgiants
--the classical, mild, and marginal Barium stars
--the Sr-Ba supergiants
--the CH and CH-like stars
--the warm carbon (early R-type) stars.

A glossary defining the major groups of Barium stars seems essential. Here, "classical Barium" star denotes the class of GK giants identified first by Bidelman and Keenan (1951) from low dispersion spectra: their "Ba II" stars, the "classical" Barium stars, show enhanced Ba II, Sr II, CH, CN, and C_2 lines. Thanks to the application of several observational techniques, one may safely assert that the enhancement of C and heavy elements shows a continuous range from no detectable enhancement in the majority of GK giants through to the substantial enhancements seen in the classical Barium stars. Such an assertion springs from the narrow-band photometry conducted by Williams (1975), the detailed spectral classifications from high-dispersion spectra reported by Wilson (1976) and Keenan and Wilson (1977), and abundance analyses (e.g. Pilachowski 1977; Sneden et al. 1981). To contend with this range, Warner (1965) introduced his subclassification Ba 1 to B5 which, in other hands, has been refined such that Ba 0.2 or even Ba 0.1 may now be appended after close scrutiny of high dispersion spectra. At the other extreme, extensions to Ba 9 have appeared -- see Catchpole et al. 1977 and Yamashita and Norimoto (1981) for catalogues of spectral types for Barium stars. Stars with a low Ba index (i.e. Ba 0 or Ba 1) are considered here to be "mild" Barium; MacConnell et al. (1972) use the term "semi-barium" stars and "marginal" Barium appears elsewhere in the literature. The distinction between "classical" and "mild" Barium stars appears, on present evidence, to be one of degree not kind.

As quantitative spectroscopy is extended to larger samples, new subgroups may emerge. Indeed, Bond {see laconic references by McClure (1984) and Pinsonneault et al. (1984)} suggests that mild Barium stars may be split into two groups. Spectra of the first group are similar to but less extreme than those of the classical Barium stars; Bond suggests that these be termed "mild" Barium stars. The second group's spectra show enhancements limited to the lighter s-process elements (e.g. Sr, Y, and Zr); Bond proposes that these stars be called "marginal" Barium stars. I am not aware of an abundance analysis for a marginal Barium star and, hence, they are excluded from this review.

Barium stars appear, in the main, to be "clump" giants (i.e. He-core burning stars) but both subgiant and supergiant variants are known (see Fig. 2). Zeta Cap (sp. type G4II - Ba 3, Yamashita and Norimoto 1981) is the prototype for the Barium supergiants. It appears to be more massive than the typical Barium star. Comparison of the compositions suggests that subgiant CH and classical/mild Barium stars belong to one group of peculiar stars. Section 4 offers a preliminary discussion of the chemical compositions in the light of the mass transfer hypothesis.

3.2 The CH Subgiants

Subgiant CH stars were isolated by Bond (1974) as a class of peculiar stars, with G-type spectra but weak metallic lines, enhanced CH and s-process lines, and absolute magnitudes ranging from the subgiant branch (M_V ~ 2) to the main sequence (M_V ~ +5); Fig. 2 shows the location of several CH subgiants on the HR diagram - see Luck and Bond (1982) for M_V and T_{eff} estimates.

Information on the chemical composition of subgiant CH stars (Table III) is drawn from 'fine' analyses by Sneden and Bond (1976), Luck and Bond (1982), Sneden (1983), and Krishnaswamy and Sneden (1984) with greater weight placed on the latter two analyses thanks to their use of higher quality spectra. A detailed analysis of the heavy elements in two stars (Krishnaswamy and Sneden 1984) is summarized in Table IV together with results for two Barium stars and KS's standard subgiant.

TABLE III

The Chemical Composition of Subgiant CH Stars

Star [HD]	T_{eff} [K]	log g [cm s^{-2}]	[Fe/H][a]	[C/Fe][b]	[O/Fe][c]	[Zr/Fe]	[La/Fe]
4395	5450	3.3	-0.3	+0.2	0.0	+0.4	+0.5
11377	6000	4.1	-0.2	-	-	+0.8	0.0
88446	6000	4.5	-0.4	+0.5	+0.2	+0.9	+0.7
89948	5950	4.1	-0.3	+0.6	+0.2	+1.7	+0.4
176021	6000	4.0	-0.4	+0.5	-	+0.4	+0.5
182274	6000	4.5	-0.4	+0.9	<+0.6	+1.0	+0.6
204613	5650	3.8	-0.4	+0.5	+0.5	+0.9	+0.9
216219	5600	3.3	-0.5	+0.3	+0.2	+0.7	+0.8

a) Metals up through the Fe-peak appear to have normal abundances i.e. [el/Fe] ≈ 0.
b) Sneden (1983) finds no evidence for a ^{13}C enrichment: $^{12}C/^{13}C > 40$ for HD 4395, HD 89948, and HD 216219. $^{12}C/^{13}C = 20$ to 60 for HD 204613. However, $^{12}C/^{13}C \simeq 10$ for HD 55496, a cooler subgiant with $T_{eff} = 4750$ K and log g = 2.5.
c) The only information on N is provided by Sneden and Bond (1976) who give [N/Fe] ~ 0.1 for HD 176021 and 204613 from CN violet system lines.

3.3. The Chemical Composition of Classical Barium Stars.

Quantitative spectroscopy of Barium stars began in the 1950's when Garstang (1952) and Greenstein (1954) showed that the heavy elements were overabundant in the atmosphere of ζ Cap. Analyses of bright giants led Burbidge and Burbidge (1957) to identify the abundance anomalies with the s-process. Warner's (1965) curve of growth analysis of 12 Barium stars (see also Danziger 1965) and Tech's (1971) thorough analysis of ζ Cap roughly define the end of the first phase of quantitative analysis. With the advent of modern detectors, a new phase has begun in the analysis of Barium and other peculiar stars.

Three points characteristic of this phase deserve emphasis:

--Improvements in spectrum quality permit the measurement of weak lines

--Analyses must contend with departures from LTE

--With access to the near-infrared and infrared, new diagnostics of nucleosynthesis and mixing are exploitable.

A marked improvement in spectrum quality with respect to S/N and resolution permits one to measure weak lines in the Barium stars and, also, to measure the same but then very weak lines in standard G or K giants. Earlier analyses were, in general, based on stronger lines in the standard giants, and hence, on rather saturated lines in the Barium stars. Tomkin and Lambert (1979, their Table 4) noted that three analyses of HR 774 based upon photographic spectra gave overabundances of 0.3 dex or less for Y, Zr, and Mo which, when high S/N spectra are analyzed, are shown to be overabundant by 0.9 dex.

Strong evidence for non-LTE effects in spectra of GK giants was presented first by Ruland et al. (1980) and confirmed by Brown et al. (1983) and Tomkin and Lambert (1983). These effects appear to be small for high excitation lines of neutral metals and lines of the ionized metals. Happily, a differential analysis of a peculiar and a standard star appears to result in a rather complete cancellation of the non-LTE effects. Table V from Tomkin and Lambert (1983) shows that the neutral and ionized lines of Y and Zr give different abundances in HR 774 and β Gem when analyzed with solar gf-values but consistent abundances in a differential analysis of HR 774 and β Gem. Clearly, this approach is an interim measure because the non-LTE effects cannot be identical in a peculiar and a standard star. Thorough non-LTE calculations are awaited with interest.

Fig. 2 - The HR diagram showing the location of representative CH subgiants, classical and mild Barium stars and Sr-Ba supergiants. Evolutionary tracks from the zero age main sequence (ZAMS) and onto the first giant branch are shown.

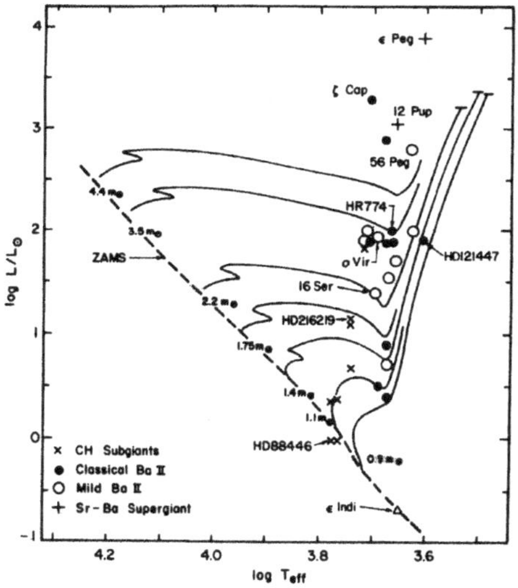

TABLE IV

Heavy Elements in Two Subgiant CH Stars

	[el/Fe]				
	Ba II stars		CH Subgiants[c]		Std. Subgiant[c]
	HR 774[a]	HD 121447[b]	HD 4395	HD 216219	94 Aqr
Cu	+0.41	-	+0.14	-	+0.07
Zn	+0.21	-	+0.39	+0.22	-0.32
Sr	+1.00	+1.22	-	+1.22	-0.22
Y	+1.17	+0.76	+1.07	+0.87	-0.11
Zr	+1.14	+0.66	+0.40	+0.74	+0.08
Ru	+0.78:	-	-	+1.00	-
Ba	+1.35	+0.57	+0.40	+1.07	+0.33
La	+1.39	+0.70	+0.46	+0.80	+0.01
Ce	+1.32	+0.73	+0.62	+0.82	+0.15
Nd	+1.49	+0.65	+0.21	+0.65	+0.15
Sm	+1.30	-	+0.22	+0.78	+0.15

a) Tomkin and Lambert (1983); b) Smith (1984); c) Krishnaswamy and Sneden (1984). The spectral type of 94 Aqr is G5 IV-V.

TABLE V

Neutral-and Ionized-Line Abundances

		[el/H]		
Species	Number of Lines	HR 774 Relative to Sun	β Gem Relative to Sun	HR 774 to β Gem
Fe I....	10	-0.44	-0.12	-0.32
Fe II...	3	-0.41	-0.09	-0.32
Y I.....	2	+0.54	-0.28	0.82
Y II....	2	+0.88	+0.01	0.87
Zr I....	7	+0.37	-0.43	0.80
Zr II...	3	+0.91	+0.05	0.86

New diagnostics drawn from the near infrared and infrared include the Rb I resonance lines and the various infrared molecular lines (CO, NH, OH, CN) which provide the important isotopic and elemental abundances of C, N, and O, the tracer of atmospheric contamination by H- and He- burning products.

TABLE VI

The Chemical Composition of Classical Barium Stars

	ζ Cap[a]		HR 774[b]		HD 121447[c]	HD 83548[d]
	New	Tech	TL	S	S	K
log ε(Li)	1.0			0.4	0.0	
log ε(C)	8.57		8.60	8.70	8.70	8.43
log ε(N)	8.13		8.31	8.33	8.40	7.85
log ε(O)	8.56		8.75	8.72	8.80	8.66
$^{12}C/^{13}C$	>40		23	15	8	
[Fe/H]	-0.11		-0.32	-0.30	+0.05	-0.33
[Cu/Fe]		+0.03	+0.41			
[Zn/Fe]		+0.09	+0.21			
[Ge/Fe]		+0.12	+0.18			
[Rb/Fe]	+0.50		+0.46			
[Sr/Fe]	+1.08	+1.01	+1.00	+1.25	+1.22	
[Y/Fe]	+0.82	+1.06	+1.17	+1.06	+0.76	+0.83
[Zr/Fe]	+0.87	+0.86	+1.14	+0.95	+0.66	+0.61
[Nb/Fe]	+0.99	+0.23	+1.19	+1.17	+0.87	
[Mo/Fe]		+0.69	+1.24	+0.85	+0.58	
[Ru/Fe]		+0.43	+0.78:			
[In/Fe]			+0.87			
[Ba/Fe]	+0.32	+1.38	+1.35	+1.03	+0.57	+0.82
[La/Fe]	+0.66	+0.93	+1.39	+0.96	+0.70	+0.26
[Ce/Fe]	+0.6:	+0.98	+1.32	+0.89	+0.73	+0.42
[Pr/Fe]		+0.88	+1.62	+0.68	+0.34	+0.38
[Nd/Fe]		+0.97	+1.49	+0.97	+0.65	
[Sm/Fe]		+0.82	+1.30			
[Eu/Fe]	-0.55	+0.10	+0.65	+0.35	+0.39	+0.05
[Gd/Fe]		+0.32	+1.08:	+0.55	+0.00	
[Dy/Fe]		+0.79	+0.91			
[Er/Fe]			+1.47:			
[Hf/Fe]			+1.09	+0.44	+0.53	
[Ta/Fe]				+0.10	+0.27	

Notes to Table VI
a) the "New" results are compiled from Smith et al. (1980), Smith and Lambert (1984a), and Pinsonneault et al. (1984). Tech's (1971) differential analyses remains the most comprehensive. The Ba abundance obtained by Smith et al. (1980) would appear to be an underestimate. Boesgaard and Fesen (1974) obtained a Tc abundance log ε(Tc) < 1.6 which is quite consistent with the abundances of adjacent elements.
b) TL denotes Tomkin and Lambert (1979, 1983). For [Fe/H] and [el/Fe], β Gem was adopted; β Gem has a solar metallicity. S denotes Smith (1984) except that the Li abundance was provided by Pinsonneault et al.
c) HD 121447 is a cool Barium star (sp. type K7-Ba3, T_{eff} = 4200K). See above note for the references to the abundances.
d) Results from Kovács (1983a). The adopted [el/Fe] values are for the ionized lines. Published results from Zr I and Zr II lines for the stellar-solar abundance difference exhibit the effect displayed in Table V.

In Table VI, I collect abundances from recently published analyses and comment below upon several key elements.

Lithium: Pinsonneault et al. (1984 PSS) report Li abundances for 13 classical Barium stars; see also Boesgaard's (1968) survey of 7 stars but note that her results expressed as a Li/Ca ratio must be adjusted because she used (apparently) gf-values from that infamous compilation by Corliss and Bozman (1962) - see remarks by Catchpole and Feast (1976). Lithium abundances in GK giants are predicted and observed (except in a few cases - Sneden et al. 1984) not to exceed a maximum value log ε(Li) ~ 1.2 (Lambert, Dominy, and Sivertsen 1980). For the observed Barium stars, the Li abundance has the same upper bound. PSS refer to "the conformity" of Li abundances in Barium stars to those in GK giants. Inspection shows that there is an apparent difference between the two data sets: the Li in Barium stars has a higher lower bound: -0.1 ≤ log ε(Li) ≤ 1.1 but -2.0 ≤ log ε(Li) ≤ 1.2 for normal GK giants.

C, N, and O: Enhanced lines of CH and C_2 define, in part, a classical Barium star. Several thorough analyses show that the C abundance in Barium stars is higher than in normal GK giants. Nitrogen and oxygen in Barium stars have normal abundances. Abundances for 8 Barium giants (Tomkin and Lambert 1979; Sneden et al. 1981; Kovács 1983a; Smith 1984) are summarized in Table VII where the star's initial el/Fe ratio is inferred from the observed Fe abundance and the abundances of C, N, and O in main sequence stars (Clegg et al. 1981). The Barium supergiant ζ Cap may be slightly deficient in oxygen: ΔC = +0.05, ΔN = +0.25, and ΔO = -0.29 from Smith, et al. (1980). One conclusion is clear. The principal contaminant of a Barium star's atmosphere is ^{12}C, a He-burning product. Exposure to the H-burning CN-cycle is at the level experienced by a typical GK giant. The limited data on the $^{12}C/^{13}C$ and $^{16}O/^{17}O/^{18}O$ (Harris et al. 1984) ratios supports these conclusions. We comment below on the correlation between the C and s-process overabundances.

TABLE VII

C, N, and O in Classical Barium Stars

Δ^a	Ba II	Normal GK[b]
C	+0.05 ± 0.12	−0.24 ± 0.13
N	+0.40 ± 0.19	+0.38 ± 0.14
O	−0.08 ± 0.06	+0.01 ± 0.10

a) $\Delta = \log \varepsilon(el)/\varepsilon(Fe)_{Ba} - \log \varepsilon(el)/\varepsilon(Fe)_{main-seq}$

b) Lambert and Ries (1981)

Na to Fe: Elements from Na through to the Fe-peak have normal abundances - see, in particular, the differential analyses presented by Smith (1984) for 7 stars. Kovács (1983a) notes that Si, Ti, and V are about 0.3 dex underabundant relative to Fe in HD 83548 but considers that non-LTE effects are responsible. He finds Na, Mg, and Si to be slightly overabundant; Smith's result provides a similar hint for Na with a mean ratio for his sample [Na/Fe] = +0.20 ± 0.16.

The $^{24}Mg/^{25}Mg/^{26}Mg$ Ratio: If the neutrons driving the s-process were released by the $^{22}Ne(\alpha,n)\ ^{25}Mg$ reaction, comparison of predicted abundances at the s-process site and the observed overabundance of s-process elements shows that ^{25}Mg and ^{26}Mg should be enhanced relative to ^{24}Mg in the atmosphere of a Barium star (Scalo 1978, Truran 1980). On the other hand, if the neutron source is the $^{13}C\ (\alpha,n)\ ^{16}O$ reaction, the heavier Mg isotopes should not be overabundant.

High resolution and high S/N spectra of MgH lines in the spectra of the classical Barium star HR 774 and the mild Barium star 16 Ser were shown by Tomkin and Lambert (1979) to provide terrestrial abundances of the Mg isotopes (Fig. 3). Since main sequence stars of comparable metallicity show a terrestrial isotopic mix, a fair inference is that the observed Barium stars are not enriched in ^{25}Mg and ^{26}Mg, and, hence, $^{22}Ne\ (\alpha,n)$ is not the neutron source. An assumption that this conclusion extends to all Barium stars deserves an observational test. The MgH lines are too weak to permit an isotopic analysis of the supergiant ζ Cap.

The Heavy Elements: A goal of the early curve of growth analyses was to prove that the s-process created the abundance anomalies seen in Barium stars. Today, spectra permit a more thorough study of the s-process. Here, I draw upon a recent analysis of HR 774 (Tomkin and Lambert 1983).

The leading contenders for the neutron source are reaction $^{22}Ne(\alpha,n)^{25}Mg$ and $^{13}C(\alpha,n)^{16}O$. The ^{22}Ne source is associated with the thermal pulses experienced on the AGB by a star of intermediate mass (Iben 1975). The ^{13}C source is presumed to operate in low-mass stars during either the thermal pulses on the AGB, the He-core flash, or quiescent He-core burning.

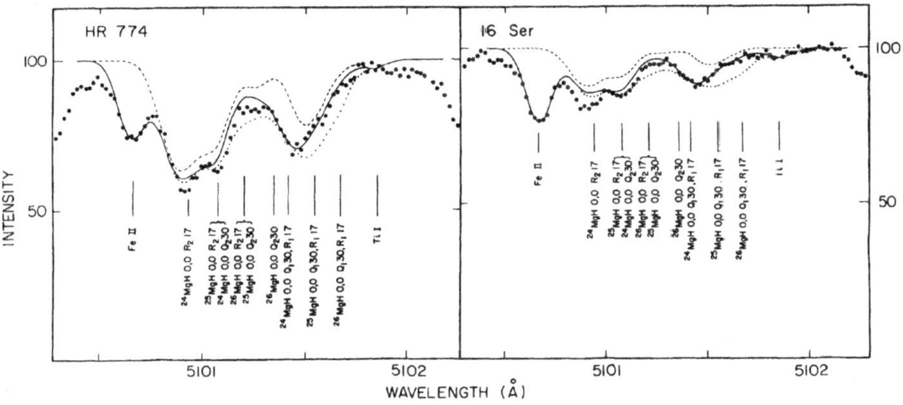

Fig. 3 - MgH in the spectra of HR 774 and 16 Ser. Dots are the data points. Three synthetic spectra, computed with C_2 lines alone (dashed line), C_2 and MgH lines with a terrestrial mixture: ^{24}Mg: ^{25}Mg: ^{26}Mg = 79:10:11 of Mg isotopes (solid line), and C^2 and MgH lines with equal abundances of Mg isotopes (dotted line). For both stars the synthetic spectra computed with terrestrial abundances of the Mg isotopes provide the best fit with the observed spectra. The Fe II and Ti I lines were included in the synthetic spectra.

Detailed predictions for the s-processing resulting from the ^{22}Ne source in thermally pulsing stars are provided by Truran and Iben (1977) and Cosner, Iben, and Truran (1980). In Figure 4, I show the predictions for the "standard" case considered by Truran and Iben and the abundance difference between this and an alternative case where the mixing fraction (1-r) is cut of 0.05 from the standard value of 0.50. This mixing fraction is "the fractional contribution to the convective shell consisting of fresh material acquired between pulses during the outward progression of the hydrogen burning shell" (Truran and Iben 1977, p. 801). These predictions do not fit the derived abundance pattern. Nonstandard cases even closer to the limit $(1-r) \rightarrow 0$ will differ insignificantly from the $(1-r) \sim 0.05$ predictions. Despain (1980) pointed out that the predicted neutron density in the He-burning shell phase of the thermal pulse is so high that the nucleosynthesis cannot follow the simple unbranched s-process chain. However, Cosner, Iben, and Truran (1980) noted that the lower neutron densities in the tail of the thermal pulse will redistribute the quasi-r-process abundances to yield a final abundance pattern not too different from their earlier calculations. The revised calculations are similar to the "standard" predictions except for a few elements associated with key branches in the s-process chain. The outstanding difference involves the Rb/Sr ratio influenced by the branch at the β-unstable ^{85}Kr.

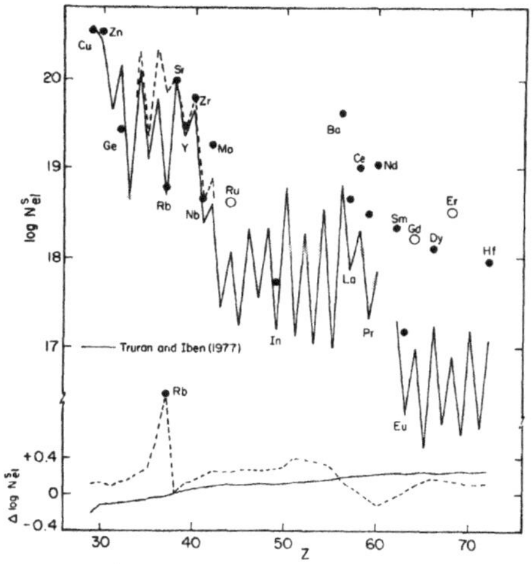

Fig. 4 - Observed and predicted abundances of heavy elements in HR 774. The elemental abundances are scaled to log N^5_{el} = 20 for Sr; open circles denote the less certain values. Predictions from Truran and Iben for the standard (r= 0.5, see text) thermal pulse model aare shown normalized at Sr. Differences between these predictions and those of a nonstandard (r=0.05) thermal pulse model and the predictions by Cosner, Iben, and Truran (1980) are shown in the bottom of the figure where Δ log N^5_{el} = log N^5_{el} (Alternate) - log N^5_{el} (standard) is plotted. The Cosner et al. predictions for elements near Rb are also shown by the broken line in the upper part of the figure.

The comparisons displayed in Figure 4 argue effectively that the ^{22}Ne(α,n)^{25} Mg reaction is not the neutron source for HR 774. It is important to stress that the predicted abundance pattern is insensitive to the detailed structure of the thermally pulsing star. Our demonstration that the ^{22}Ne(α,N)^{25} Mg reaction is incapable of accounting for the composition of HR 774 confirms Truran's (1980) contention that the observed Ba/(Sr, Y, Zr) ratios for Barium stars exceed those expected for the ^{22}Ne source. In addition, models for thermally pulsing intermediate mass stars predict similar surface enhancements of carbon and the s-process elements. With strong s-process but mild carbon enhancements, the Barium stars again fail this prediction. The conclusion is clear: the s-process in HR 774 (and other classical Barium stars) was not driven by the ^{22}Ne neutron source.

If ^{13}C(α,n)^{16}O is the neutron source, very few neutrons are captured by the ^{16}O nuclei and the existing light elements. This is in sharp contract to the situation prevailing with the ^{22}Ne(α,n)^{25}Mg source. (Other neutron sources -- $^{17}O(\alpha, n)^{20}Ne$, for example--behave like the ^{13}C source but are unlikely to provide sufficient neutrons.) Cowley and Downs (1980) predicted s-process abundances for the standard s-process chain beginning with ^{56}Fe and two types of neutron exposures--a single burst and an exponential distribution function.

The s-processing is characterized by the neutron exposure

$$\tau = \int N(n)\, v\, dt$$

where N(n) is the neutron number density, v is the mean relative velocity of neutrons and seed nuclei, and the integral is taken over the duration of the event; τ is commonly given in the unit millibarn^{-1}. The distribution function $\exp(-\tau/\tau_0)$, is attractive not only because the exact solution for the s-processed abundances is known, but because mixing in real stars may simulate this distribution of exposures rather than a single burst (Ulrich 1973, Clayton and Ward 1974). Our analysis of the abundance pro- file for HR 774 showed that the predictions for an exponential distribution function with τ_0 = 0.6 are an excellent fit (Fig. 5). No equally satisfactory fit was achieved with predictions for a single burst. The τ_c = 0.6 predictions match the HR 774 abundances extremely well, and except for a striking discrepancy (Cu and Zn) and a minor disagreement (Mo), the mean difference between the observed and predicted abundance is less than 0.1 dex. Smith (1984) obtained a lower Mo abundance. Cu and Zn appear to be overabundant relative to the predictions. We suspect that either the observed abundances are in error (both elemental - abundances are based on a single line, or the atmosphere contains a small amount of very lightly exposed material). Since Smith obtained lower abundances for Ba the rare-earths, his τ_0's are smaller: τ_0 ~ 0.3 to 0.4 for 7 stars.

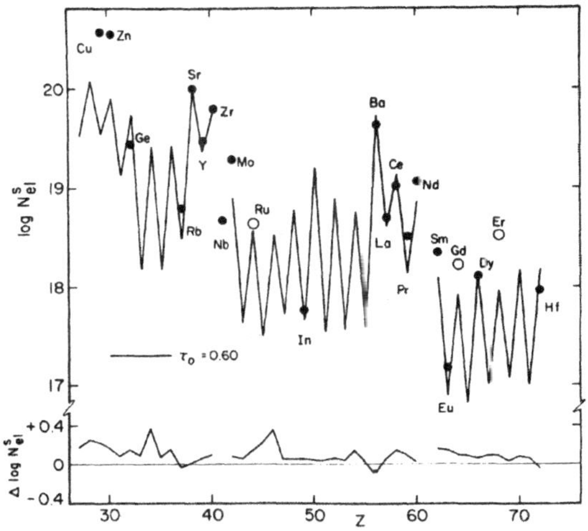

Fig. 5 - Observed and predicted abundances of heavy elements in HR 774. Predictions correspond to a distribution function exp $(-\tau/\tau_0)$ of neutron exposures with τ_0 = 0.6 and neutron capture cross sections taken from Käppeler et al. (1982). Newman's (1978) recommended cross sections lead to a slightly different set of predicted abundances: see $\Delta\log N^5_{el} = \log N^5_{el}$ (Newman) - $\log N^5_{el}$ (Käppeler) given at the bottom of the figure.

With the increasing completeness and precision of the abundance determinations, more detailed answers to key questions about the s-process environment are obtainable: What was the neutron density? The temperature? The duration of the active phase of s-processing? How long ago did processing cease? Did the neutron captures follow a classical s-process? Abundance predictions such as those presented in Fig. 4 and 5 are based on an unbranched s-process chain. A few relatively long-lived radioactive nuclei are links in this chain. At low neutron densities, such nuclei decay rather than capture a neutron. At high density, a neutron capture will be more probable, and so the s-process chain takes a branch not opened at low densities. The branching may also be influenced by the temperature because excited states populated at high temperature can have quite different, usually shorter, lifetimes than the ground state. A branch leads to neutron density-dependent and temperature-dependent abundance predictions. Examination of potential branches in the s-process chain shows that the branch at ^{85}Kr through its influence on the Rb/Sr ratio is probably the only indicator of neutron density accessible to stellar spectroscopists - see discussion by Tomkin and Lambert (1983). The abundance of Rb changes by a factor of about 25 between the low and high neutron density limis. At low densities, ^{85}Rb is the dominant isotope. ^{87}Rb is dominant at high densities. Unfortunately, the isotopic ratio is effectively unobtainable from the profiles of the Rb I resonance lines (Lambert and Luck 1976). For HR 774, we measured the Rb abundance relative to Sr, Y, and Zr and found ratios just in excess of the low density limit: $N(n) = (5\pm3) \times 10^6$ cm^{-3}. This estimate is a mean value for the closing phases of the s-processing. No information is provided on the distribution function of neutron density; for example, we cannot exclude the extreme but simple case in which a small amount of material exposed to a high neutron density ($N(n) > 10^8$ cm^{-3} and high Rb content) is mixed with a larger amount of material exposed to a low neutron density ($N(n) < 10^6$ cm^{-3} and a low Rb content).

The duration of active processing is obviously calculable given the neutron flux (Φ) and the capture cross-sections down the chain - see calibration provided by Ward et al. (1976). For HR 774, we estimated $7000 \lesssim \langle t\rangle \lesssim 7 \times 10^4$ yr. Total neutron consumption ($\Phi \langle t\rangle$) is readily estimated to correspond to about 20 neutron captures per Fe seed nucleus. Production must exceed this level because several light elements act as poisons by absorbing neutrons.

The temperature at the s-process site is not obtainable directly from the observations. Observations show that the ^{22}Ne$(\alpha,n)^{25}$Mg source is not ignited; then, examination of the rates suggests that $T_8 \lesssim 3$. A lower bound to T_8 ($\gtrsim 1.2$) is provided by equating $\langle t\rangle$ with the lifetime of ^{13}C against α-capture ($T_8 = 10^{-8}$ T[K]).

Technetium is the classical s-process chronometer for red giants. Unsuccessful searches for Tc in Barium stars have been reported; e.g. Smith's (1984) limit to the Tc abundance just exceeds the predicted abundance at the termination of s-processing. Niobium represented by a single stable isotope, ^{93}Nb, is also a potential chronometer. In the standard s-process chain, ^{93}Nb is bypassed because the unstable, long-lived isotope ^{93}Zr captures a neutron to form stable ^{94}Zr. Hence, the Nb abundance at the s-process site is negligibly small; the neutron density must be very low, $N(n) \lesssim 100$ cm^{-3}, before the ^{93}Zr nuclei β^--decay instead of capturing a neutron. Anders (1958) pointed out that the Nb abundance is then a monitor of the elapsed time (t_{el}) from the cessation of the s-process in the case of a single episode of s-processing. An equivalent definition of t_{el} pertains to the case of repeated episodes. With the complete decay of ^{93}Zr, the Nb abundance is constant. Nb is a monitor of t_{el} as long as $t_{el} \lesssim 2\ \tau_{1/2}$ where $\tau^{1}/_{2}$ is the half-life of ^{93}Zr, which, at low temperatures $T \lesssim 2 \times 10^8$K, is 1.5×10^6 yrs. Nb complements Tc because ^{99}Tc has a half-life at low temperatures of just 10^5 yrs decreasing to 44 yrs at $T = 2 \times 10^8$K (Cosner and Truran 1981; Schatz 1983). Examination of Table VI shows that Tech's Nb abundance for ζ Cap is strikingly low and perhaps $t_{el} < 10^6$ yrs. However, our new spectra (Smith and Lambert 1984a) lead to a higher Nb abundance; the Nb chronometer in ζ Cap has run down and $t_{el} \gtrsim 3 \times 10^6$ yrs.

At the s-process site, the heavy element to iron ratio must be high. Then, the modest ratio in the atmosphere implies that the total amount of s-processed material mixed into the atmosphere is small: a mixing fraction $f \sim 10^{-3}$ to 10^{-4} is suggested (Warner 1965; Tomkin and Lambert 1979).

3.4 The Chemical Composition of Mild Barium Stars.

The lack of an accurate quantitative survey of mild Barium stars must be recognized. Nonetheless, interesting parallels between these groups of Barium stars may be sketched.

Lithium: PSS determined the Li abundance for 10 mild Barium stars; I exclude v^2 Cas (a normal giant?) and 55 Cam (a Sr-Ba supergiant?). The abundances range from $\log \varepsilon(\mathrm{Li}) = +0.6$ to an upper limit $\log \varepsilon(\mathrm{Li}) < -1.2$. Relative to the results for classical Barium star, the distribution for mild Barium stars extends to lower Li abundances. The two samples may also differ at the other extreme; however, only 2 of the 13 classical Barium stars have an abundance exceeding that of the most Li-rich mild Barium star in the (small) sample of 10 stars.

C, N, and O: Mild Barium stars like their classical cousins and have similar N and O abundances to the normal GK giants. As noted by Sneden et al. (1981), mild Barium stars show only a slight enhancement of C. In Fig. 6, I compare the s-process and C enhancements for mild and classical Barium stars for which an abundance analysis was completed recently using model atmospheres and quite similar techniques. Although additional analyses of mild Barium stars are desirable, a correlation between [s/Fe] and [C/Fe] is indicated. The solid line shows how the composition of a GK giant would change if material from a 'typical' cool carbon star were added in increasing amounts; the fraction of added material is indicated along the line.

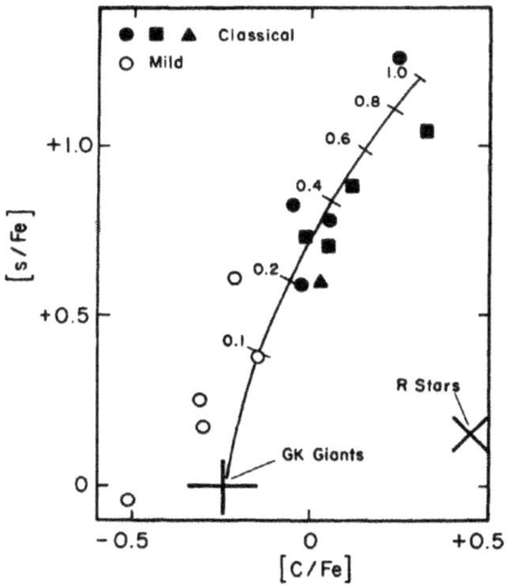

Fig. 6 - The abundance of s-process elements and carbon in mild and classical Barium stars. Data for mild Ba stars are from Sneden et al. (1981) and for classical Ba stars from: circles - Sneden et al. (1981) Tomkin and Lambert (1979), squares - Smith (1984), and triangle - Kovács (1983a). Points for GK Giants and R stars are taken from Lambert and Ries (1981) and Dominy (1984) respectively.

Na to Fe: A thorough element-by-element analysis of digital spectra is as yet unavailable but abundance anomalies for this group are neither expected nor reported.

TABLE VIII

Heavy Elements in the Mild Barium Star o Vir

Lines	[el/Fe]		
	o Vir[a]	HR 774-mod[b]	HR 774[c]
1 Y II, 2 Zr II	+0.6	+0.60	+1.15
1 Ba II, 1 Ce I, 2 Ce II 1 Pr II, 2 Nd II	+0.8	+0.84	+1.45
1 Eu II	+0.4	+0.25	+0.68

a) Abundances relative to ε Vir as the standard star. [Fe/H] = -0.4.
b) s-processed material with the inferred composition of that present in HR 774 is added to material with a solar composition such that [el/Fe] = +0.60 for Y and Zr.
c) Tomkin and Lambert (1983)

Heavy Elements: Published analyses based on photographic spectra suggest that the relative abundances of the heavy elements are similar to those seen in the classical Barium stars (e.g. Helfer and Wallerstein 1968, Chromey et al. 1969, Williams 1972, Pilachowski 1977, Smith and Harmer 1982). Preliminary results from our current analysis of oVir and 16 Ser confirm the suggestion - see Table VIII. Elements are collected into groups to provide more accurate mean abundances, but the results from individual lines are quite uniform. The comparison with the classical Barium star HR 774 shows that the contaminant in oVir is s-processed material with a similar exposure but the mile Barium star is less contaminated: 20% of the amount present in HR 774 - see Table VIII. This mixing fraction is consistent with that required to explain the C abundance of oVir (Fig. 6).

In summary, the chemical composition of classical and mild Barium stars suggest strongly that Barium stars are formed when material enriched in carbon, the heavy elements and, possibly lithium is mixed with material having a normal composition. When a large amount of the processed material is added, a classical Barium star results. When smaller amounts are added, a mild Barium star results. Both the quantitative analyses and qualitative measures such as spectral classification show classical and mild Barium stars to belong to a single family of peculiar giants.

3.5 The Chemical Composition of the Sr-Ba Supergiants.

When Williams (Warren and Williams 1970, Williams 1971) proposed that some unremarkable G and K supergiants were enriched in Ba, he sparked a continuing debate. Recent developments deserve comment in this review. I refer to these stars as Sr-Ba supergiants and include two on the HR-diagram in Fig. 2: εPeg (Warren and Peat 1972, Kovács 1983b) and 12 Pup (Kovács 1983a). These supergiants should not be confused with mild Barium stars such as oVir and 16 Ser.

Kovács (1983a,b) and Holweger and Kovács (1984) observe that the supergiants are enriched in Sr and Ba but not in the other heavy elements customarily associated with the s-process; e.g. 12 Pup has [Sr/Fe] = 1.0 and [Ba/Fe] = 0.7 but [el/Fe] ~ 0 to -0.3 for Y, Zr, La, Ce, and Nd (also, Eu). This odd pattern was suspected earlier for Hyland and Mould (1974), wrote "rare-earths do not show the expected enhancements were the s-process operative". The new piece of the pattern is the Sr abundance which comes from the Sr II 10036 and 10327 Å lines. Recently, Holweger and Kovács (1984) showed that the abundance pattern could be attributed to a s-process. For very light neutron exposures ($\tau \sim 0.04$ mbn^{-1}), the Ba abundance increases by 0.5 dex as the nearby lighter elements Sn, Te, and Xe are converted to Ba. A similar overabundance of Sr is created from Ge, Se, and Kr. For higher exposures, the relative abundances of heavy elements begin to approach the levels predicted for the classical s-process and the local enhancements of Sr and Ba are less spectacular. Since the predicted Sr and Ba enhancements are close to observed levels, Holweger and Kovács noted that a large fraction of the outer envelope of these supergiants must have undergone the light s-processing. The fact that the ^{13}C abundance for εPeg is low ($^{12}C/^{13}C = 5.1 \pm 0.5$, Tomkin and Lambert 1974) is, perhaps, additional evidence for extensive processing (Kovács 1983b). Although the novel suggestion of a light (or modified) s-process resolves the nucleosynthesis of the Sr-Ba supergiants, the assignment of the episode of nucleosynthesis to a point in the life of the supergiants is quite unresolved.

In the debate initiated by William's identification of the Sr-Ba supergiants, several adverse criticisms were directed at the analyses and specifically at the derivation of the Ba abundance. One fact cannot be over emphasized: The Ba and Sr abundances depend or one or two very strong lines; e.g. two Ba II lines, W_λ = 211mÅ(5854Å) and 315 mÅ(6496Å), were available for 12 Pup (Kovács 1983a), Other heavy elements are represented by weaker lines; the next strongest line in Kovács's list is the Y II line at 5200Å with W_λ = 137mÅ but Y II is represented also by 3 weaker lines. Lines as strong as the Ba II and Sr II lines are sensitive to the adopted microturbulence (an increase by about 0.5 kms^{-1} eliminates the Ba and Sr overabundances), the structure of the upper photosphere, and to non-LTE effects.

Hyland and Mould (1974) sought to determine the Ba abundances for εPeg and another supergiant from weak Ba I lines at 5777, 5971, 6341, and 6595Å. A differential analysis relative to αBoo gave [Ba/Fe] = 0.0 and a reasonable value, [Ba/Fe] = 1.0 for the classical Barium star HR5058; a fine analysis of αBoo gives [Ba/Fe] = -0.25 relative to the sun (Mäckle et al. 1975). This analysis of Ba I lines, as emphasized by Hyland and Mould, suggests that Ba is not overabundant in two Sr-Ba supergiants. Some reservations must be noted. Tomkin and Lambert (1983) suggested that two of the Ba I lines (5971 and 6595Å) may be blended. Unfortunately, the 6595Å line was the only line measured in αBoo (and εPeg). However, the fact that the differential analysis gave a plausible result for HR5058 implies that the result for the supergiants is probably not too much in error; [Ba/Fe] = +0.75 which was obtained by Kovács (1983b) for εPeg would appear to be excluded. Departures from LTE affecting the Ba I lines must be another concern. A new investigation of the Ba I lines using high S/N spectra is underway (Smith and Lambert 1984b). Sr-Ba supergiants and normal supergiants are being compared in an attempt to reduce systematic errors. A search for weak near-infrared Ba II in the warmer supergiants is also being conducted. Our preliminary conclusion is that the case for a Sr and Ba enhancement in the Sr-Ba supergiants is unconvincing.

The challenge to translate the concept of light s-processing to a detailed description of a phase of nucleosynthesis and mixing in a star is exceedingly stiff. Recall that the inferred enhancement of Sr and Ba is close to the maximum permitted by a light s-process and, therefore, a large fraction of the atmosphere of a Sr-Ba supergiant would appear to have been exposed to neutrons; by contrast, the equivalent fraction for a classical Barium star is only 10^{-3} to 10^{-4}. A supergiant's atmosphere is surely part of the deep convective envelope extending over 50% by mass of the star. Then, a large amount of material was exposed to a neutron source and so to a high temperature $T \gtrsim 1.5 \times 10^8$K. Surely, this exposure must result in additional nucleosynthesis and observable consequences. Yet, the composition of the Sr-Ba supergiants is strikingly normal apart from Sr and Ba, and, perhaps, ^{13}C. In particular, since neutron capture cross-sections vary slowly with atomic number, one expects the light s-process to create abnormal ratios of certain light elements; standard abundance ratios Ca/Sc = 1800, Ca/Ti = 25, and Sc/V = 210 suggest that Ca is a likely seed for a severe overproduction of Sc, Ti, and V (Cowley 1984). Published analyses provide rather normal abundances for Na through to Fe. These objections might be avoided if the light s-processing were to be applied to material previous exposed to the standard (say τ_0 ~ 0.6) s-process.

In summary, the evidence linking the Sr-Ba supergiants to the Barium star family (the subgiant CH and giant CH stars, the mild and classical Barium stars) is less than convincing. My suspicion is that Sr-Ba supergiants will be shown to possess normal abundances of both elements. If their status as peculiar supergiants is confirmed, a remarkable episode of nucleosynthesis and mixing within a star of moderate mass must be missing from contemporary calculations of stellar structure and evolution prior to the thermal pulsing state on the AGB.

3.6 The Chemical Composition of CH Stars

As members of pop. II, the CH stars fall outside the defined scope of the review but their claim to the title pop. II Barium stars' justifies some brief remarks. CH stars were classified first by Keenan (1942): enhanced bands of CH, CN, and C_2 and enhanced lines of the heavy elements together with a high space velocity are the defining characteristics.

Quantitative analysis has been reported for just 3 field CH stars. It is no surprise that they are metal poor and C-rich: [Fe/H] = -0.4 for HD 26 (Luck and Bond 1982) to -1.5 for HD 201626 (Wallerstein and Greenstein 1964); [C/Fe] $\simeq$ 1.2 for HD201626 according to Schadee (1968). Several stars appear to be ^{13}C-rich: e.g. $^{12}C/^{13}C$ = 6 for HD26 (Dominy and Lambert, unpubl. results). Determinations of the N and O abundances have yet to be made. Heavy-element abundances relative to iron reach the levels achieved by classical Barium star: [el/Fe] $\simeq$ +1.2 for Ba to Nd in HD26, 201626, and 209621 (Wallerstein 1969). The lighter elements - Y and Zr - may be less overabundant. If to these estimates of the composition we add the fact that all CH stars may belong to binary systems (McClure 1985), the parallels with the Barium stars are obvious.

Yamashita (1972, 1975) introduced the "CH-like" stars: low velocity stars with the spectral characteristics of the CH stars. He lists 16 CH-like stars (Yamashita 1975) but I can find a quantitative analysis for only one star: HD16115, one of the early R stars analyzed by Dominy (1984), has [Fe/H] = -0.2, [s/Fe] = +0.3 and the C and N enhancements of a typical early R star. Low dispersion spectra suggest that the Ba II 4554A line is greatly enhanced in some CH-like stars and so, perhaps, the list contains some CH stars too. It is now imperative that CH-like stars be analysed to isolate the distinctive characteristics (if any) that set them apart from either the early R and CH stars.

When Bond (1974) introduced the subgiant CH stars, he suggested that they were the progenitors of Keenan's giant CH stars. However, the kinematics suggests that the class is a mixture of halo and disk stars. As Table III shows, the metal abundance would permit many subgiant CH stars to be the progenitors of classical or mild Barium stars which are often slightly metal-poor. Bond's suggestion might be broadened: some subgiant CH stars will evolve into CH stars and others will evolve into mild or classical Barium stars.

3.7 The Chemical Composition of the Early R Stars

In the initial search for relatives of the classical Barium stars, the early R (warm carbon) stars are leading candidates. With the trend highlighted in Fig. 6, one expects the C-rich (i.e. C is more abundant than O) R stars to be also enriched in the s-process elements.

The remarkable fact is that the R stars are not enriched in the heavy elements. This conclusion was obvious from even the early attempts at quantitative analyzes (Buscombe 1955, Warner 1962, Gordon 1968, Greene et al. 1973). Dominy (1984) in a comprehensive analysis fully confirms this result (see, in particular, his Fig. 2). The point in Fig. 6 representing the early R stars is computed from Dominy's results for the 3 stars - HD113801, BD+33° 2399, and HD156074 - for which his analysis was the most reliable. Other differences with the classical Barium stars are found in the C and O abundances. The R stars are, of course, more C-rich: [C/Fe] = +0.5 with [C/Fe] = +0.1 for the Barium stars. Dominy found a [N/Fe] = +0.8 for the R stars but [N/Fe] = +0.4 for normal GK giants and the Barium stars. This N enhancement with the higher ^{13}C abundance ($^{12}C/^{13}C < 10$, in general) shows that extensive CN-cycling has occurred. Finally, McClure (1985) finds that early R stars, unlike the classical and mild Barium stars, do not show a high frequency of binaries.

In summary, the early R stars and the Barium stars appear to be unrelated.

4. MASS TRANSFER?

With the discovery (McClure et al. 1980; McClure 1983) that all Barium stars may belong to binary systems, production of a Barium star by mass transfer across a binary system began to replace the He-core flash as the leading explanation for these peculiar stars. McClure (1985) reports the following classes of peculiar stars to belong exclusively in binary systems: the classical and mild Barium stars and the subgiant and giant CH stars. I close this review by comparing the chemical composition of Barium stars and the qualitative predictions of the mass transfer hypothesis.

A brief sketch of the hypothesis must precede the comparison. Consider a wide binary in which mass transfer via Roche lobe overflow cannot occur until the primary has evolved to the AGB and dredged up C and the s-process elements to its surface. Through the mass transfer from the AGB star, the companion is converted to a peculiar or Barium star. I assume that the transferred mass is mixed with the outer envelope of the companion which will, in general, retain the appearance of a main sequence star. On evolution to the red giant branch, the convective envelope should thoroughly erase composition gradients and, as in normal stars, dredge up material lightly exposed to the CN cycle. I adopt the plausible assumption that all of the transferred mass is provided by the AGB star's envelope and, hence, its composition is that of the atmosphere of the cool carbon stars. Mass transfer in a few cases may occur before the AGB star has completed its conversion from an O-rich to a C-rich star. To an observer, the mass transfer hypothesis has three attractive features: (i) AGB stars are predicted and observed to be enriched in C and s-process elements, (ii) mass transfer is inevitable for wide binaries and, (iii) the chemical composition of Barium stars is, in principle, predictable given the composition of the AGB stars.

A close affinity between Barium and cool carbon stars is suggested by the abundance ratio of C to s-process elements. For extreme classical Barium stars (Fig. 6), [s/Fe] ~ 1.3 at C/O ~ 1. Utsumi's (1985) results for Zr and Sr in 12 carbon stars (^{13}C-rich stars are excluded-see below) provide a mean [s/Ti] = 1.3. A preliminary result of an analysis of infrared molecular lines (Eriksson et al. 1984) is that C/O ratios in carbon stars cluster between 1 and 1.5 with very few stars having a ratio greater than 1.5. The correspondence between the extreme Barium and cool carbon stars is surely not a coincidence. The mass transfer hypothesis accounts simply and naturally for the correspondence and, also, for the existence of subgiant CH, CH, mild and classical Barium giant and supergiant (i.e. ζCap but not ε Peg) stars showing differing levels of C and s-process enrichment. The ratio of C to s-process abundances is set by structural details rather than intrinsic properties of the nuclear reaction network. Hence, it would be a remarkable coincidence if a He-core flash in a low mass star were to provide at the surface the same mix of C and s-process elements as is observed in cool carbon stars where thermal pulses create the enrichment. The chemical composition of classical Barium stars indicates that the AGB companion be a low mass star with ^{13}C (α,n) as the neutron source. The 'marginal' Barium stars with a less pronounced enchancement of Ba and the rare-earths may result from mass transfer from an intermediate mass AGB star with ^{22}Ne (α,n) as the neutron source.

The mass transfer hypothesis may also explain why C-rich Barium stars appear to be rare. Since the C/O ratio is close to unity in the majority of carbon stars, a slight mixing of the transferred materials with the envelope of the companion for which C/O < 0.6 will ensure that C/O $\lesssim$ 1 for the Barium star. Nonetheless, some C-rich Barium stars should result. The CH-like stars are candidates for this class. Note also that the first dredge-up will reduce the C/O ratio of the Barium giants and convert some C-rich subgiant CH stars to O-rich Barium stars. The ^{13}C-rich cool carbon stars are not enriched in the heavy elements (Utsumi 1985) and, hence, are not progenitors of C-rich Barium stars. Early R stars may result from mass transfer in a binary system containing a ^{13}C-rich cool carbon star. Calculations show that a low mass AGB star may dredge up carbon without accompanying s-process elements (Wood 1985) and so produce the ^{13}C-rich cool carbon stars. An objection to this explanation of early R stars is McClure's (1985) claim that these stars show a normal frequency of binary stars. I shall not be surprised if further observations show that all R stars belong to binary systems!

The typical mild or classical Barium star and, in particular, the subgiant CH stars are slightly metal-poor stars (Iben 1983). Our incomplete analysis of carbon stars offers a tantalizing hint that they are metal-poor too. Dredge-up on the AGB is sensitive to the metallicity (Wood 1981) and, in particular, the dredge-up of C and s-process elements in low mass stars is predicted to be enhanced for metal-poor stars. Although detailed analyses and calculations must be completed, the mass transfer hypothesis may explain the metal deficiency of Barium stars.

The typical cool carbon star has a rather low abundance of ^{13}C: $^{12}C/^{13}C \sim 40$ to 70 (Eriksson et al. 1984). Then, subgiant CH stars are expected on the mass transfer hypothesis to have $40 < {}^{12}C/^{13}C < 90$, where I assume $^{12}C/^{13}C = 90$ for the companion prior to mass transfer. Sneden (1983) reported values of > 40, > 40, 20 - 60, and > 40 for 4 subgiant CH stars. Again, the mass transfer predictions are supported. The $^{12}C/^{13}C$ ratio for giant stars are affected strongly by the first dredge-up and no simple predictions are available for testing.

Predictions for Li even in main sequence peculiar stars must be uncertain because Li is so readily destroyed. There is a positive correlation between ^{13}C and Li in cool carbon stars (Gordon 1971). The ^{13}C-rich, Li-rich stars are not enhanced in the s-process elements (Utsumi 1985). However, Li-rich S stars (Boesgaard 1968; Catchpole and Feast 1976) are presumably enhanced in the s-process elements. Lithium in other carbon stars spans a range from log ε(Li) ~ 1.8 to a detection limit log $\varepsilon(Li) < 0.5$ (Wallerstein and Conti 1969). The Li abundance in the thin outer shell of the main sequence companion prior to receipt of mass from the carbon star companion will not exceed the cosmic abundance, log $\varepsilon(Li) = 3.0$, and may be substantially less. The Li abundance immediately following mass transfer is obviously dependent on the mixing which may occur. Depletion of Li is likely to occur for the lower mass Barium stars. Dilution by a factor of 20 to 60 may be expected for a red giant. Perhaps, the only clear prediction for the mass-transfer hypothesis is that it is not surprising that Li is present with roughly normal abundances in Barium stars. One interesting observation is suggested. The abundance log $\varepsilon(Li) \sim 1.0$ seen in several Barium giants implies that some of the subgiant and main sequence CH progenitors must show a Li abundance log $\varepsilon(Li) \sim 2.3$ to 3.0.

This commentary shows that the mass transfer hypothesis with the observed composition of carbon stars accounts qualitatively for the chemical composition of the several kinds of Barium stars. The presence of white dwarf companions to some Barium stars (Böhm-Vitense 1980; Dominy and Lambert 1983, Böhm-Vitense et al. 1984) is also consistent with the hypothesis. The paucity of Barium main sequence stars remains to be explained; the subgiant CH stars seem too few in number to be the sole main sequence product of mass transfer. Much remains to be done!

I thank M. J. Harris, K. Krishnaswamy, C. Sneden, V. V. Smith, and J. Tomkin for helpful discussions and for providing results in advance of publication. I apologize to J. F. Dominy, B. Gustafsson, and A. Zook who supplied unpublished results and interesting ideas on AGB stars which I included in my talk at Stasbourg but have omitted here for lack of space. This research has been supported in part by the U. S. National Science Foundation (grant AST 83-16635) and the Robert A. Welch Foundation.

REFERENCES

Adelman, S. J. 1973, Ap. J., 183, 95.
Anders, E. 1958, Ap. J., 127, 355.
Anders, E. and Ebihara, M. 1982, Geochim. Cosmochim. Acta, 43, 2363.
Biémont, E., Karner, C., Meyer, G., Träger, F., and zu Putlitz, G.1982, Astr. Ap., 107, 166.
Bidelman, W. P. and Keenan, P. C. 1951, Ap. J., 114, 473.
Boesgaard, A. M. 1968, Ap. J., 161, 1003.
Boesgaard, A. M. and Fesen, R. A. 1974, Pub. A. S. P., 86, 76.
Böhm-Vitense, E. 1980, Ap. J. (Letters), 239, L79.
Böhm-Vitense, E., Nemec, J., and Proffitt, C. 1984, Ap. J., 278, 726.
Bond, H. E. 1974, Ap. J., 194, 95.
Boyarchuk, A. A. and Eglitis, I. 1984, Bull. Crimean Ap. Obs., 64, 11.
Brown, J. A., Tomkin, J., and Lambert, D. L. 1983, Ap. J. Letters, 265, L93.
Burbidge, E. M. and Burbidge, G. R. 1957, Ap. J., 26, 357.
Buscombe, W. 1955, Ap. J., 121, 312.
Catchpole, R. M. and Feast, M. W. 1976, M.N.R.A.S., 175, 501.
Catchpole, R. M., Robertson, B. S. C., and Warren, P. R. 1977, M.N.R.A.S., 181, 391.
Clayton, D. D. and Ward, R. A. 1974, Ap. J., 193, 397.
Clegg, R. E. S., Lambert, D. L., and Tomkin, J. 1981, Ap. J., 250, 262.
Corliss, C. H. and Bozman, W. R. 1962, N.B.S. Monograph No. 53.
Cosner, K., Iben, I., Jr., and Truran, J. W. 1980, Ap. J. Letters, 238, L91.
Cosner, K. and Truran, J. W. 1981, Ap. Space Sci., 78, 85.
Cowley, C. R. 1984, private communication.
Cowley, C. R. and Downs, P. L. 1980, Ap. J., 236, 648.
Danziger, I. J. 1965, M.N.R.A.S., 131, 51.
Despain, K. H. 1980, Ap. J. (Letters), 236, L165.
__________ 1982, Ap. J., 253, 811.
Deupree, R. G. 1984, Ap. J., 282, 274.
Deupree, R. G. and Cole, P. W. 1983, Ap. J., 269, 676.
Dominy, J. F. 1984, Ap. J. Suppl., 55, 27.
Dominy, J. F. and Lambert, D. L. 1983, Ap. J., 270, 180.
Duncan, D. K. 1981, Ap. J., 248, 651.

Eriksson, K., Gustafsson, B., Hinkle, K. H., and Lambert, D. L. 1984, in preparation.
Garstang, R. H. 1952, Pub. A.S.P., 64, 227.
Gordon, C. P. 1968, Ap. J., 153, 915.
___________ 1971, Pub. A.S.P., 83, 667.
Greene, T. F., Perry, J., Snow, T. P., and Wallerstein, G. 1973, Astr. Ap., 22, 393.
Greenstein, J. L. 1954, Mém. Soc. Roy. des Sci. Liège, 14, 402.
Harris, M. J., Lambert, D. L., and Smith, V. V. 1984, in preparation.
Herbig, G. H. 1965, Ap. J., 141, 588.
Holweger, H. and Kovács, N. 1984, Astr. Ap., 132, L5.
Hyland, A. R. and Mould, J. R. 1974, Ap. J., 187, 277.
Iben, I., Jr. 1975, Ap. J., 196, 525 and 549.
___________ 1983, Ap. J. (Letters), 275, L65.
Iben, I., Jr. and Renzini, A. 1984, Phys. Repts., in press.
Käppeler, F., Beer, H., Wisshak, K., Clayton, D. D., Mäckle, R. L., and Ward, R. A. 1982, Ap. J., 257, 821.
Keenan, P. C. 1942, Ap. J., 96, 101.
Keenan, P. C. and Wilson, O. C. 1977, Ap. J., 214, 399.
Kollatschny, W. 1980, Astr. Ap., 86, 308.
Kovács, N. 1983a, Astr. Ap., 124, 63.
_________ 1983b, Astr. Ap., 120, 21.
Krishnaswamy, K. and Sneden, C. 1984, in preparation.
Kurtz, D. W. 1976, Ap. J. Suppl., 32, 651.
Kwiatkowski, M., Zimmermann, P., Biémont, E., and Grevesse, N. 1982, Astr. Ap., 112, 337.
Lambert, D. L., Dominy, J. F., and Sivertsen, S. 1980, Ap. J., 235, 114.
Lambert, D. L. and Luck, R. E. 1976, Observatory, 96, 100.
Lambert, D. L. and Ries, L. M. 1981, Ap. J., 248, 228.
Luck, R. G. and Bond, H. E. 1982, Ap. J., 259, 792.
MacConnell, D. J., Frye, R. L., and Upgren, A. R. 1972, A. J., 77, 384.
Mäckle, R. Holweger, H., Griffing, R., and Griffin, R. 1975, Astr. Ap., 38, 239.
McClure, R. D. 1983, Ap. J., 268, 264.
____________ 1984, Pub. A.S.P., 96, 117.
____________ 1985, this volume.
McClure, R. D., Fletcher, J. M., and Nemco, J. M. 1980, Ap. J. (Letters), 238, L35.
Mengel, J. G. and Sweigart, A. V. 1981, in Astrophysical Parameters for Globular Clusters, ed. A. G. D. Phillip, 277.
Moore, C. E., Minnaert, M. G. J., and Houtgast, J. 1966, NBS Monograph No. 61.
Newman, M. J. 1978, Ap. J., 219, 676.
Paczyńi, B. and Tremaine, S. 1977, Ap. J., 216, 57.
Pilachowski, C. A. 1977, Astr. Ap., 54, 465.
Pinsonneault, M. H., Sneden, C., and Smith, V. V. 1984, Pub. A.S.P., 96, 239.
Roby, S. W. 1984, private communication.

Ruland, F., Holweger, H., Griffin, R., Griffin, R., and Biehl, D. 1980, Astr. Ap., 92, 70.
Scalo, J. M. 1978, Ap. J., 221, 627.
Schadee, A. 1968, Ap. J., 151, 239.
Schatz, G. 1983, Astr. Ap., 122, 327.
Smith, M. A. 1971, Astr. Ap., 11, 325.
Smith, V. V. 1984, Astr. Ap., 132, 326.
Smith, V. V. and Lambert, D. L. 1984a, Pub. A.S.P., 96, 226.
______________________________ 1984b, in preparation.
Smith, V. V., Sneden, C., and Pilachowski, C. A. 1980, Pub. A.S.P., 92, 809.
Sneden, C. 1983, Pub. A.S.P., 95, 745.
Sneden, C. and Bond, H. E. 1976, Ap. J., 204, 810.
Sneden, C., Brown, J. A., Dutchover, E., Jr., Lambert, D. L. 1984, B.A.A.S., 16, 491.
Sneden, C., Lambert, D. L., and Pilachowski, C. A. 1981, Ap. J., 247, 1052.
Tech, J. L. 1971, N.B.S. Monograph No. 119.
Tomkin, J. and Lambert, D. L. 1974, Ap. J., 194, L89.
______________________________ 1979, Ap. J., 227, 209.
______________________________ 1983, Ap. J., 273, 722.
Truran, J. W. 1980, Nucleonika, 25, 1463.
Truran, J. W. and Iben, I., Jr. 1977, Ap. J., 216, 797.
Ulrich, R. K. 1973, in Explosive Nucleosynthesis, eds. D. N. Schramm and W. D. Arnett, 139.
Utsumi, K. 1985, this volume.
van den Heuvel, E. P. J. 1968a, Bull. Astr. Inst. Netherlands, 19, 309 and 326.
Wallerstein, G. 1969, Ap. J., 158, 607.
Wallerstein, G. and Conti, P. S. 1969, Ann. Rev. Astr. Ap., 7, 69.
Wallerstein, G. and Greenstein, J. L. 1964, Ap. J., 139, 1163.
Ward, R. A., Newman, M. J., and Clayton, D. D. 1976, Ap. J. Suppl., 31, 33.
Warner, B. 1962, Observatory, 82, 164.
__________ 1965, M.N.R.A.S., 129, 263.
Warren, P. R. and Peat, D. W. 1972, Astr. Ap., 17, 450.
Warren, P. R. and Williams, P. M. 1970, Observatory, 90, 115.
Williams, P. M. 1971, Observatory, 91, 37.
________________ 1975, M.N.R.A.S., 170, 343.
Wilson, O. C. 1976, Ap. J., 205, 823.
Wood, P. R. 1981, in Physical Processes in Red Giants, eds. I. Iben Jr. and A. Renzini, 135.
Yamashita, Y. 1972, Ann. Yokyo. Astr. Obs., 13, 169.
_____________ 1975, Pub. A.S.J., 27, 325.
Yamashita, Y. and Norimoto, Y. 1981, Ann. Tokyo Astr. Obs., 18, 125.

DISCUSSION

MALANEY - May I make two comments. Firstly with regard to the overabundance of Eu which you have remarked upon. This element is not produced in the classical s-process as you have stated. It is more associated with the r-process. However is now clear that the neutron fluxes associated with helium shell flashes for low mass stars on the AGB are typically ∿ 7 orders of magnitude greater than those considered for a classical s-process flux. Such a high flux can still produce many classical s-process elements, as well as the occasional r-process element. It may be then that this Eu overabundance observed is an indication of processing under a higher neutron flux than anticipated. However I know of no mechanism to produce such a high flux for single stars on or near the main sequence. So perhaps these observations of Eu may be an indications for mass transfer in binaries.

Secondly, I think it is dangerous to extrapolate from a value of the neutron exposure τ, which gives best agreement with the observations, to a value for the neutron flux and its time of duration. That is, similar observed abundances can be produced by low neutron fluxes on a long time scale and higher fluxes on shorter time scales. Such extrapolations may lead to wrong conclusions.

LAMBERT - I agree, the observational data is usually not sufficient to warrant an extrapolation for some stars.

CAYREL - Your abundance analysis for stars with T_{eff} below 4000 K are based on which grids of models ?

LAMBERT - For the O- rich stars, we have used models computed by Johnson and his colleagues. For the C - rich stars, Gustafsson and their colleagues have computed a large number of models.

DAVIS LOCANTHI - Did you include in chart of Zr isotopes the work by Schadee and Davis. This paper was ommitted from Zook's bibliography.

LAMBERT - I know about it, but I did not have it on chart.

RICHER - Do you have any evidence of the Ne^{22} neutron source operating in the C - stars ?

LAMBERT - Not as yet. However the pattern of the s-process elements is unknown in detail. Searches for the Mg isotopes must contend with severe blending with C_2 and CN lines.

BIDELMAN - There are a very few Ba stars that have considerably stronger C_2 bands than the majority. Have any of them been analyzed ? They might be transition cases to a star like V Cr B.

LAMBERT - Yes, Smith (1984) analyzed two of these cool Ba stars.

BARBUY - What lines did you use for determining the oxygen isotopes ?

LAMBERT - CO lines at $\lambda \sim 2.3\ \mu$.

JOHNSON - Have you searched your spectra of C stars for the (1 - 0) S (1) line of H_2, as well as the line you showed ? It strikes me that there is not a good correlation between the strength of the S (1) line we looked at and the line of H_2 you studied. I have in mind, for example, WZ Cas (where we found a strong line) and UU Aur (where we found no line).

LAMBERT - We did not study the S (1) line although I did check a few lines and thought I saw a correlation. We checked carefully the line we used for possible contaminents.

TSUJI - Do you have any result on C, N and O abundances in C stars ?

LAMBERT - In collaboration with K.H. Hinckle (KPNO), B. Gustafsson (Stockholm) and K. Eriksson (Uppsala), I am analyzing infrared spectra of 30 cool carbon stars. We hope to conclude this analysis shortly.

JASCHEK - What do you expect a carbon star to become after evolution goes on ?

LAMBERT - Some theoretical work suggests that a very luminous star carbon star may develop a hot bottom convective envelope and CN cycling will then convert the envelope to a N - rich and O - rich (relative to C) composition. Some C stars may shed their envelope and become C - rich planetary nebulae before this reconversion to a C - star can occur.

RARE EARTH PATTERNS IN RED GIANTS AND BLUE DWARFS

Charles R. Cowley and Robert C. Dempsey
Department of Astronomy, University of Michigan
Ann Arbor, MI 48109-1090, USA

ABSTRACT. Lanthanides are intercompared in normal and chemically peculiar red giants and A dwarfs. There is a surprising similarity, much of which may be accounted for in terms of the observational techniques, the properties of the atomic spectra, and the column densities above optical depth unity. There are important differences between the CP lanthanide anomalies and those of the red giants, especially for the elements praseodymium and europium.

1. LANTHANIDES: RED GIANTS AND BLUE DWARFS

It is an observational fact (Kurtz 1976) that the abundances of the elements beyond the iron peak show remarkably similar patterns in chemically peculiar red giants and blue dwarfs. This is perhaps surprising in view of our current ideas of how the anomalies were produced.

Abundance or identification studies in stars typically find the lighter, even-Z lanthanides, Ce, Nd, Sm, and Gd (Z=58-64). This is true both for the Ap and Am stars (CP stars), and for G and K stars both normal and chemically peculiar. The odd-Z elements behave more capriciously. For the most part, our information comes from observations of the second spectra. The heaviest lanthanide included in the Cayrel's (1963) paradigm study of ϵ Vir was gadolinium (Z=64), while Tech's (1971) identifications of lanthanides in ζ Cap ended with dysprosium (Z=66).

The strength of Pr II and Eu II in the peculiar A and Barium stars presents an interesting contrast. Both spectra are broadened markedly by hyperfine structure, which distorted early abundance determinations (Hartoog, Cowley, and Adelman 1974, Allen and Cowley 1974). While Eu II is often extremely strong in the early, chemically peculiar dwarfs, the *weakness* of this spectrum in barium stars was an inportant confirmation of the relevance of the s-process. Pr II is typically weak or absent, especially in the hot CP

M. Jaschek and P. C. Keenan (eds.), Cool Stars with Excesses of Heavy Elements, 225–229.

stars, but it is often very strong in the barium stars.

The odd-Z lanthanides beyond europium (Z=63), terbium, holmium, thulium, and lutetium (Z=65-71), are only sporadically identified in stellar spectra, and their abundances are rarely determined.

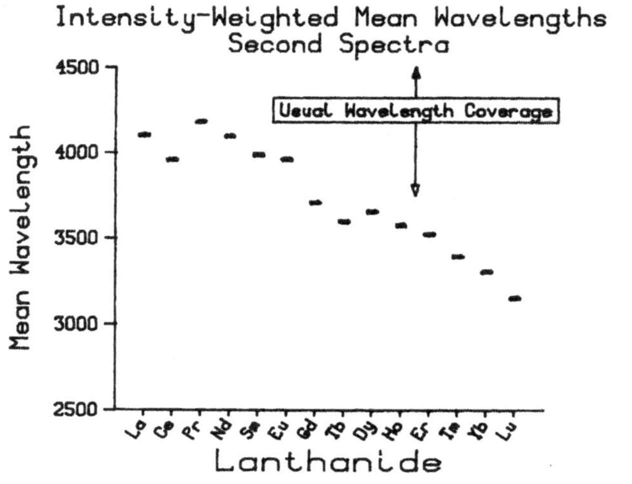

Figure 1. Mean Lanthanide Wavelengths.

Element identifications by means of wavelength coincidence statistics (WCS) have now been carried out for a variety of CP stars on the one hand, and F, G, and K giants (and dwarfs) on the other. The lanthanide-WCS parameters for the cool stars were surprisingly similar to those of the CP stars, in many cases (Cowley 1983). Cluster analysis of the lanthanide parameters usually separated the low-dispersion types, but in a few cases, the chemically peculiar dwarfs seemed to be more closely related to the late-type giants than to their CP congeners. For example, the Am stars 32 Aqr, τ UMa, and HR 178 clustered with the Ap star γ Equ, the mild barium star o Vir, and the G8 III η Dra.

2. SYNTHESIS

Figure 1 is a plot of data from Meggers, Corliss, and Scribner (1975). One sees immediately the powerful selection effect which makes the lighter lanthanides more accessible.

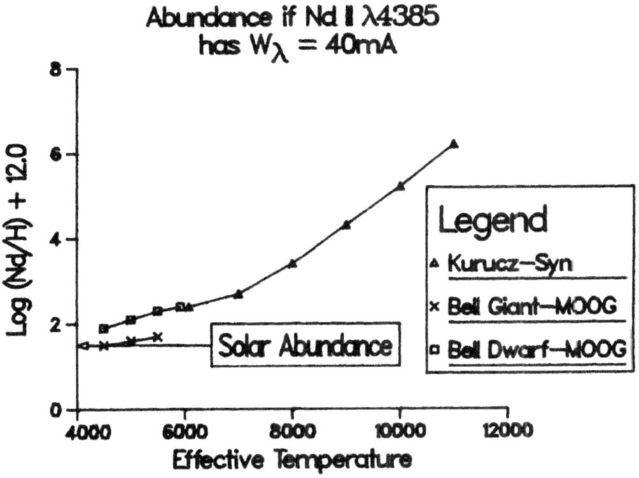

Figure 2. Model atmosphere calculations using Kurucz (1979), Bell, et al. (1976), and Bell (1984). MOOG and Syn are programs which calculate stellar absorption spectra.

In Figure 2 we provide the basis for an understanding of how the lanthanide WCS patterns are similar in the cool stars with normal or moderately enhanced (~ 1 dex) abundances and the CP dwarfs with abundance excesses that are often orders of magnitude higher. The 0.2eV Nd II line λ4385 has been chosen as "representative" of the lanthanide spectra. All calculations assume a microturbulent velocity of 1 km/sec. One sees that a chemically normal giant (Te=4500K) and a dwarf with Te=8000K and an excess of about 2 dex would both have a 40mA Nd II λ4385 line. The very large excesses quoted for some of the hotter CP stars will inevitably arise if the second spectra are detectible.

3. GOODNESS OF FIT TESTS

Cowley and Downs (1980) applied the chi-squared methodology to discriminate among hypotheses for the presence of heavy element excesses in evolved stars. Ideally, one would like to reject all but one hypothesis with $\geq$ 95% confidence. Table 1 shows the results of tests made for Barium and Am stars. The procedures and notation are those of Cowley and Downs; the estimated error is in column 4. The best fits for the Am stars require large values of the τ-zero parameter, but they cannot be rejected at 95% confidence unless σ is better than ca. 0.3. A different distribution of exposures from the exponential of Clayton and Ward would surely improve the fit.

Table 1 Clayton-Ward Fits for Sr - REE

Star	df	D	$\hat{\sigma}$	τ_o	Ref
HD116713	13	1.53	0.34	0.5	Warner(1965)
ζ Cap	13	0.34	0.16	0.3	Tech(1971)
HR774	12	0.24	0.14	0.6	Tom&Lam(1983)
HR774	14	1.54	0.33	0.3	Smith(1984)
τ Uma	9	1.5	0.41	>10	Smith(1973)
63 Tau	8	1.6	0.45	>10	Hundt(1972)

For the Barium stars, analyses such as those of Tech (1971) or Tomkin and Lambert (1983) provide remarkably good fits and critical decisions, perhaps arguably, may be made now.

It is a pleasure to thank Dr. Roger Bell for use of his model atmospheres, and Dr. John Laird for help with the programs.

4. REFERENCES

Allen, M. S., and Cowley, C. R. 1974, *Ap. J.*, **190**, 601.
Bell, R. A. 1984, *unpublished dwarf models.*
Bell, R. A., Erickson, K., Gustafsson, B., and Nordlund, A. 1976, *Astron. and Ap. Suppl.*, **23**, 37.
Cayrel, R., and Cayrel, G. 1963, *Ap. J.*, **137**, 431.
Clayton, D. D., and Ward, R. A. 1974, *Ap. J.*, **193**, 397.
Cowley, C. R. 1983, in *Statistical Methods in Astronomy*,
Cowley, C. R., and Downs, P. L. 1980, *Ap. J.*, **236**, 648.
comp. by E. J. Rolfe, ESA SP-201, p. 153.
Hartoog, M. R., Cowley, C. R., and Adelman, S. J. 1974, *Ap. J.*, **187**, 551.
Hundt, E. 1972, *Astron. and Ap.*, **21**, 413.
Kurtz, D. W. 1976, *Ap. J. Suppl.*, **32**, 651.
Kurucz, R. L. 1979, *Ap. J. Suppl.*, **40**, 1.
Meggers, W. F., Corliss, C. H., and Scribner, B. F. 1975,
Smith, M. 1973, *Ap. J. Suppl.*, **25**, 277. *N. B. S. Monograph*, **145**.
Smith, V. V. 1984, *Astron. and Ap.*, **132**, 326.
Tech, J. L. 1971, *N. B. S. Monograph*, **119**.
Tomkin, J., and Lambert, D. L. 1983, *Ap. J.*, **273**, 722.
Warner, B. 1965, *M. N. R. A. S.*, **129**, 263.

DISCUSSION

JASCHEK - But Am stars cannot have gone through s-process because they are young. On the other hand why does one not invoke for Ba stars the hypothesis of gravitational separation, as for Am stars ?

COWLEY - First if (N.B.) the s-process has any relevance to the Am stars, the nuclear processing must have gone on in a companion. The fact that Am's like Ba stars are all binaries should be noted in this regard. In answer to the second question, a deep convective envelope of the cool giants is generally taken to preclude a separation of the chemical elements by diffusion.

RICHER - The s-process distribution which you have shown, is it at all like the solar system distribution ?

COWLEY - The neutron exposures are considerably higher.

WING - The shift of the spectra of heavier rare earths toward shorter wavelengths will increase the blending with lines of iron-peak elements. This will also cause discrimination against measuring the heavier elements, in addition to the instrumental effects you mentioned.

COWLEY - Yes. The point is that the absence of the heavy lanthanides in abundance studies of most Ba and Am stars is not to be interpreted as a sign that the abundance patterns are necessarily the same.

SCHATZ - If I understood the last transparency right, you compare nine abundance data to a theory with eleven parameters (five τ's, five weights and δ).

COWLEY - We assumed a binomial distribution of exposures with N values, at equal steps up to τ_0. So the exposure model has three free parameters, τ_0, N, and the probability p. This model is a null hypothesis, and is offered in this sense. Certainly the fact that one cannot reject a null hypothesis can in no way be taken as proof the null hypothesis is correct. What we really need are a few more abundances, and to improve the accuracy for those elements that are plotted.
(Note : this binomial distribution of exposures is not discussed in the published paper).

STATISTICAL EQUILIBRIUM OF FE I/FE II IN COOL STARS

W. Steenbock
Institut für Theoretische Physik
und Sternwarte der Universität
2300 Kiel
Federal Republic of Germany

ABSTRACT. We investigate the statistical equilibrium of Fe I/Fe II in the sun and in the red giant Pollux (KO III). Special attention is given to a realistic treatment of the ionizing ultraviolet radiation field including absorption lines, to the thermalizing effect of inelastic collisions with hydrogen atoms, and to the influence of high-lying levels near the ionization limit.

The solar iron abundance is hardly affected by departures from LTE. In Pollux, non-LTE abundance corrections are 0.1 to 0.2 dex and depend on equivalent width and excitation potential.

I. INTRODUCTION

Knowledge of population numbers of atomic states is essential for the quantitative analysis of stellar spectra. In late-type stars usually the assumption of LTE is made. However, improved methods of observation and model-atmosphere analysis recently have revealed inconsistencies in LTE analyses: abundances derived from lines of high and low excitation disagree, and ionization equilibria cannot be reproduced consistently. In particular the iron-group elements, which play a mayor role in the analysis of cool stars, show these effects, as demonstrated in the LTE -analysis of Pollux by Ruland et al. (1980). Recently Brown et al. (1983) found anomalous Ti/Zr ratios in 33 red giants, which these authors attribute to non-LTE effects.

The statistical equilibrium of Fe I in the solar atmosphere has been studied by Athay and Lites (1972) with a 15 level Fe I model atom. Lites and Cowley (1974) extended this investigation to other cool stars. As Ruland et al. (1980) point out, there is no obvious way to explain the observed non-LTE effects in Pollux on the basis of these pioneering studies.

In an attempt to improve our theoretical understanding of departures from LTE in cool stars we have developed a non-LTE code with the intention to avoid certain simplifications made in earlier studies. The most critical of these simplifications appeared to us: (1) Neglect of inelastic collisions with hydrogen atoms; (2) Crude approximation of the uv

M. Jaschek and P. C. Keenan (eds.), Cool Stars with Excesses of Heavy Elements, 231–235.

radiation field which governs the ionization balance; (3) The model atom is too simple. In particular, highly excited levels which may couple to the continuum are neglected.

II. COMPUTATIONAL METHOD

A short description of the non-LTE code used in the present work is given by Steenbock and Holweger (1984, paper I). The solar model atmosphere used here is that of Holweger and Müller (1974); for test purposes we have also employed the HSRA (Gingerich et al. 1971). For the red giant Pollux we adopted the model of Ruland et al. (1980).

Our model atom involves 49 terms of Fe I and 20 terms of Fe II plus continuum. The Fe I model atom contains about 1/3 of the terms listed in the Atomic Energy levels. The coverage is complete in the lower part of the term diagram; at higher energies those terms have been preferred which are involved in strong radiative transitions. The 17 uppermost terms of Fe I were assumed to be coupled with the other terms only through collisions. Terms were represented by single levels. The strongest transition of each multiplet was selected to represent the radiative coupling between the various terms. The number of bound -bound transitions treated explicitly was 38 for Fe I plus 23 for Fe II. Line profiles were assumed to be Voigt profiles, with typically 27 frequency points per (half)profile. Transition probabilities are taken from various sources, preferring the f-values from the Oxford group.

Photo-ionization and recombination rates were computed assuming a LTE-radiation field where line opacities were taken into account following the distribution function scheme of Kurucz (1979) (see paper I). Cross sections were taken from Lites (1972), otherwise hydrogenic approximation were used. Excitation and ionization by collisions with electrons are treated following the semiempirical scheme of Drawin (1966). For transition with known large f-values we used the dipole approximation. For all other we have set the cross sections equal to πa_0^2. As was shown in paper I the very large N_H/N_e ratio in cool stars causes neutral-atom collisions to become the dominant thermalizing mechanism. Collisional excitation and ionization has therefore been included in our calculations as outlined in paper I; for transitions with unknown f-values the respective Q factor was set equal to one.

III. RESULTS

a) The solar non-LTE abundance corrections

The result is shown in Fig. 1 in terms of logarithmic abundance corrections, $\Delta\log\varepsilon$ being that quantity which must be added to the LTE abundance inferred from individual lines to correct for non-LTE effects. The sample comprises 101 lines of Fe I and 30 of Fe II out of the Ruland et al. (1980) list. Thermal excitation was assumed for fine structure levels of a multiplet.

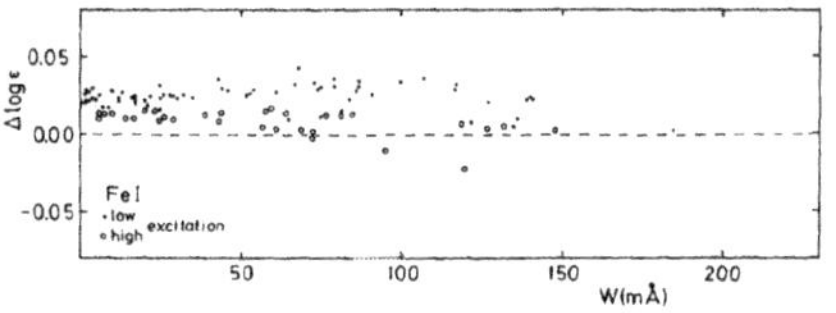

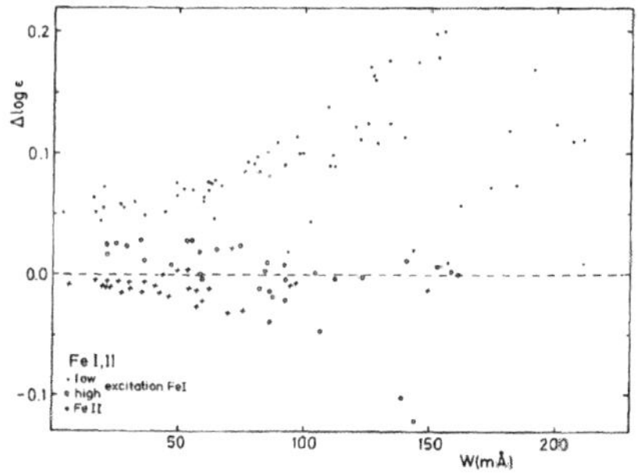

Fig. 1 Non-LTE abundance correction in the sun derived from individual lines versus equivalent width. Corrections for Fe II are zero (not shown).

Fig. 2 Same as Fig. 1 in Pollux

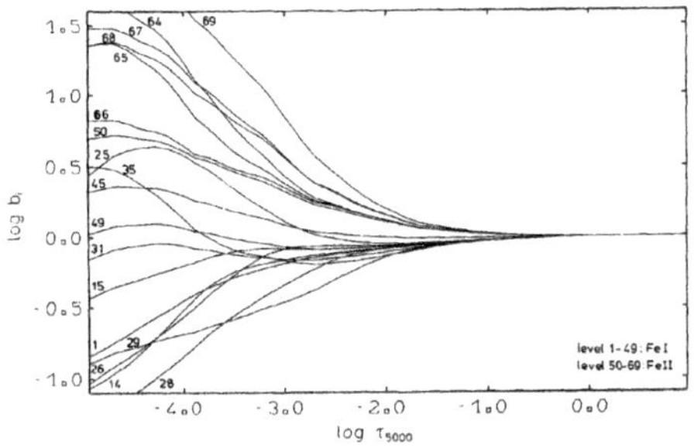

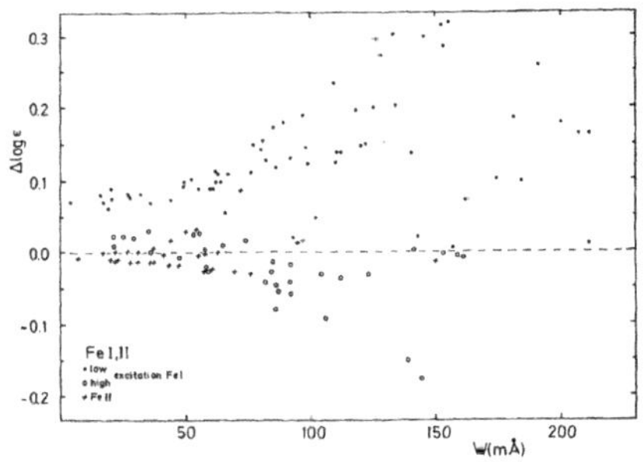

Fig. 3 Departure coefficients log b_i = log(n_i (non-LTE)/n_i (LTE)) for selected levels of Fe I/Fe II versus optical depth in the continuum.

Fig. 4 Same as Fig. 2, but neutral-atom collisions reduced by a factor of 3.

The non-LTE corrections of Fe I in the sun, although small, show a dependence on the lower excitation potential. As Ruland et al. did in the case of Pollux, we have grouped the lines into lines of low and high excitation, the intersection being made at 3.5 eV. Test calculations show that the dependence of $\Delta \log \epsilon$ on excitation results from the general overionization of Fe I and simple reflects the different height of formation of both groups of lines. Departures from thermal excitation are negligible. The low-lying Fe II levels of even parity are in LTE; Fe II levels with odd parity are overpopulated in high photospheric layers, but are in LTE in deeper layers where equivalent widths are formed.

The non-LTE corrections are largely independent of the $T_e(\tau)$ model used; the HSRA gives only slightly (~0.02 dex) larger $\Delta\log\varepsilon$ values. If the uppermost 17 terms of Fe I are omitted in the model atom, the solar $\Delta\log\varepsilon$ values increase to ~0.05 dex. Much larger departures occur if neutral -atom collisions are suppressed: the $\Delta\log\varepsilon$ values then scatter over more than 0.2 dex.

b) Abundance corrections for the red giant Pollux

Abundance corrections for Pollux (T_{eff} =4850 K, log g = 2.4, [M/H]= -0.1) are shown in Fig. 2, whereas Fig. 3 display the departure coefficients of representative terms. Compared to the sun the departures from LTE are larger. Fe II is no longer the dominant ionization stage of iron, and an overpopulation of the low-lying Fe II levels with even parity occurs. Because Fe II lines originate in the lower photosphere where the departures are small, the Fe II abundance corrections remain also quite small.

Abundance corrections for Fe I derived from low excitation lines, in particular those on the flat portion of the curve of growth(100-150 mÅ), fall systematically above the high excitation lines. The reason for this dependence on excitation is the same as in the sun. The larger scatter reflects the fact that departures from LTE become more important at lower gravities because collisions are less frequent. Neutral-atom collisions dominate the collisions by electrons. That collisions are still important in Pollux is demonstrated by Fig. 4, where neutral-atom collision rates have been reduced to 1/3 of their former value. Comparison of Fig.2 or Fig.4 with Fig.4 of Ruland et al. (1980) shows that our non-LTE calculations explain the main empirical results: the dependence on excitation potential and equivalent width.

Interestingly, those six lines in Fig.2 that show the largest $\Delta\log\varepsilon$ values all belong to multiplet no. 62 : these are the 2.2 eV lines which, according to Blackwell et al.(1984), show an abundance anomaly in the sun if analyzed assuming LTE.

REFERENCES

Athay,R.G.,Lites,B.W.:1972,Astrophys.J.176,809
Blackwell,D. E. Booth,A. J.,Petford,A. D.:1984,Astron.Astrophys.132,236
Brown,J.A.,Tomkin,J.,Lambert,D.L.:1983,Astrophys.J.265,L93
Drawin,H.W.:1966,Collision and Transport Cross-Sections,Report EUR-CEA-FC-383,Association Euratom-C.E.A.,Fontenay-aux-Roses(1967, revised)
Gingerich,O.,Noyes,R.W.,Kalkofen,W.,Cuny,Y.:1971,Solar Phys.18,347
Holweger,H.,Müller,E.A.:1974,Solar Phys.39,19
Kurucz,R.L.:1979,Astrophys.J.Suppl.40,1
Lites,B.W.:1972,NCAR Cooperative Thesis No.28,University of Colorado
Lites,B.W.,Cowley,C.R.:1974,Astron.Astrophys.31,361
Ruland,F.,Holweger,H.,Griffin,R.,Griffin,R.,Biehl,D.:1980,Astron.Astrophys.92,70
Steenbock,W.,Holweger,H.:1984,Astron.Astrophys.130,319 (paper I)

DISCUSSION

CHMIELEWSKI - It has been often remarked that the temperature structure in the Holweger - Mueller solar model atmosphere is rather an average excitation temperature structure, than a true a kinetic temperature. Could you comment on the effect this would have on your conclusions ?

STEENBOCK - These calculations with a HSRA model show that on the average abundance corrections for FeI increase only by less than 0.02 dex with a slightly larger scatter in the corrections. Fe II abundance corrections inferred from lines with equivalent widths greater than 100 mA may decrease somewhat more ($\sim$0.03 dex).

CAYREL - Can you tell us what would become of the Δ log ε NLTE corrections you have shown us for a solar model, if instead of a solar model, the computations would have been done with the same T_{eff} and the same log g, but with M/H ratio 0.1 or 0.2 dex above the solar value ?

JOHNSON - What values did you use for the radiation fields at the wavelenghts of the bound-free transitions ?

STEENBOCK - We used theoretical values from the models. Photoionization rates were computed assuming a LTE radiation field. Because the ionizing UV radiation field is strongly influenced by absorption and, possibly, re-emission of spectrum lines, line opacities were taken into account. In suitable spectral intervals the line opacities can be represented by a distribution function, which has a much smoother variation with wavelength. Our approach makes use of the statistical distribution function representation of line opacities worked out by Kurucz (1969) for use in his ATLAS 6 Code (see paper I).

CHEMICAL COMPOSITION OF BARIUM STARS

N. Kovacs
Institut für Theoretische Physik
und Sternwarte der Universität Kiel
Olshausenstraße
2300 Kiel
Germany (F. R.)

ABSTRACT. A model atmosphere analysis of seven barium stars has been carried out, based on high-dispersion spectra.

The s-process elements are found to be enhanced relative to iron by factors between 2 (HD 139195) and 30 (HD 92626).

HD 65699 shows a special abundance pattern insofar as out of the s-process elements only strontium and barium are enhanced. A "mild" s-process is proposed in order to explain this result.

For four stars CNO abundances could be determined. In one star (HD 83548) carbon is slightly enhanced with respect to normal red giants, in two other stars (HD 46407 and HD 116713) carbon is enhanced by a factor 2.5 to 3. No carbon enhancement was found for HD 65699.

1. INTRODUCTION

The discovery by McClure et al. (1980) and by McClure (1983) that all barium stars appear to be members of binary systems where the companions might be white dwarfs, and the detection of the postulated white dwarfs through their UV-radiation in some cases (e.g. Böhm-Vitense, 1980; or Böhm-Vitense et al., 1983) has re-animated the discussion about the origin of the abundance pecularities in these cool giants of Population I, which form about 1-2% of all red giants.

Any theory clearly needs support by model-atmosphere analyses based on high-resolution spectra.

Here it is advantageous to have atmospheric parameters and determinations of the chemical composition for whole samples of barium stars at hand in order to allow the study e.g. of relative abundances. One needs data which are not biased by the fact that a comparison has to be made between different authors who used spectral material of different quality and who employed different model atmosphere codes or different comparison stars.

Comparable studies have recently been performed e.g. by Pilachowski (1977) for mild barium stars or by Smith (1984) for cool barium stars.

The author has started his survey with the analyses of two barium

M. Jaschek and P. C. Keenan (eds.), Cool Stars with Excesses of Heavy Elements, 237–241.

stars, HD 65699 and HD 83548 (Kovacs, 1983 b); now this survey has been extended to a whole of seven stars.

It is hoped that this work which is based on a homogeneous set of high-resolution spectra of some of the brightest southern barium stars will provide additional facts about the chemical composition of these stars.

2. OBSERVATIONS AND REDUCTIONS

For all seven stars Reticon spectra have been recorded by H. Holweger with the ESO Coudé Echelle Spectrograph, fed by the 1.4 m Coudé Auxiliary Telescope.

Up to ten spectral regions have been registered per star. The 1872 Reticon channels covered wavelength intervals which went from a width of 43 A at 5160 A to 83 A at 10330 A.

The resolving power is 100 000.

The reductions have been carried out by H. Holweger and the author at ESO, Garching with the IHAP image processing system.

3. MODEL-ATMOSPHERE ANALYSIS

The abundances of the chemical elements have been computed with a program which calculates the product $gf\varepsilon$ of oscillator strength f (times statistical weight of the lower level of the transition, g) and abundance ε for a given equivalent width.

Abundances were derived relative to the Sun, in this way errors in the oscillator strengths cancel out.

The atmosphere models have been scaled from the empirically derived $T(\tau)$ relationship for Pollux (Ruland et al., 1980). With an effective temperature of 4850 K for Pollux, the maximum difference to the temperatures of the seven barium stars is 225 K, but in most cases this difference is much smaller. Gravity, metallicity, and turbulent velocities of Pollux are also very similar to the respective parameters for the barium stars of our sample.

For the calculation of the effective temperatures we used near-infrared broad-band colours and the Johnson (1966) calibration.

This method is sensitive to the infrared excess of the barium stars which makes the more pronounced barium stars look brighter in the near infrared and consequently leads to a lower effective temperature, but an estimation shows that the maximum effect is less than 100 K. Even though the excess radiation, which probably is related to the carbon overabundances of many barium stars, influences the gravity determination, our abundances are not affected significantly by these possible temperature errors.

We did not use excitation temperatures because we found that our barium stars were subject to non-LTE effects which lead to a split between the abundances derived from high-excitation lines and from low-excitation lines, respectively, of the order of 0.12 dex by which low-excitation abundances e.g. of iron were lower than high-excitation abundances in our barium stars. The reality of this split has been verified by statistical equilibrium calculations (Steenbock, 1985) for a 50-level iron model atom.

TABLE I

Star	T_{eff} [K]	log g	$[Fe/H]^{*}_{\odot}$	MICRO	MACRO	[Ba/Fe]	[R.E./Fe]
HD 139195	5075	2.6	-0.15	1.5	3.4	0.50	0.27
HD 111315	4870	1.9	-0.25	2.2	5.6	0.64	0.28
HD 65699	4850	1.5	-0.20	2.0	9.5	0.70	-0.28
HD 83548	4930	2.0	-0.33	1.6	5.5	0.82	0.35
HD 116713	4720	2.4	-0.29	1.3	3.9	1.24	1.11
HD 46407	4800	2.4	-0.42	1.7	4.9	1.39	1.08
HD 92626	4670	1.7	-0.35	2.2	4.2	1.62	1.50

Physical parameters of the barium stars. T_{eff} : effective temperature; g : gravity; $[Fe/H]^{*}_{\odot}$: iron abundance relative to the Sun; MICRO : microturbulent velocity in km/sec; MACRO : macroturbulent velocity in km/sec; [Ba/Fe] : Ba/Fe ratio relative to the Sun; [R.E./Fe] : Rare Earths/Fe ratio relative to the Sun, where R.E. average of La II, Ce II, and Nd II.

The gravity was determined from the ionization equilibrium of iron, the element with the largest number of lines of both ionization stages in this temperature range.

The error in the gravity determination is of the order of 0.3 dex.

The microturbulence was derived from the usual assumption that abundances must be independent of equivalent widths, and the macroturbulence was found by fitting theoretical line profiles for different turbulent velocities to a composed line consisting of eight clean Fe I lines.

The physical parameters of the barium stars are shown in TABLE I.

4. RESULTS AND DISCUSSION

Whenever it was possible, we derived the abundances of the chemical elements from high-excitation lines, which are less influenced by departures from LTE than low-excitation lines (c.f. Ruland et al., 1980).

For four stars, namely HD 46407, HD 65699, HD 83548, and HD 116713, we could determine CNO abundances. In all four stars the oxygen abundance is nearly unchanged and nitrogen is enhanced compared to the solar value, but close to the average nitrogen abundance in K-giants (Sneden et al., 1981).

While the carbon abundance in HD 65699, adjusted to the iron deficiency, also has a value which is common for K-giants, we obtain slight overabundances in HD 83548 and significant overabundances in HD 46407 and HD 116713, but still C/O less than 1.

For the elements of the iron group, and especially for iron itself, we obtain moderate inderabundances compared to the Sun. Danziger (1965) has found similar underabundances of iron in two stars of our sample, HD 83548 and HD 116713. According to Tomkin and Lambert (1983), the barium star HR 774 also has a moderate iron deficiency.

For the heavy elements from strontium on we restricted ourself to ionic lines for the abundance determination because Brown et al. (1983) have shown that for field giants the use of neutral lines of zirconium results in an apparent underabundance of this element when we compare the Ti/Zr ratio to its solar value.

This finding has some relevance to the classification e.g. of ε Peg as a mild barium star. Hyland and Mould (1974) have questioned this classification because they found solar barium abundances from neutral low-excitation lines, while Kovacs (1983 a) derived an enhanced barium abundance for this star from an ionic line.

The overabundances of the rare earths vary between a factor 2 in HD 139195 and a factor 30 in HD 92626.

HD 65699 shows a special abundance pattern insofar as out of the s-process elements only strontium and barium are enriched. According to Holweger and Kovacs (1984), this is probably due to the especially small neutron capture cross sections of ^{88}Sr and ^{138}Ba, in combination with a "mild" s-process which shifted pre-existing heavy nuclei by some places in the mass scale and piled up excess abundances at the "bottle-necks" strontium and barium. A neutron exposure $\tau \sim 0.04$ mb^{-1} suffices to increase the abundances of strontium and barium by a factor three without significantly affecting the abundances of the other s-process elements.

The abundances of barium and the rare earths lanthanium, cerium, and neodymium (average value) are also contained in TABLE I.

REFERENCES

Böhm-Vitense, E. : 1980, Astrophys. J. Letters 239, L79

Böhm-Vitense, E., Nemec, J., Proffitt, C. : 1983, publication of the University of Washington, Department of Astronomy

Brown, J.A., Tomkin, J., Lambert, D.L. : 1983, astrophys. J. Letters 265, L93

Danziger, I.J. : 1965, Monthly Notices Roy. Astron. Soc. 131, 51

Holweger, H., Kovacs, N. : 1984, Astron. Astrophys. 132, L5

Hyland, A.R., Mould, J.R. : 1974, Astrophys. J. 187, 277

Kovacs, N. : 1983 a, Astron. Astrophys. 120, 21

Kovacs, N. : 1983 b, Astron. Astrophys. 124, 63

McClure, R.D., Fletcher, J.M., Nemec, J. : 1980, Astrophys. J. Letters 238, L35

McClure, R.D. : 1983, Astrophys. J. 268, 264

Pilachowski, C.A. : 1977, Astron. Astrophys. 54, 465

Ruland, F., Holweger, H., Griffin, R., Griffin, R. : 1980, Astron. Astrophys. 92, 70

Smith, V.V. : 1984, Astron. Astrophys. 132, 326

Sneden, C., Lambert, D.L., Pilachowski, C.A. : 1981, Astrophys. J. 247, 1052

Steenbock, W. : 1985, this conference

Tomkin, J., Lambert, D.L. : 1983, Astrophys. J. 273, 722

DISCUSSION

CATCHPOLE - Do you consider the Fe/H deficiencies you show are real ?

KOVACS - I think that part of the deficiencies could be caused by the way in which I determined the effective temperatures. Due to the excess radiation in the near infrared of the more pronounced Ba stars (like e.g. HD 46407 or HD 92626) my effective temperatures for these stars, as already mentioned, might be somewhat too low. If we would raise the effective temperature by say 100 K, the iron abundances would go up as well by about 0.1 to 0.2 des.

However, this mechanism does not apply to the milder Ba stars without any noticeable infrared excess. Therefore I think that the underabundance of iron of the seven Ba stars of my sample is real, even though it might be exaggerated.

RICHER - In your summary table it appears that there is a correlation between derived Ba abundance and temperature. Is this of any evolutionary significance ?

KOVACS - The answer is similar to the one to Dr Catchpole : the infrared excess being stronger in the stars with larger Ba overabundances, I also obtained lower effective temperatures for these stars than for the stars with only moderate overabundances and no infrared excess. Probably this is the reason for the apparent correlation mentioned by you.

COWLEY - This is a general comment, and perhaps a question for the theoretician. The solar abundance of scandium is very low, comparable to that of yttrium, and I would expect it to be raised if part of the star had been exposed to neutrons and mixed into the visible atmosphere. It seems to me this could be a difficulty for the "dredge-up" hypothesis, in view of your results that scandium is not enhanced, but actually somewhat deficient.

KOVACS - Sorry, I do not feel competent to answer your question. I can only state that in the four stars where I was able to determine its abundances, scandium in no way stands out from the other elements of the iron group.

ABUNDANCE ANALYSIS OF COOL CARBON STARS

Kazuhiko Utsumi
Department of Astronomy,
Faculty of Integrated Arts and Sciences,
Hiroshima University, Hiroshima, Japan.

ABSTRACT. Curve-of-growth analyses for thirty cool carbon stars have been made, and physical parameters and relative abundances for fourteen elements have been determined. In J-type cool carbon stars, abundances of s-process elements with respect to Fe are nearly normal. In normal carbon stars of C5∿C8, heavy metals are overabundant by factors of 10∿ 100, and rare-earth elements are overabundant by a factor of about 10.

1. INTRODUCTION

Identification of spectral lines in the visual region for two cool carbon stars U Hya (C6,3) and Y CVn (C5,5) was made by Utsumi (1967), and it was found that the effect of molecular bands is not serious in the spectral regions from 4750 A to 4900 A, and from 4400 A to 4500 A, and physical parameters and abundances of the elements can be obtained from the atomic lines in these regions. Curve- of-growth analyses for 22 cool carbon stars were made by Utsumi (1970), and it was found that in J-type cool carbon stars, Sr, Y, Zr, and Ba show nearly normal abundances, while rare-earth elements are overabundant by factors of 10∿ 100, and in normal cool carbon stars,all s-process elements are overabundant by factors of 10∿100. These results were obtained with the use of the absolute gf-values by Corliss and Bozman (1962), and by Corliss and Warner (1964), and the cosmic abundance by Aller (1961) as a standard. The gf-values by Corliss and Bozman and those by Corliss and Warner were found to contain large systematic errors, and as a result, the solar photospheric abundances were revised. Our previous result, therefore, must be revised by using new gf-values, and abundance analyses will be extended for as many carbon stars as possible.

2. OBSERVATIONS

We have obtained about two hundred spectrograms of sixty-two carbon stars (forty-eight stars later than C4, fourteen stars earlier than C3) in the spectral region from 4400 to 5000 A with a dispersion of 13.7

M. Jaschek and P. C. Keenan (eds.), Cool Stars with Excesses of Heavy Elements, 243–247.

A/mm, with the coude spectrogragh attached to the 188 cm reflector of the Okayama Astrophysical Observatory. Some of the spectrograms were obtained by Prof. Y. Yamashita. Logarithmic intensity tracings were made with a Nalumi microphotometer of the Department of Astronomy, University of Tokyo by Prof. Y. Yamashita, and with a Nalumi microphotometer of the Department of Astronomy, Hiroshima University by Utsumi. We have made analyses of thirty cool carbon stars which include eight stars of C4∿C5 in addition to twenty-two stars studied before.

3. CURVE OF GROWTH

The differential curve-of-growth method is improper for cool carbon stars because the spectral region which is almost free from molecular absorption is different for M-type stars and for cool carbon stars, and appropriate atomic lines of cool carbon stars in common with those of K-type or Ba II stars are very few. For this reason, we constructed empirical curves of growth for fourteen elements using the equivalent widths and absolute gf-values. Absolute gf-values were mainly taken from the Table by Kurucz and Peytremann (1975). Atmospheric parameters were derived from the plots for Fe I and Ti I lines so as to show the best fitting to Unsöld's (1955) theoretical curve of growth.

4. RESULTS

Excitation temperatures of thirty cool carbon stars thus derived are given in Table 1, together with the spectral types by Yamashita (1972, 1975).

TABLE 1.

Star	Spectral Type	θ_{ex}	Star	Spectral Type	θ_{ex}
HD 79319	R4; C4, 4J	1.6	U Hya	N2; C6, 3	1.8
HD 19557	R5; C4, 5J	1.6	VY UMa	N0; C6, 3	1.8
HD 52432	R5; C4, 5J	1.6	DS Peg	N1; C6, 3	1.8
HD 168227	R5; C4, 5J	1.7	Y Tau	N2; C6, 4	1.8
UV Cam	R8; C4, 5J	1.6	TU Gem	N3; C6, 4	1.8
RY Dra	N4; C4, 5J	1.8	UU Aur	N3; C6, 4	1.8
Y CVn	N3; C5, 5J	1.8	S Sct	N3; C6, 4	1.8
RV Mon	Nb; C4, 4	1.8	V Aql	N6; C6, 4	1.8
AQ And	Nb; C5, 4	1.8	HD 180953	N2; C6, 4	1.8
ST Cam	Nb; C5, 4	1.8	Z Psc	N0; C7, 2	1.8
W Ori	N5; C5, 4	1.8	19 Psc	Nb; C7, 2	1.8
X Cnc	N3; C5, 4	1.8	UX Dra	N0; C7, 3	1.8
SS Vir	Np; C5, 3e	1.8	RR Her	K5ep; C8, 1e	2.0
HD 59643	N9; C6, 2	1.8	RS Cyg	N0pe; C8, 2	2.0
BL Ori	Nb; C6, 3	1.8	WZ Cas	N1p; C9, 2J	2.0

Most of normal cool carbon stars indicate $\theta_{ex}\sim 1.8$, and R-type stars of C4J∿C5J $\theta_{ex}\sim 1.6$. Excitation temperatures of N-type stars of C4J∿C5J (Y CVn and RY Dra) are difficult to determine, because lines of 4400 A∿ 4500 A region cannot be used in curve-of-growth analysis due to strong molecular bands and UV-depression. However, from intensities of lines in 4400 A∿ 4500 A region of Y CVn and RY Dra, $\theta_{ex}\sim 1.8$ can be indicated. In the present study, ionization degree was determined by using Ti I and Ti II lines, since most of Y I and Y II lines are weak and blended. Thus, we have determined relative abundances of the following elements; Ca, Sc, Ti, V, Cr, Mn, Fe, Sr, Y, Zr, Ba, La, Nd, and Sm for thirty cool carbon stars. On account of strong molecular bands and blending of both molecular and atomic lines, the accuracy of the result is different from element to element. Since our main interest is abundance anomaly in s-process elements, [Y/Ti], [Zr/Ti], [Ba/Ti], [La/Ti], [Nd/Ti], and [Sm/Ti] for twenty stars are given in Table 2. Here, the notation [M/Ti] denotes the logarithmic relative abundance with respect to Ti, normalized by the solar abundance by Allen (1976). The accuracy is estimated to be about 0.4 dex.

TABLE 2.

Star	Spectral Type	[Y/Ti]	[Zr/Ti]	[Ba/Ti]	[La/Ti]	[Nd/Ti]	[Sm/Ti]
AQ And	C5, 4	1.7	1.9	0.6	1.6	0.8	1.0
ST Cam	C5, 4	1.7	1.8	0.9	1.2	0.5	0.6
X Cnc	C5, 4	2.3	1.7	2.1	1.9	1.2	1.1
BL Ori	C6, 3	1.6	1.5	1.2	1.3	1.1	1.2
U Hya	C6, 3	1.8	1.5	1.2	1.0	0.8	1.2
VY UMa	C6, 3	1.9	1.7	1.4	1.0	0.7	0.8
Y Tau	C6, 4	1.5	1.4	0.4	—	1.5	0.6
TU Gem	C6, 4	1.1	2.2	1.4	—	1.1	1.3
UU Aur	C6, 4	1.7	1.5	1.3	1.0	0.8	0.9
Z Psc	C7, 2	2.3	1.8	2.0	1.1	1.3	1.7
19 Psc	C7, 2	1.8	1.4	2.2	0.9	0.9	1.0
RS Cyg	C8, 2	2.5	1.9	1.8	0.6	0.6	0.6
HD 79319	C4, 4J	<0.8	0.0	0.5	<1.3	0.3	0.3
HD 19557	C4, 5J	<0.8	0.0	0.5	<1.3	0.3	0.3
HD 52432	C4, 5J	—	0.5	0.0	<1.5	0.8	0.5
HD 168227	C4, 5J	-0.2	0.1	-0.3	<1.1	0.6	0.2
UV Cam	C4, 5J	0.4	0.1	-0.2	—	0.5	0.3
RY Dra	C4, 4J	<0.7	0.1	0.1	—	0.1	-0.4
Y CVn	C5, 5J	0.5	0.2	0.0	—	0.5	0.5
WZ Cas	C9, 2J	<1.2	0.5	0.1	0.0	0.0	0.4

In J-type cool carbon stars of both R- and N-type, abundances of s-process elements with respect to Fe are nearly normal. In normal carbon stars of C5∿C8, heavy metals are overabundant by factors of 10∿100, and rare-earth elements are overabundant by a factor of about 10.

5. DISCUSSION

In connection with heavy metals relative to Ti, the present results are not so different as from the previous ones.

In connection with abundances of rare-earth elements relative to Ti, the present results show the values less than the previous results by factor of 10∿100.

Such discrepancies are believed to be

(1) due to new gf-values adopted .... about 0.5 dex,

(2) due to difference of ionization degree .... about 0.3 dex,

(3) due to the effect of blending in J-type cool carbon stars .... about 0.5 dex.

REFERENCES

Allen, C. W. 1976, Astrophysical Quantities p. 31∿ p. 33 3rd Ed. Athlone Press.

Carliss, C. H., and Bozman, W. R. 1962, Experimental Transition Probabilities for Spectral Lines of Seventy Elements, NBS Monograph 53.

Corliss, C. H., and Warner, B. 1964, Astrophys. J. Suppl., 8, 395.

Kurucz, R. L., and Peytremann, E. 1975, Smithonian Astrophys. Obs. Spec. Rep., No. 362.

Unsöld, A. 1955, Physik der Sternatmosphären, 2te Aufl. (Springer)

Utsumi, K. 1967, Publ. Astr. Soc. Japan, 19, 342.

Utsumi, K. 1970, Publ. Astr. Soc. Japan, 22, 93.

Yamashita, Y. 1972, Ann. Tokyo Astr. Obs. 2nd Series, 13, No. 3.

Yamashita, Y. 1975, Ann. Tokyo Astr. Obs. 2nd Series, 15, No. 1.

DISCUSSION

PEERY - Do you and Utsumi attempt to get any quantitative information from molecular bands in the blue and violet ? They are found throughout the wavelengths considered here, specially metallic hybrides.

FUJITA - They have not yet been taken into account, but they should, and perhaps we will.

LLOYD EVANS - We should congratulate Dr Utsumi on this monumental study. It would be valuable to obtain similar results for more cool J stars, because we expect (from theory and MC observations) that such stars with enhanced s-process elements should also occur.

FUJITA - As Utsumi obtained the spectrograms of more than sixty carbon stars, I am sure that will continue his work on other thirty of them.

JASCHEK - I would like to offer to all people having large collection of wavelengths and/or equivalent widths to send them to any one of the data centers for publication, rather than to send them to journals where the chance of being published is rather slim.

CALCIUM CHLORIDE IN THE IUE SPECTRA OF CARBON STARS

P. D. Bennett
University of British Columbia
Vancouver, B.C. V6T 1W5
Canada

H. R. Johnson
Indiana University
Bloomington, IN 47402
U.S.A.

ABSTRACT. An identification of CaCl in the IUE spectra of N-type carbon stars is proposed. This is based primarily on the coincidence of several strong and previously unidentified stellar features with the strongest band heads of the D-X system of CaCl. It is possible that these lines carry information about the upper photosphere and perhaps even the temperature minimum.

1. INTRODUCTION

The identification of absorption bands in the spectra of carbon stars was first made by Sanford in 1942. These bands of the red system, $A^2\Pi$-$X^2\Sigma$, at 6200 Å and the orange system, $B^2\Sigma$-$X^2\Sigma$ at 5800 Å can become strong enough to dominate the optical spectrum of certain C and SC stars, e.g. VX Aql (Wyckoff and Wehinger, 1976) and WZ Cas (Keenan and Bidelman, 1979). It has been suggested by Clegg and Wyckoff (1977) that these CaCl bands occur commonly in the spectra of carbon stars.

2. THE CALCIUM CHLORIDE D-X BAND SYSTEM

The bands of the $D^2\Sigma$-$X^2\Sigma$ system under consideration in this investigation lie in the wavelength range of 3000-3300 Å, which appears to be a spectral region of relatively low opacity in carbon stars. The closed nature of the CaCl bands render resolution of their rotational structure difficult, if not impossible and hindered analysis for many years. Recently, both laser spectroscopy (L.-E. Berg et.al., 1978) and traditional techniques (L.-E. Berg et.al., 1981) have yielded complete analyses of the D-X system. Their spectroscopic constants (L.-E. Berg, 1981) have been used in this work and are listed in Table I, where the conventional notation of molecular spectroscopy (e.g. Herzberg, 1950) has been followed.

M. Jaschek and P. C. Keenan (eds.), Cool Stars with Excesses of Heavy Elements, 249–254.

TABLE I. Molecular constants in cm^{-1} for $Ca^{35}Cl$

	$X^2\Sigma$	$D^2\Sigma$
T_e	0.0	31110.934
ω_e	370.201	423.254
$\omega_e x_e$	1.3732	1.581
$\omega_e y_e$	9.3×10^{-3}	1.6×10^{-3}
$\omega_e z_e$	-4.8×10^{-4}	-2.9×10^{-4}
B_e	0.152233	0.162975
α_e	7.92×10^{-4}	7.968×10^{-4}
D_e	1.068×10^{-7}	1.0090×10^{-7}
γ_e	1.38×10^{-3}	1.2×10^{-3}

The term values for each of the spin-split $^2\Sigma$ states involved were evaluated with the standard model:

$$T_1(N) = T_v + B_v N(N+1) - D_v N^2(N+1)^2 + \tfrac{1}{2}\gamma N$$

where $N = J - \frac{1}{2}$, and

$$T_2(N) = T_v + B_v N(N+1) - D_v N^2(N+1)^2 - \tfrac{1}{2}\gamma(N+1)$$

where $N = J+\frac{1}{2}$, and J is the total angular momentum quantum number. The constants for the vibration level v were obtained in turn from the following formulae:

$$T_v = T_e + \omega_e(v+\tfrac{1}{2}) - \omega_e x_e(v+\tfrac{1}{2})^2 + \omega_e y_e(v+\tfrac{1}{2})^3 + \omega_e z_e(v+\tfrac{1}{2})^4$$

$$B_v = B_e - \alpha_e(v+\tfrac{1}{2})$$

$$D_v = D_e$$

Rotational line strengths S_{HL} (Hönl-London factors) were calculated for each of the six branches using the appropriate formulae from Kovacs (1969):

$$P_1(J) = P_2(J) = \frac{4J^2-1}{4J}$$

$$R_1(J) = R_2(J) = \frac{4(J+1)^2-1}{4(J+1)}$$

$$^{P}Q_{12}(J) = {}^{R}Q_{21}(J) = \frac{2J+1}{4J(J+1)}$$

where J refers to the lower (ground) state rotational level.

Franck-Condon factors $q_{v'v''}$ (vibrational band strengths) remain unavailable for this system, and so were estimated on the basis of the visibility of band heads in the laboratory spectra. The values assumed for this work are given in Table II.

TABLE II. Assumed CaCl D←X Franck-Condon factors

V'V" →	0	1	2	3	4	5	6
0	0.2	0.4	0.3	0.1	-	-	-
1	0.4	-	0.2	0.3	0.1	-	-
2	0.3	0.2	-	0.1	0.3	0.1	-
3	0.1	0.2	0.1	-	-	0.2	0.1
4	-	0.1	0.2	-	-	-	0.1
5	-	0.1	0.1	0.2	0.1	-	-
6	-	-	0.1	0.1	0.2	-	0.2
7	-	-	-	0.1	0.1	0.2	-
8	-	-	-	-	0.1	0.1	-

Electronic transition moments (or equivalently, radiative lifetimes) needed to establish an absolute absorption coefficient are also unknown at present. It is hoped that both the Franck-Condon factors and radiative lifetimes for the CaCl D←X system will be available soon (L.-E. Berg, 1984).

For now, relative line strengths α_i for a transition i can still be estimated by

$$\int \alpha_i(\nu)d\nu \sim q_{v'v''}\ S_{HL}\ \exp(-hcT_i/kT)$$

where T_i is the lower state term value in cm^{-1}, and T is the gas temperature. This strength was evaluated for approximately 50000 lines of the CaCl D-X system, with each line strength subsequently binned by wavelength to generate a "continuous" relative absorption coefficient over the spectral region 3000-3400 Å. The result for a gas temperature of 3000K is shown in Figure 3.

3. OBSERVATIONAL EVIDENCE FOR CALCIUM CHLORIDE

The case for CaCl in the ultraviolet spectra of N-type carbon stars rests primarily on the coincidence between several moderately strong unidentified absorption features in these spectra with strong band heads of the CaCl D-X system. It can be seen from Figure 3 that the 2-0, 1-0, 0-1 and 0-2 heads at 3127, 3168, 3249 and 3288 Å respectively dominate the absorption spectrum, with the 4-2, 2-1, 0-0 and 1-2 heads at 3118, 3163, 3211 and 3243 Å also conspicuous. When compared to typical carbon spectra, e.g. BL Ori (Figure 1) or TX Psc (Figure 2) as recorded with the LWR camera of the IUE satellite (Johnson and O'Brien, 1983, 1984), strong stellar features at 3163, 3168, 3249 and 3288 Å for which no other identification is available match the strongest of the CaCl band heads. Significantly, these features are also strong in the

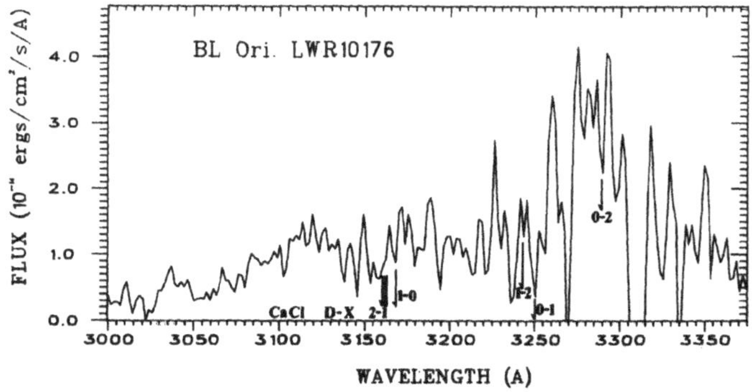

Figure 1

The spectrum of the N-type carbon star BL Ori obtained with the IUE LWR camera, with the positions of the principal CaCl features indicated.

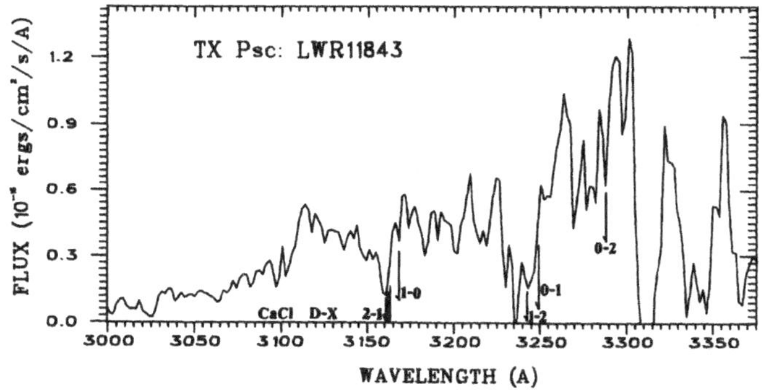

Figure 2

The spectrum of the N-type carbon star Psc obtained with the IUE LWR camera, with the positions of the principal CaCl features indicated.

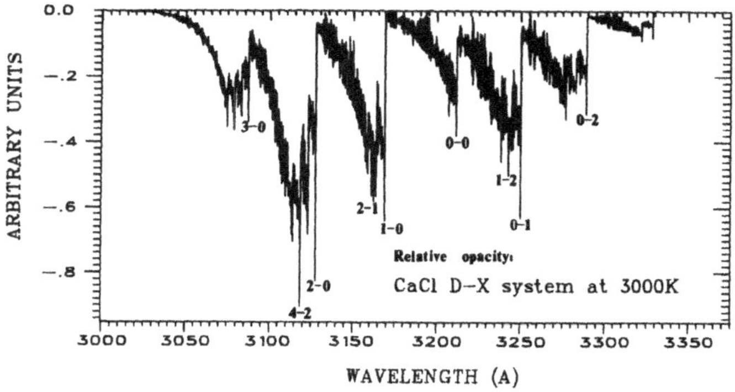

Figure 3

The theoretically generated absorption spectrum of D-X system of CaCl at a temperature of 3000 K, with the major band heads labelled.

spectrum of the late M giant star RZ Ari. This is consistent with the proposed identification since the CaCl abundance does not depend significantly upon the C/O ratio. Finally, the inclusion of TX Psc by Clegg and Wyckoff (1977) on their list of CaCl stars, along with their assertion that CaCl bands occur commonly on the spectra of carbon stars further strengthens this identification.

4. ASTROPHYSICAL IMPLICATIONS

The abundance of calcium chloride is very temperature sensitive (Clegg and Wyckoff, 1977) and so the presence of the molecule should serve as a useful thermometer in the establishment of a temperature scale for the N-type carbon stars. These lines of the D-X system of CaCl presumably arise from the cooler upper photosphere and so may carry information about this region and perhaps even about the temperature minimum. Further work must wait until an absolute opacity for CaCl can be obtained and a full model atmosphere calculated.

ACKNOWLEDGEMENTS

We are grateful to L.-E. Berg for supplying information about the calcium chloride molecule. PDB wishes to acknolwedge the support of a UBC University Graduate Fellowship.

REFERENCES

Berg, L.-E., Klynning, L. and Martin, H. 1978, Physica Scripta **18**, 61.
Berg, L.-E., Klynning, L., Martin, H., Pereira, A. and Royen, P. 1981, Physica Scripta **24**, 23.
Berg, L.-E. 1984, private communication.
Clegg, R. and Wyckoff, S. 1977, Mon. Nat. R. Astr. Soc. **179**, 417.
Herzberg, G. 1950, Molecular Spectra and Molecular Structure, Vol. I, van Nostrand Reinhold.
Johnson, H.R. and O'Brien, G.T. 1983, Ap. J. **265**, 952.
——— 1984, in preparation.
Keenan, P.C. and Bidelman, W.P. 1979, Publ. Astr. Soc. Pacific **91**, 365.
Kovacs, I. 1969, Rotational Structure in the Spectra of Diatomic Molecules, pp. 125-127, American Elsevier.
Sanford, R.F. 1942, Publ. Astr. Soc. Pacific **54**, 158.
Wyckoff, S. and Wehinger, P.A. 1976, Mon. Not. R. Astr. Soc. **175**, 587.

DISCUSSION

QUERCI - It seems necessary to verify your identification through model atmosphere calculations and synthetic spectra using some approximate values for the Franck - Condon factors and for the oscillator strength of the electronic transitions. This region is rich in Fe I v. 7, v. 8 absorption lines in the carbon stars.

BENNETT - I agree, and this work so far was intended as a preliminary step in this direction.

BIDELMAN - Could we estimate a chlorine abundance from the visual region bands ? Also some years ago I obtained a Lick 120 - inch plate of TX Psc extending to your ultraviolet region, so useful information can also be obtained from the ground.

BENNETT - Aside from the complexity of the C-star spectra in the visual region, abundance determinations using molecular features require accurate oscillator strengths for the electronic transitions involved, which are not available presently for Ca Cl. A much more direct approach to obtain a Cl abundance would be to observe H Cl lines in the infrared.

CARBON ISOTOPE RATIOS IN OXYGEN RICH MIRA AND SRa VARIABLES

Kenneth H. Hinkle
Kitt Peak National Observatory
National Optical Astronomy Observatories[1]
P. O. Box 26732
Tucson, Arizona, U.S.A. 85726

Werner W. G. Scharlach
National Radio Astronomy Observatory[2]
2010 North Forbes Blvd.
Tucson, Arizona, U.S.A. 85726

ABSTRACT. Carbon isotope ratios have been measured in oxygen rich Mira and SRa variables brighter than K=+0.6. Results are now available for 34 stars (25 Miras and 9 SRa variables). The longer period Miras have a $^{12}C/^{13}C$ ratio about three times larger than the shorter period Miras. For Miras, period, Tc abundance and $^{12}C/^{13}C$ are shown to be functionally related. This is interpreted in terms of a period-mass relation.

1. INTRODUCTION

Mira variables present an opportunity to select a sample of stars from the field that is in a homogeneous evolutionary state. The high luminosities of the Miras demand that these stars be double shell burning asymptotic giant branch stars (Scalo 1976). The progenitors are generally considered to be 1 to 2 solar mass stars when on the main sequence (Feast 1963; Wood and Cahn 1977). The Miras thus form a group of stars of both known evolutionary stage and known evolutionary history. This is in contrast to, for instance, the field K giants where a wider range of masses ($\sim$1 to 4 $M_\odot$) are present and the stars may be in either the hydrogen shell or helium core burning stage.

Recent observational results (Dominy, Wallerstein and Suntzeff 1984; Hall and Hinkle 1984) show that the surface gravity of Miras is nearly constant over a large range of period. Dominy, Wallerstein and Suntzeff argue that constant surface gravity combined with a mass-radius relation implies that the mass of the Miras increases

[1]Operated by the Association of Universities for Research in Astronomy, Inc., under contract with the National Science Foundation.
[2]Operated by Associated Universities, Inc. under contract with the National Science Foundation.

M. Jaschek and P. C. Keenan (eds.), Cool Stars with Excesses of Heavy Elements, 255–259.

with period. A similar conclusion has been reached independently by Wood and Cahn (1977). By combining evolutionary tracks and the known distribution of Miras with period, Wood and Cahn found that Miras were limited to a wedge-shaped area on the H-R diagram. In this model, period is a function of progenitor mass and age. The argument that the mass of the Miras is an increasing function with period is strengthened by Dominy, Wallerstein and Suntzeff and Wood and Cahn both reaching this conclusion through independent arguments and from different pulsation models for the Miras (fundamental pulsation for the former, first overtone for the latter).

2. OBSERVATIONS AND ANALYSIS

Spectra of 38 oxygen rich Mira and SRa variables were obtained in the 1.5 to 2.5 μm region. All observations were made with the Kitt Peak 4-meter telescope and the coude Fourier transform spectrometer (Hall et al. 1979). The stars are part of a 39-member set of all Mira and SRa variables brighter than m_K=+0.6 and further north than -30°. The limiting magnitude was chosen to permit high resolution (0.07 cm^{-1}), high signal-to-noise ratio (∿100:1) and broad wavenumber coverage (∿3000 cm^{-1}) in integration times of order one hour. Approximately one-third of the stars were observed more than once.

The $^{12}C/^{13}C$ was measured from the CO second overtone. Details of the method and results for individual stars may be found in Hinkle and Scharlach (1984). At this time results are available for 34 stars. The typical uncertainty on a ratio derived from a single spectrum is 30%. In cases where the star was observed more than once, the average ratio is used. For 12 stars, spectra were obtained throughout the light cycle. In no case was $^{12}C/^{13}C$ found to be a function of phase or dependent on the excitation temperature.

3. DISCUSSION

While our sample of SRa stars is meager, it does not appear that these stars have the same distribution as the Miras. Hinkle, Scharlach and Hall (1984) demonstrated that one SRa star has a different velocity curve from Miras. This implies that these stars are in a different pulsation mode and have a different mass-radius-period relation from the Miras. In the following discussion, only the Miras will be considered.

Among the Miras, no Mira with $^{12}C/^{13}C > 25$ can be considered a normal Mira variable. The Miras with $^{12}C/^{13}C > 25$ are either S-type Miras or T Cep-type Miras. Hinkle, Scharlach and Hall identified the T Cep-type Miras as Miras with a pronounced bump on the rising light part of the visual light curve and doubled CO lines in the infrared at phases as early as 0.6. The $^{12}C/^{13}C$ of the normal Miras form a tight distribution with mean = 14.

The mean $^{12}C/^{13}C$ ratio for the "normal" Miras is slightly smaller than for an "average" K giant (Tomkin, Luck and Lambert, 1976). The agreement of the Mira and K giant ratios indicates that for the normal Miras only CNO processed material formed during the star's main sequence lifetime has been mixed to the surface. This is in agreement with theory (Iben and Renzini, 1983) which predicts that mixing of processed material to the surface should not occur for low mass asymptotic giant branch stars. The $^{12}C/^{13}C$ results also seem to confirm the suggestion of Lambert and Ries (1981) that low mass main sequence stars (the stars that evolve into Miras) become ^{13}C-rich giants.

In Figure 1, $^{12}C/^{13}C$ is seen as a function of period for the Miras. At a period near 380 days, the $^{12}C/^{13}C$ ratio changes from

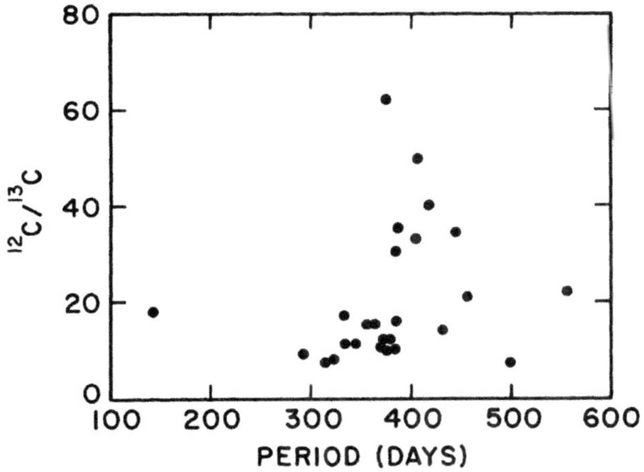

Figure 1. $^{12}C/^{13}C$ as a function of period for the Miras (SRa variables are not plotted).

a mean of 12 to a mean of 32. IK Tau, period 500 (?) days, stands out as an exceptionally low $^{12}C/^{13}C$ star with a long period. This star has a peculiar spectrum (weak lines) at the phases we have observed and we consider the current $^{12}C/^{13}C$ ratio (7) very uncertain.

Variations in the initial (main sequence) abundances can complicate the interpretation. Kinematics indicate that Miras with periods less than 200 days are mainly population II (Feast, 1963). Globular cluster Miras, which always have short periods (Frogel, 1983), support this argument. However, all the stars in our sample are bright and since Miras do not exhibit a great range of intrinsic luminosity the sample stars must be at distances of only a few hundred parsec. All the sample stars also have low radial velocities. We assume that they are uniformly population I.

As discussed in the introduction, there are several indications that the longer period Miras are more massive than the shorter period Miras. If this is so, then Figure 1 indicates that mixing of 3α-processed material to the surface of Miras occurs only for Miras more massive than the mass at a 380-day period. 3α-process ^{12}C is mixed to the surface in these stars producing large ratios of $^{12}C/^{13}C$. One scenario to explain the differences between the S-type stars and the T Cep stars involves the amount of CNO-tricycle material present. The CNO-tricycle can operate in tandem with the 3α-process by C-rich 3α-processed material undergoing abundance changes via the CNO-tricycle as the material is convected to the surface (Iben and Renzini, 1983). Alternately, the atmosphere may already be rich in CNO-tricycle products before the 3α-processed material is mixed out. In the S-type stars CNO-tricycling may reduce the O so the 3α material pushes C/O near unity. The T Cep stars seem to have C added without the reduction of O.

s-process material also has been mixed out in the large $^{12}C/^{13}C$ Miras. The radioactive element technetium is known to exist in the atmospheres of many Mira variables. Technetium in late type stars is formed in the s-process and indicates mixing from the interior more recently than 2 X 10^5 years ago (Iben and Renzini, 1983). A strong correlation exists between our $^{12}C/^{13}C$ results and the Tc results of Little-Marenin and Little (1979). Tc is generally not present in the normal Miras (P < 380 days, $^{12}C/^{13}C$ $\sim$12), but the long period, $^{12}C/^{13}C$ $\sim$32 Miras have strong Tc lines. This is further evidence that extensive mixing to the surface of s-processed and ^{12}C rich material is happening in these stars.

4. CONCLUSIONS

$^{12}C/^{13}C$ ratios demonstrate that the longer period, population I Miras are undergoing extensive mixing of material to the surface that has been processed by He-burning in their interiors. The mixing enriches the surface in Tc and ^{12}C. The shorter period Miras are not undergoing mixing.

REFERENCES

Dominy, J., Wallerstein, G., and Suntzeff, N. 1984, preprint.
Feast, M. W., 1963, Ap.J. **137**, 342.
Frogel, J. A., 1983, Ap.J. **272**, 167.
Hall, D. N. B. and Hinkle, K. H., 1984, in preparation.
Hall, D. N. B., Ridgway, S. T., Bell, E. A. and Yarborough, J. M., 1979, Proc. Soc. Photo-Opt. Instrum. Eng. **172**, 121.
Hinkle, K. H. and Scharlach, W. W. G., 1984, in preparation.
Hinkle, K. H., Scharlach, W. W. G., and Hall, D. N. B., 1984, Ap.J. Suppl., in press.
Iben, I. Jr. and Renzini, A. 1983, Ann. Rev. Astron. Astrophys. **21**, 271.
Lambert, D. L. and Ries, L. M., 1981, Ap.J. **248**, 228.
Little-Marenin, I. R. and Little, S. J., 1979, A.J. **84**, 1374.
Scalo, J. M., 1976, Ap.J. **206**, 474.
Tomkin, J., Luck, R. E., and Lambert, D. L., 1976, Ap.J. **210**, 694.
Wood, P. R. and Cahn, J. H., 1977, Ap.J. **211**, 499.

DISCUSSION

QUERCI - Have you seen S shaped C/Oline radial velocities in the SRa stars ?

HINCKLE - At this time one SRa star (X Oph) has a measured radial velocity curve at 1.6 μm. It is similar to the Miras, but lacks the discontinuity near maximum light.

BIDELMAN - Have you observed the K giant companion of X Oph for C^{12}/C^{13} ?

HINCKLE - Not yet.

RICHER - Just a comment. We tend to forget what the prototype of the Miras is really like ; a late type star together with a very hot companion with evidence that the two are strongly interacting.

HINCKLE - There are few Miras with companions. o Cet is unusual in having a companion. Infrared spectra show no evidence that the companion of o Cet has an influence in the photosphere. Recent observations of circumstellar CO in the microwave region (Knapp, 1984) show that the companion has only a small impact on the circumstellar shell.

WING - The two M stars which are up among the S stars with high C^{12}/C^{13} ratios -T Cep and T Cas- do show moderately ZrO at λ 9300 (Spinrad et al, Ap. J. **43**, 291 - 1961)

JASCHEK - Can one estimate, independently from Tc, the circulation time (interior to surface) in Miras ? How does that compare with the duration of the Mira episode in giants ?

HINCKLE - The circulation time is certainly much less than the duration of the Mira phase. Tc^{99} and Nb^{93} are the two well known chronometers. I know of only qualitative Tc line strength estimates in the Miras. Quantitative abundance work on Tc, Nb and Zr is badly needed.

NIOBIUM IN R AND (S6, 6e) and HR 1105 (S5,3)

D. N. Davis
Jet Propulsion Laboratory
4800 Oak Grove Drive
Pasadena, California, U.S.A.

ABSTRACT. Lines of the first multiplet of niobium are strong in R And and HR 1105. These lines are also present in other S stars: HR 8714, R Cam, V Cnc, R CMi, and T Sgr. They are also visible in the M stars, β Peg and μ UMa. An approximation to the abundance ratio, Nb/Fe, has been deduced from pairs of lines having nearly equal intensity. In R And the ratio is about 200 times the solar value.

It is hoped that good plates will soon be obtained for the near infrared region, so that the significant Nb/Rb abundance ratio may be determined.

The importance of determining the Nb/Rb abundance ratio in BaII stars has been shown by Smith and Lambert (1984). For their analysis they measured only two unassigned Nb lines, λ4606.77Å (120) and λ 5344.17Å (22)*. Of even greater interest would be the value of the Nb/Rb ratio in S type stars, where the s-process is actively producing peculiar abundance ratios for heavy elements.

The first multiplet of niobium appears in the accessible blue region ca. λ4160Å, for which there are excellent Palomar plates with dispersion from 2.2 to 4.5Å/mm. On microphotometer tracings high resolution has been attained at 0.025 to 0.0544Å/mm. Niobium lines of the first multiplet are strong in the spectra of R And, HR 1105, R Cam, T Sgr, V Cnc, and other S stars. These lines are also present in μ UMa (M0). Unfortunately, most of the niobium lines are blended with other strong lines. This is shown in Fig. 1, where four niobium lines are shown. Only one Nb line, λ4168.10Å, is relatively free of blending in R And.

Table I gives equivalent widths measured for the best (i.e. least blended) Nb lines. Niobium lines appear to be strongest in HR 1105, a non-variable binary S star, the spectrum of which does not show TcI lines. Niobium lines are fairly strong in the M star, μ UMa. The higher level lines at λ4606Å and λ5344Å are suprisingly strong in both S stars.

* Laboratory intensities are from Meggers, Corliss, and Scribner (1961).

M. Jaschek and P. C. Keenan (eds.), Cool Stars with Excesses of Heavy Elements, 261–266.

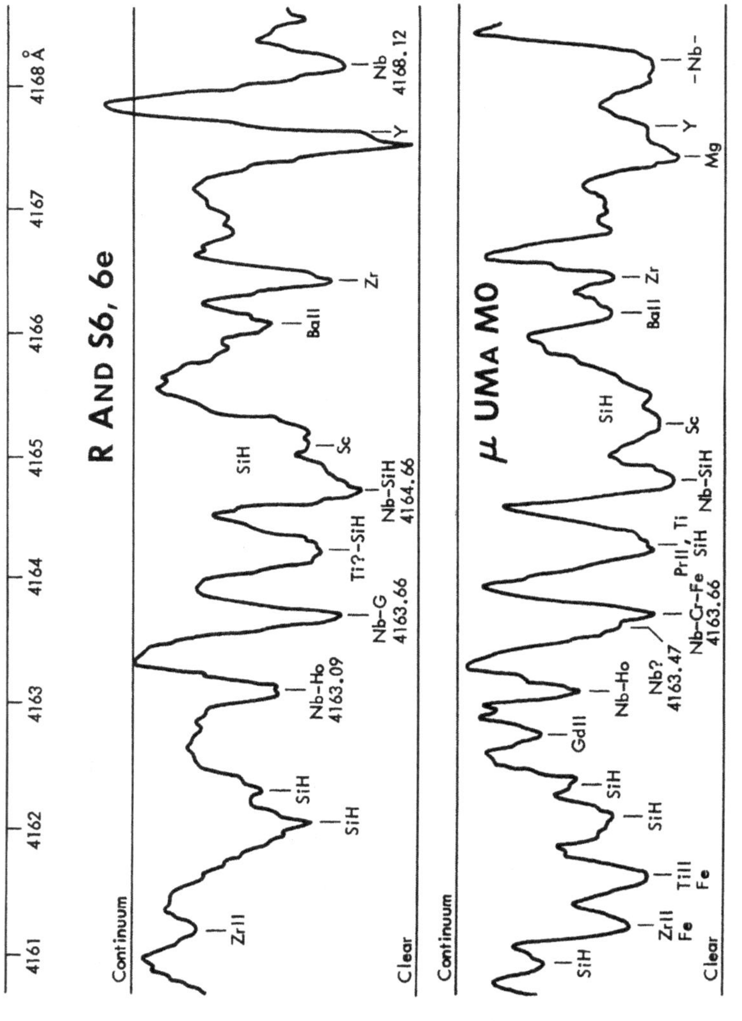

Figure 1. Comparison of Nb lines at λ4163.0Å, λ4163.66Å, λ4164.66Å and λ4168.12Å, in R And (S6,6e) and μ UMa (MO).

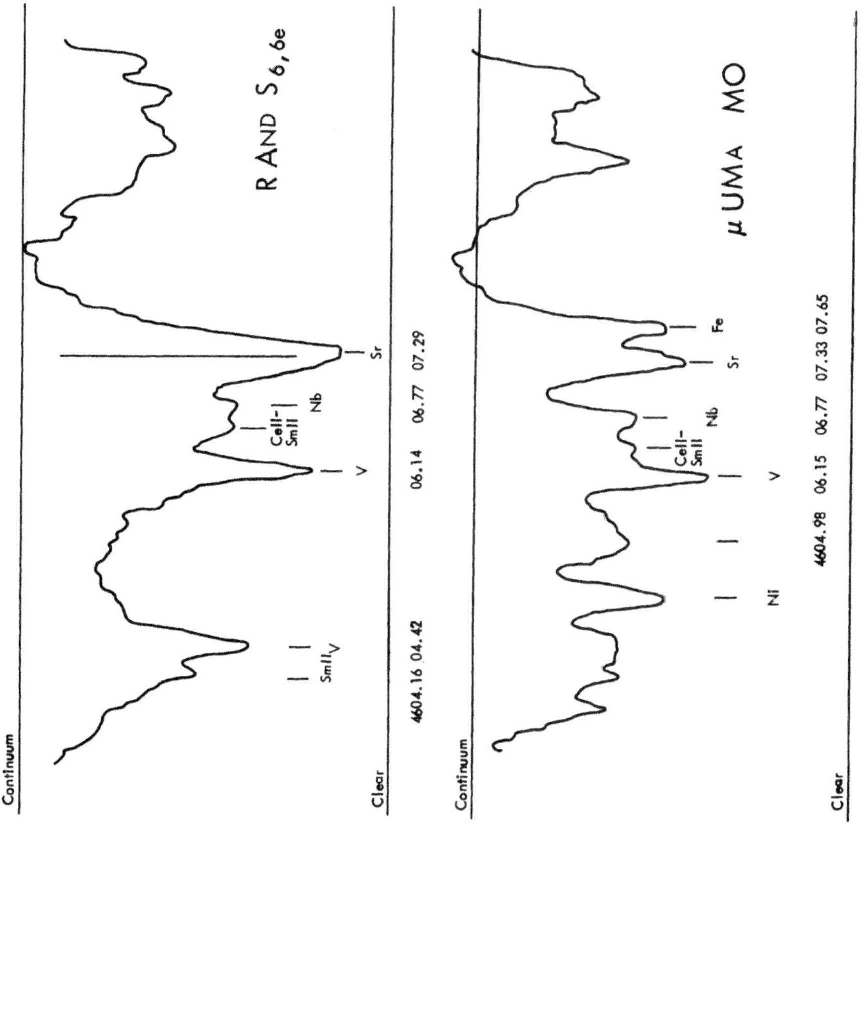

Figure 2. Comparison of Nb line at λ4606.77Å in R And (S6,6e) and μ UMa (MO).

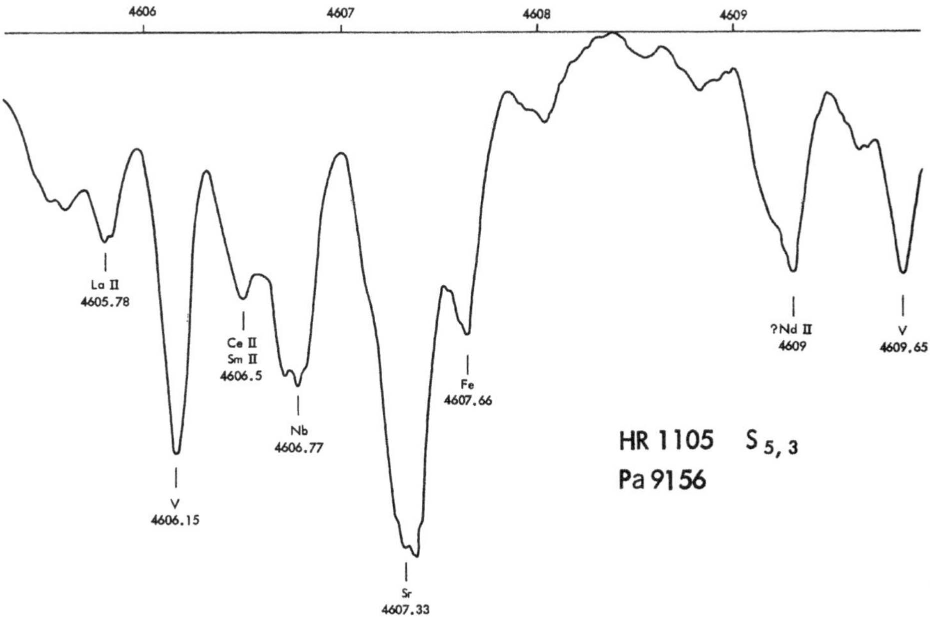

Figure 3. λ4606.77Å line of Nb in HR 1105 (S5,3). Note that the Nb line is relatively stronger than the CeII-SmII line in HR 115, as compared with R And and μ UMa.

Although the significant Nb/Rb abundance ratio cannot yet be determined for these S stars, the Nb/Fe ratio could be calculated from pairs of Nb and Fe lines having nearly equal intensities, using a method developed earlier (Davis, 1978). Because of the scarcity of good Fe lines and the prevalence of blending only two barely acceptable pairs could be found: Nb(λ4079 Å)/Fe(λ 4271 Å) and Nb(λ 4163 Å)/Fe(λ 4204 Å), which yield values of −2.60 dex. and −2.14 dex., resp. The average of these logarithmic ratios indicates that the Nb/Fe abundance ratio for R And is approximately 200 times the solar value, −4.69 (GMA, 1960). Some of the difference may be due to the apparent underabundance of iron in S stars.

Table I. Equivalent Widths of Niobium Lines in R And, HR 1105, and μ UMa.

λ Å	E. P. e.V.	log gf	RAND S6,6e Å	HR 1105 S3, 5 Å	μ UMa MO Å	Notes
4079.73	0.08	0.37	0.262:	0.216:	0.183:	a
4137.09	0.00	-0.44	0.191	0.314:	0.205	b
4163.66	0.02	-0.15	0.149	0.186:	0.138	c
4168.12	0.00	-0.28	0.168	0.334	0.170	d
4606.77	0.34	-0.56	0.208:	0.274	0.111:	e
5344.17	0.34	-1.47	0.122:	0.130	0.165:	

Notes:

a. In wing of saturated SrII line λ4077.71Å .
b. Continuum depressed by Si H.
c. Wing depressed on longward side.
d. Continuum depressed by SiN (1,1).
e. Nb line resolved only in HR 1105.

ACKNOWLEDGEMENTS

I am grateful to Dr. J. L. Greenstein for the privilege of examining his Palomar plates and microphotometer tracings. I appreciate also the assistance from Dr. S. Adelman in making tracings for HR 1105 and HR 8714.

REFERENCES

Davis, D. N., 1978 Comm. XXII COLL. INT'L, D'ASTROPHYSIQUE, LIEGE.
Goldberg, L., Muller, E., and Aller, L. H., 1960, Ap. J. Suppl, 5, No. 45.
Humphreys, C. J. and W. F. Meggers, 1945, NBS J. Res., 34, 477.
Meggers, W. F., C. H. Corliss, B. F. Scribner, 1961, NBS Mono. #32.
Meggers, W. F., C. H. Corliss, B. F. Scribner, 1975, NBS Mono. #145.
Smith, V. V., and D. L. Lambert, 1984, P.A.S.P., 96, 226.
Tompkin, J., and D. L. Lambert, 1983, Ap. J., 273, 722.

COMMENTS AND QUESTIONS

Dr. Catchpole: I was able to determine [Nb/Rb] = $0.6^{+.4}_{-.6}$ in UY Cen, using a line at λ5819Å .

Dr. Davis-Locanthi: Very interesting! I note, however, that your niobium line is very weak having an intensity of only 5.

Dr. Wing: Do any of the other niobium lines have the same upper state as the line at λ4101Å? If so, the coincidence with Hδ might affect their intensities in R And or even cause them to go into emission.

Dr. Davis-Locanthi: I don't know. This would be worth looking into.

Dr. Peery : Have you been able yet to estimate the Nb abundance in HR 1105 ? This would be interesting because, since it results from decay of Zr^{93}, it may tell us something of the past of these curious stars.

Dr. Davis-Locanthi : No, I have not yet, but I intend to. Nb appears strong in HR 1105.

EXCITATION OF SOME La II, Gd II and V I LINES BY THE FLUORESCENCE MECHANISM IN THE SPECTRA OF THE LONG-PERIOD VARIABLE o CETI

S. Grudzinška
Institute of Astronomy, Nicolaus Copernicus University
ul Chopina 12 - 18, 87 - 100 Torun, Poland

ABSTRACT. In the near infrared spectrum of the star o Ceti some emission lines of La II and Gd II were found. It is possible that some lines of La II are excited by Hβ and some of Gd II by Hε. The lines of V I from the upper level $z^4\ P^o{}_{1/2}$ are excited by the line 4412.25 of Cr I.

For the first time the occurence of a series of emission lines from the visible range in late-type variables on the ground of the excitation by strong, bright lines from the far ultraviolet was given by Thackeray (1935, 1937). A series of titanium, vanadium, indium and especially iron lines can be excited by Mg II, Fe II and Si I.

The emission lines in the visual range of long-period variables have low intensity with exception of hydrogen and the manganese triplet lines. Some of the emission lines excited by the hydrogen ones were found in the near infrared spectrum (5000 - 8800) of o Ceti obtained by Ch. Fehrenbach at the phase 0.05 in the Haute Provence Observatory (Grudzinska, 1978). Another spectrum of this star, but with lower dispersion was obtained at the phase 0.13.

For Gd II a radiative excitation by the Hε line takes place upward to the excited level $z^8\ D_{4\ 1/2}$ (3.1 eV). In the observed spectral range three lines from that level can exist. One of them (5393.659 multiplet 100) is in the region of the V0 absorption band [A ($^2\Delta$) - (X) ($^2\Delta$)]. The two other lines are λλ 6004.57 and 5897.62, both from the multiplet (112). The observed intensity ratio 4 : 3 differ a little from the laboratory one 5 : 2 (Mocre, 1945) . Probably the explanation is that both lines are also in the region of absorption bands.

The Hβ fluorescence is the excitation mechanism that produces the La II lines from the upper level $z^3\ D^o_3$ (2.78 eV). Three lines from that level can exist in the observed region. The line 5183.42 (multiplet 36) is not observed because of the overlapping of the TiO band [$C^3\Pi$ - $X^3\Pi$(0.0)]. The observed intensity ratio of the two other lines (6129.57 multiplet 47 and 6626.53 mutiplet 61) is 9 : 6 and that from the laboratory 50 : 5. But the intensity of the lines may be affected by the absorption of the TiO band [a ---> $X^3\ \Pi$(0,1)].

M. Jaschek and P. C. Keenan (eds.), Cool Stars with Excesses of Heavy Elements, 267–268.

In the case of the V I lines, the atoms could be excited to the upper level $z^4P^o{}_{1/2}$ by the absorption in the line 4412.14 which correspond closely to the Cr I line 4412.25. All the lines of neutral vanadium originating from the level $z^4P^o{}_{1/2}$ exist in the emission of the observed spectrum and their intensity is higher than the intensity of the other lines of that multiplet.

TABLE I

Lines of V I from the level $z^4\ P^o{}_{1/2}$

λ lab	Multiplet	Intensity (lab)
6135.36	34	15
6111.62	34	25
6565.88	48	3
6605.98	48	10

So the excitation of some of emission lines of the star o Ceti can be explained by the fluorescence mechanism. In particular there is no doubt that this mechanism is responsible for the excitation of the lines of neutral vanadium originating from the level $z^4\ P^o{}_{1/2}$.

REFERENCES

Grudzinška S. : 1978, Studia Soc. Sc. Torun, section F, vol. VI, 7
Moore Ch. : 1945, Contr. Princeton Univ. Obs. N° 20
Thackeray A.D. : 1935, Astrophys. J. 81, 467
Thackeray A.D. : 1937, Astrophys. J. 86, 499

DISCUSSION

M. QUERCI - Have you radial velocity of these fluorescent lines ?

GRUDZINSKA - The radial velocities shows, of course, different values for different elements, as compared to the mean radial velocity of the star determined from emission lines.

BIDELMAN - The line energized by Hβ emission should, I would think, be considerably stronger in the S and C stars than in the M's, because of the much greater strenght of Hβ emission in the former stars, at least from outside the star.

GRUDZINSKA - I agree, but I had not the spectra of C or S stars.

V. - MODEL ATMOSPHERES FOR THE PECULIAR STARS

MODEL ATMOSPHERES FOR PECULIAR RED GIANT STARS

Hollis R. Johnson, Astronomy Department, Swain West 319
Indiana University, Bloomington, IN 47405 U.S.A.

ABSTRACT. From the pragmatic point of view of the maker and user of model atmospheres--really model photospheres, we provide an overview of those available for peculiar red giant (PRG) stars. After listing the assumptions of the classic stellar atmosphere, we examine the prospects for generalizing each of these. Except for convection and spherical geometry, these generalizations will likely not appear soon. An examination of the parameter space in effective temperature, surface gravity, and chemical composition covered by published atmospheres demonstrates the sketchiness of that coverage. Tests of the adequacy of current models as regards both self-consistency and agreement with stellar observations are discussed. Key observations against which to test models are scarce, particularly because essentially all red giants are variable. Models for warm and intermediate M giants and probably also the warmer S stars appear to be quite accurate. Models for cooler M and S stars and for N-type carbon stars must be regarded as tentative. Models for R stars may also be accurate, but very little testing has been done. Several applications of models, especially in calibrating the temperature scale, are described. We are reminded that photospheres exist in a wider world of stellar interiors and of stellar chromospheres, circumstellar shells, and mass loss. Spherical models and chromospheres are discussed.

I. INTRODUCTION

Although atmospheric models for red giant stars have been very slow in arriving, sufficient are now available to permit some systematic analysis. A detailed, comprehensive review of certain aspects of model photospheres for stars of intermediate (mostly) and low temperature has been given by Carbon (1979) and a topical review focussed upon the interplay of observations and models has been given by Gustafsson (1981). Our perspective here is the practical view of the maker and user of atmospheres. What is currently being done? How well does it work? What are the results?

M. Jaschek and P. C. Keenan (eds.), Cool Stars with Excesses of Heavy Elements, 271–292.

Our starting point (Sect. II) is a description of the assumptions normally employed in calculating models and a discussion of current and possible future generalizations. We then (Sect. III) examine the results -- the atmospheric models -- as they are distributed over parameter space in effective temperature, surface gravity, and chemical composition. Perhaps the toughest part of our discussion (Sect. IV) regards tests to measure their validity in representing real stars and some applications of the models to the solution of astrophysical problems. Stellar atmospheres are then viewed (Sect. V) in the wider context of their relation both to stellar interiors and to mass loss and stellar chromospheres. Finally we touch on several special topics (Sect. VI). At this conference we are interested in model atmospheres for red giant stars, particularly those of peculiar chemical composition: the Ba stars--real and marginal, R stars, N stars, S stars, CH stars, and other lesser known and perhaps even stranger objects. It would be wise to keep in mind that these are among the coolest of all stars and their photospheres are likely the most difficult to model.

II. THE ASSUMPTIONS

The physical principles assumed are those adopted from calculations for hotter, main-sequence stars. Originally justified on the basis of computational tractability, these assumptions were simply carried over, and only later have we worried about their applicability. The standard assumptions are: hydrostatic equilibrium, radiative equilibrium, and local thermodynamic equilibrium. Equations incorporating these are then solved in a plane-parallel, horizontally homogeneous geometry. Together these form the classic problem of stellar atmospheres, and the equations describing the system are well known and have been thoroughly treated in several articles and books (see f. ex., Gray 1976; Mihalas 1978; Baschek and Scholz 1982). The whole system is determined by specifying, in addition to the chemical composition, the effective temperature and the surface gravity.

Certain additional auxiliary equations are necessary to connect the phenomenological or macroscopic quantities involved in the above equations to the microscopic properties of atoms and molecules -- the radiative opacity as a function of temperature, density, and frequency and the equation of state connecting electron pressure, gas pressure, temperature and density, which must include the solution of the Saha equation for the ionization fraction of every element considered and the equation of molecular dissociative equilibrium for every molecule. In all of these calculations, LTE is assumed. Grain formation is normally not included in these equations but is treated as a separate phenomena occurring well outside the photosphere.

At several institutions, computer codes are now available to find numerical solutions to the above equations and thus construct model atmospheres for PRG stars. The efficient solution of these equations, especially for the temperatures and chemical compositions found in the atmospheres of PRG stars is an important topic in its own right, and several descriptions are available: Kurucz (1970, 1979); Gustafsson and Nissen (1972); Querci et al. (1974); Gustafsson et al. (1975); Tsuji (1978); Mihalas (1978); Johnson et al. (1980); Kalkofen (1984); and references therein.

Two generalizations of the classic physical principles -- the inclusion of convective energy transport and the extension to spherical geometry -- are currently available or on the horizon. Other generalizations will not come quickly.

Although the deep hydrogen-helium convection zone controls the thermal structure of the stellar envelope, the photosphere of a cool PRG star rides above this layer. When treated by the usual local mixing-length theory (Bohm-Vitense 1958; Henyey et al. 1965) without overshoot, therefore, convection has only a small effect upon the structure of red giant photospheres (Auman 1969; Johnson 1974; Gustafsson 1981). In fact, convection has been entirely neglected in many current model atmospheres. When included, convection has in every case been treated by the standard local mixing length theory.

Yet convection cannot so easily be dismissed. It lurks just below the photosphere, where a rapid transition to a convective regime takes place. Convection, and possible associated inhomogeneities, may make itself evident in several ways. It is vital to the structure of stellar envelopes, especially in red giant stars. Convection may influence the temperature gradient in the deep photosphere of PRG stars, which could in turn affect both the ionization zone and the matching of observed colors and luminosities to theoretical evolutionary tracks. Convection may also become more important in possible inhomogeneous atmospheres. Mechanical energy associated with convection is generally thought to be responsible for chromospheres, transition regions, and coronae, (see, f. ex., Linsky 1980, 1982; Praderie, et al. 1981; Brown 1984) although both observation and theory of chromospheres for red giants has been directed almost exclusively toward G, K and early M giants. Studies of chromospheres for PRG stars are still in a rather primitive state (a review is given by de la Reza 1984).

Lastly, while the inadequacies of the local mixing length theory are well known (see, f. ex., Nordlund 1974; Ulrich 1976; Latour et al. 1976; Marcus, et al. 1983), the development of a better theory has been remarkably slow. While we wait, we note some progress -- both in theoretical development (Nordlund 1976, 1980, 1982; Latour, et al. 1981; Chan, et al. 1982; Marcus et al. 1983) and in deducing observational constraints on the present theory (Nelson 1980; Bohm-Vitense 1982; Lester, et al. 1982). Unfortunately, these studies are either very general or specifically aimed at hotter, main sequence stars. Applications of any improved theory to atmospheres of PRG stars are definitely yet future.

Atmospheres with spherical geometry constitute such a noteworthy generalization of the classic stellar atmosphere that they will be described later in a separate section (Sect. VI).

At least for the foreseeable future hydrostatic equilibrium will almost surely continue to be used in models for stellar photospheres. Yet the photosphere is part of the moving atmosphere and dynamics must eventually be included, at least for certain variable stars. Well developed theories of mass flow in regions relatively far from the photosphere already exist, including mechanisms for grain nucleation and growth, wind dynamics, mass flow, and evolutionary implications for mass loss (see, f. ex., papers at this meeting; proceedings of recent conferences on mass loss; Thomas 1983; Goldberg 1984b). Whether grain formation and consequent mass loss by radiative pressure may begin in the stellar photosphere is still an open question. Theoretical calculations (Deguchi 1980; Draine 1981; Woodrow and Auman 1982) allow that possibility, but none of these include the effects of non-radiative heating in the outer photosphere, and the answer could depend sensitively on the thermal structure there.

Although chromospheres of warmer stars have been extensively studied, research has barely begun on the structure and energy balance of the outer photosphere and chromosphere of cool PRG stars where departures from radiative equilibrium occur. Observations of emission lines of Fe II, Ca II, Mg II, and Hα signal the presence of a warm chromosphere, but not a transition region or corona, in many PRG stars. Theoretical studies lag badly. The ultimate goal, once we also understand the formation of chromospheres (see f. ex. Schmitz and Ulmschneider 1981) in RG stars and the detailed interaction of shock waves in the outer atmospheres (Willson and Hill 1979), is to combine the photosphere, chromosphere, and outer atmosphere self-consistently into a fully dynamic theory of the whole stellar atmosphere. That goal could be a decade away. Current work on chromospheres is discussed under special topics (Sect. VI).

Investigations of departures from LTE in photospheres of PRG stars are also scarce. Most research has been directed to the circumstellar environment, where LTE ceases to exist and atomic and molecular concentrations are fixed by reaction rates (see, f. ex., Zuckerman 1980; Glassgold and Huggins 1984). Extreme examples of non-LTE effects are the H_2O and OH maser lines often seen in the circumstellar shells of oxygen-rich mira variable stars. The recognition (Elitzur 1980) that SiO masers might be located in the upper photosphere of the star may unite the heretofore separate fields of photospheric and circumstellar research. In the photosphere itself very little work has been done; we recall the non-LTE calculations of electron donors in M stars (Auman and Woodrow 1975) and of Li in a carbon star (de la Reza and Querci 1978). That's all. To say that further study is needed is a wild understatement. Prudence suggests we begin with H I, H_2, and H^- since these control the atmospheric thermodynamics, and H_2 quadropole lines appear to be weaker than expected in cool non-Mira stars of both oxygen-rich (Tsuji 1983) and carbon-rich (Johnson et al. 1983) composition (but see Eriksson et al. 1984).

Mounting evidence hints that inhomogeneities -- perhaps on several scales -- are common in red-giant atmospheres. In Arcturus, for example, there is observational evidence for inhomogeneities (Chiu et al. 1977; Heasley et al. 1978). Existence of inhomogeneities might have profound consequences for formation of molecules and perhaps even grains, interpretation of spectral lines, small-amplitude stellar variability, and chromospheric heating. The (possibly) drastically changed molecular concentrations in such an atmosphere might produce positive feedback on the atmospheric structure (as suggested by Gustafsson 1981) or perhaps even instability (as suggested by Ayres 1981 and Kneer 1984). Yet even a simple, two-component model atmosphere challenges the model-makers with considerable difficulties (in determining the relative ratios of surface areas covered), but provides such great power to match observations that other important effects might be covered up.

One of the toughest problems in constructing cool-star atmospheres is the treatment of the radiative extinction due to millions of molecular lines. Fortunately, several statistical methods for treating such a multitude of lines exist. Here we simply list briefly these methods -- in order that the reader be reminded but not bored -- and then restrict our discussion of atmospheres to those computed with reasonably accurate opacity treatments. There are three such methods: the opacity distribution function (ODF), opacity sampling (OS), and the band model (VAEBM). In applying an ODF, one rearranges the absorption in a selected spectral interval to resemble one giant spectral line and then approximates this by a series of pickets (Strom and Kurucz 1966; Gustafsson et al. 1975; Kurucz 1979; Carbon 1984). In opacity sampling (Peytremann 1974; Johnson et al. 1980 -- henceforth JBK; Carbon 1984), one treats the opacities as exactly as desired at each selected frequency, and chooses a sufficient number of frequencies to assure accuracy in frequency-integrated quantities. The VAEBM is an approximate ODF in which lines in a molecular band are assumed to be equally spaced and of equal strength (Tsuji 1976). A cruder mean opacity sometimes used is the straight-mean (SM), in which the integrated absorption coefficient of all lines in a chosen spectral interval is simply averaged across the interval; in a sense, the SM is equivalent to a one-picket ODF. General discussions of the mathematical formulation and accuracy of these methods may be found in several reviews (Tsuji 1978; Mihalas 1978; Carbon 1979, 1984; Gustafsson 1981; Johnson 1984). Especially enlightening is a penetrating analysis of the OS method by a Monte Carlo approach and its comparison to the ODF (Carbon 1984). A recent breakthrough in formulating the ODF allows partial ODF's, computed for each molecule, to be combined correctly to yield the same result as the standard ODF even in spectral regions of overlapping molecular bands (Saxner and Gustafsson 1984). This greatly increases the flexibility of the ODF method.

Which molecular opacities must be included in models of PRG photospheres? Most models of oxygen-rich composition include at least some band systems of CH, NH, OH, CN, CO, MgH, TiO, and H_2O, the last often being treated less accurately than the others. For carbon-rich stars the list is similar except that C_2 replaces OH, TiO, and H_2O. Other molecules -- such as VO in the coolest M stars and ZrO in S stars -- ought to be added. The H_2O opacity should be treated much more carefully in oxygen-rich stars with $T_{eff} \leq 3200$ K. Finally, the importance of polyatomic molecules -- especially HCN and C_2H_2 -- in carbon stars has recently been emphasized by several workers, though to somewhat different degrees (Eriksson et al. 1984; Jorgensen, this meeting; Tsuji 1984).

Another problem in the construction of very cool models is their sensitivity against small perturbations in the temperature stratification and the abundances. This results from the strong temperature dependence of all equilibrium constants and the effective interconnection of the species via molecule formation (Scholz and Wehrse 1982; Eriksson et al. 1984).

III. THE RESULTS

To present as concisely and comprehensibly as possible an overview of atmospheres computed to date, we resort to pictures. Figure 1 displays, in effective temperature (Teff) and surface gravity (log g) space, the models with solar composition and plane-parallel geometry (PPG) computed to date. Here BEGN denotes Gustafsson et al. (1975) and Bell et al. (1976); T denotes Tsuji (1976, 1978), and J denotes JBK. It would appear that at least some models are available for almost all parts -- even the coolest regions -- of the diagram. While this is technically true, all models are not equally appropriate. The BEGN models, which are excellent for warmer stars, fail for the cool PRG stars because they do not include TiO opacity. Worse yet, all cool models shown treat H_2O with a mean opacity, and these may therefore not be highly accurate (JBK 1980). In fact, one of the barriers limiting calculations at present is the difficulty of treating the opacities of polyatomic molecules -- especially H_2O and HCN -- accurately. For stars of solar composition, this limits us to effective temperatures above about 3200 K, below which water vapor becomes a dominant opacity in the outer atmosphere. (Projects presently underway may allow the H_2O opacity to be statistically incorporated into the OS or ODF treatments: Alexander, 1984, private communication; Bennett, 1984, private communication.) In addition, the low gravity of several of these models suggests that, if the corresponding stellar masses are near 1 solar mass as expected (Scalo and Miller 1979), sphericity effects may become important.

Much less of the total parameter space in chemical composition has been explored than in other parameters. The coverage is also more difficult to describe simply, for chemical composition is not one additional parameter, but several -- Fe/H, CNO/H, C/H, N/C, or C/O. Figure 2 displays the temperature and C/O space and outlines the regions covered to the present. Even including all the models, it is obvious there is a long way to go. Furthermore, it is not yet clear that models by different investigators agree in the regions of overlap, nor is it yet established how accurately these models represent real stars. It is perhaps ironic that the only published models for S stars are those of Piccirillo (1980), computed with SM opacities, and of Johnson (1982), computed with OS opacities; yet S stars can perhaps be better represented by current models than can the carbon stars or the coolest M giants because polyatomic molecules (H_2O, HCN, C_2H_2) are unimportant in S stars, and they are also likely better represented by plane-parallel atmospheres.

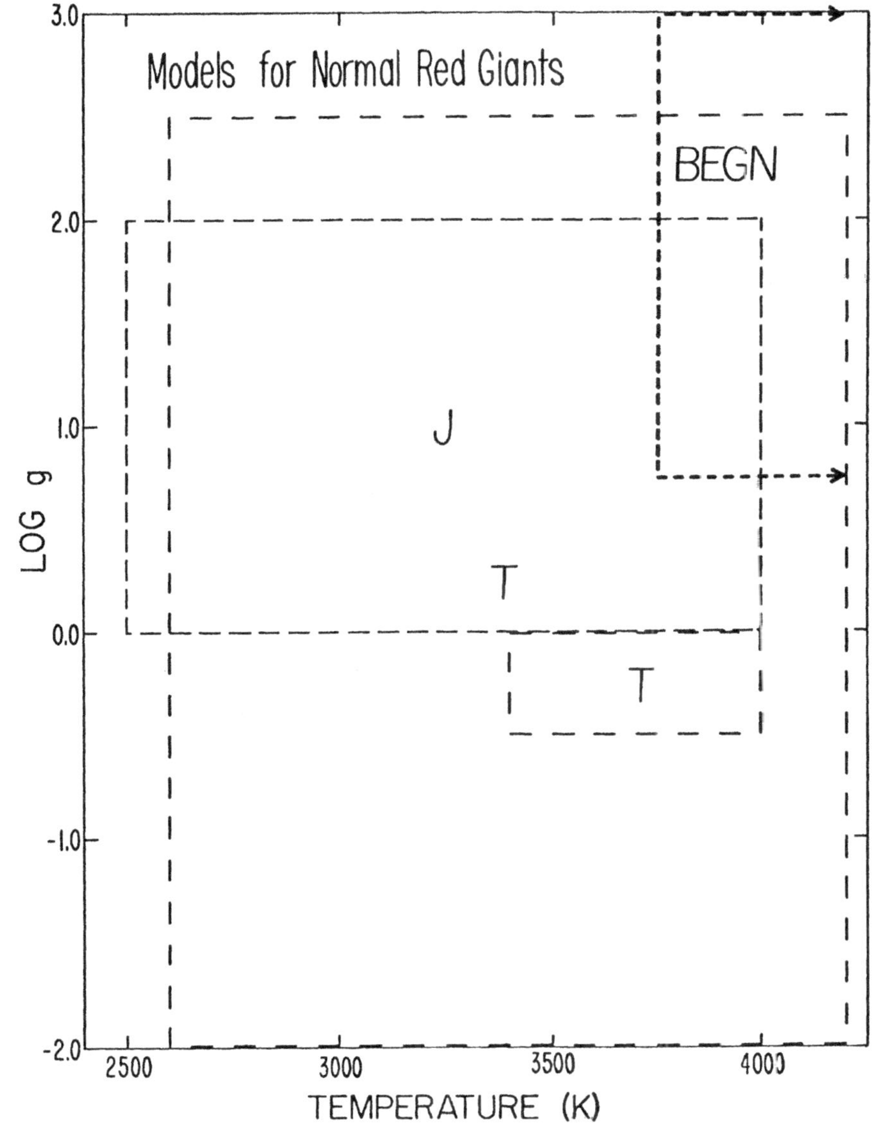

Figure 1. Published atmospheric models for red giant stars of solar composition displayed in effective temperature and surface gravity (log g) space. References are given in the text.

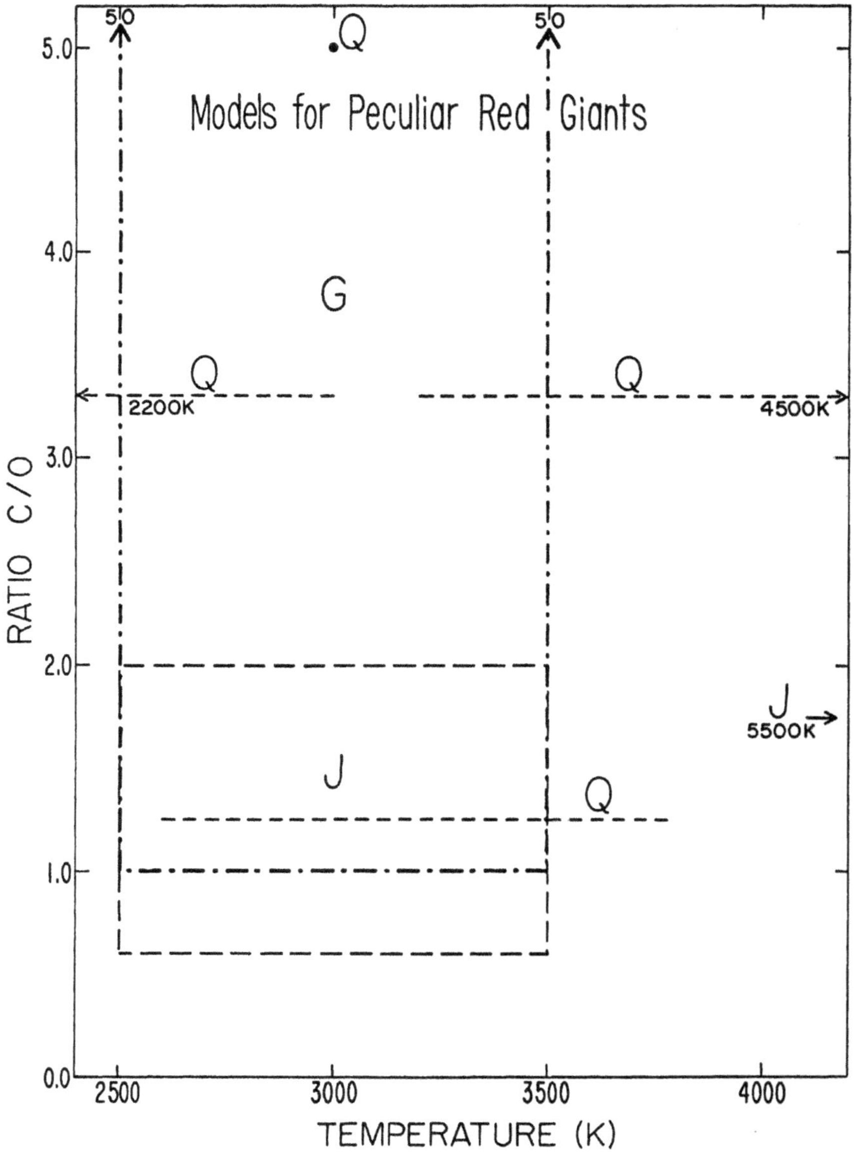

Figure 2. Atmospheric models for red-giant stars with non-solar chemical compositions displayed in effective temperature and C/O space (C/O = 0.6 for the sun). References are given in the text.

Note that models, such as those of Q (an abbreviation for Querci et al. 1974 and Querci and Querci 1975), which define a line in the Teff - C/O space, are simply projections onto this plane of a series of models varying in some other aspect of chemical composition -- in this case N/C. In the same way the abundance of the metals (as in CH stars and stars in the Large and Small Magellanic Clouds) or of hydrogen (as in hydrogen deficient carbon stars) would define a new dimension which is not shown on Fig. 2. Although we cannot easily display, for a wide variety of abundance variations, the range of surface gravity, several of the computations cover a substantial range in surface gravity. For some of these, spherical effects may well be important.

The set of models labelled G comprises several hundred models computed by the Swedish group but not yet published (Eriksson, et al. 1985, henceforth EGJN; Gustafsson 1984, private communication). These are computed by a new ODF treatment which accounts for changes with depth of the principal absorbers (Saxner and Gustafsson 1984) and include the opacity of HCN, which becomes extremely important for cooler models (Teff < 3200 K).

IV. TESTS AND APPLICATIONS

How good are the models? The first test is that of self-consistency. Are the equations correct? Are the best current values of atomic and molecular parameters used? Do calculations by different workers agree where they overlap? Although the relatively few groups computing atmospheres for RG stars tend to work independently, a few comparisons have been made. The results are mixed. For K giant stars, excellent accord in the structure is obtained between models by BEGN and JBK (Johnson and Krupp 1976), and for M giants the structure of models by Tsuji and JBK agree well (JBK 1980) for $T_{eff} \geq 3400$ K. Below this temperature H_2O opacity becomes increasingly important, and it is approximated rather crudely.

For the computational challenge of a cool carbon star, the thermal structures ($T(\tau)$) of models computed by the Swedish group (EGJN 1985; Gustafsson 1984, private communication) using ODF opacities and by the Indiana group (Johnson 1982) using OS opacities agree very well (easily within 100 K), and the pressure structures ($P(\tau)$) also agree quite well when comparable diatomic opacities are used. When polyatomic opacities (HCN and C_2H_2) are added to the EGJN models, the thermal structure is only very slightly changed, but the pressure is drastically decreased (by a factor 10-100) in the outer photosphere; only in the deepest photosphere do the pressures agree. Since the codes, molecular data, and opacity treatments are independent, the relatively good agreement is reassuring. Fair agreement is also obtained between the structures of the EGJN models and the Q models (Gustafsson 1984, private communication). Yet much work lies ahead, for adding each new molecular opacity (see Jorgensen, this conference on HCN, for example) is a tedious task.

More central is a second question. How well do predictions from the models compare with observations of real stars? Unfortunately, little has yet been done for PRG stars. We must regard atmospheric models for PRG stars computed to the present as generally exploratory, but with increasing reality as time passes. What are the key observations which models must match? The most important observations are: (1) colors of stars, or, better, complete spectrophotometry; (2) molecular band strengths; and (3) profiles of individual spectral lines. Detailed discussions of the use of model atmospheres in the inference of effective temperatures, surface gravities, chemical compositions, and turbulent velocities from observations have been given by Pecker (1973), Gray (1976), Gustafsson (1981), and Johnson (1984). Because the determination of elemental and isotopic abundances in the PRG stars is well covered by another review (Lambert, this conference), we do not discuss this important use of model atmospheres, but stress temperature determinations.

A few empirical confrontations have been arranged. Although the BEGN models are specifically relevant at considerably higher temperatures (Fig. 1), emergent fluxes from the coolest of these models have been compared (Manduca et al. 1981) to infrared observations of late K and early M giants (Strecker et al. 1979) with quite satisfactory results. While the lack of molecular opacities in these models is obvious in the calculated fluxes, the overall fit is nevertheless sufficient to allow a determination of the effective temperature for a star as late as M0.

In a classic piece of research, Ridgway et al. (1980, henceforth RJWW) collected all available angular-diameter measurements and used these to define a new temperature scale for K and M giant stars. This semi-empirical relation between effective temperature and color temperature or spectral type (TiO band strengths) provides a marvelous opportunity to test predictions of models. For K and early M giants, the relation between the Wing color temperature and the effective temperature of the star given empirically by RJWW is correctly matched by predictions of the models by BEGN, JBK, and Tsuji (Piccirillo et al. 1981), and this agreement constitutes one of the strongest evidences to date that model atmospheres for K and early M giant stars from several sources are basically correct. These models are admittedly easier to construct than models for some of the PRG stars which are the primary focus of this conference since these are oxygen-rich stars whose composition, effective temperature, and gravity are presumably well determined.

The use of model atmospheres for red-giant (RG) stars in a comprehensive way is nicely illusrated by the attempt to calibrate the effective temperature scale for K and M giants. Using VAEBM opacities for CO, CN, OH, CH, TiO, MgH, CaH, SiH, and H_2O, Tsuji (1978) constructed a grid of models for K and M giants over the range of parameters shown in Figure 1. From these models he calculated crude spectral energy flux distributions, and chose for each star the effective temperature of that model which best matched the slope of the observed infrared broad-band colors. In this way he derived a temperature scale in fair agreement with that obtained from angular diameter measurements and complete energy flux curves. Aware of possible deficiencies in his models, Tsuji (1981a) then used the method of infrared photometry (Blackwell-Shallis method: Blackwell and Shallis 1977; Blackwell et al. 1980) to derive a new, independent scale of effective temperature for K and M giants. Because this method is only weakly model dependent, it is preferred at the present stage of model development. This scale is generally in good agreement with that of RJWW although the temperature plateau from M0-M2 obtained by Tsuji appears not to be real (Keenan 1982). Steiman-Cameron and Johnson (1984) compare these scales with the purely theoretical relation between T_{eff} and TiO band strengths computed from the JBK models. The theoretical scale appears to match the observations as well as does the "empirical" scale, at least for the warmer stars. Below about 3400 K, however, the JBK models predict strengths of TiO bands, on which the spectral classification is based, which are too weak. No hypothesis provides a convincing explanation for the above discrepancy. Adoption of spherical models would tend to strengthen the predicted TiO bands and produce better agreement, but it seems strange that the sphericity effects in M giants should be that important. A possible explanation based on departures from LTE appears to fail because the work of Auman and Woodrow (1975) predicts, if anything, an overionization of Ti I, which would weaken the TiO bands in the star and yield a discrepancy in the direction opposite to that observed. Uncertainties in f-values seem insufficient to account for the effect. The overestimation of H_2O opacities (by the use of SM opacities) is unlikely to be important at 3400 K, and the effect goes in the wrong direction. A final untested explanation attributes the discrepancy to chromospheric heating in the upper photosphere; that is, if two otherwise identical M giants have differing amounts of chromospheric heating, the TiO band strengths of the two might be different, and they might be assigned to different spectral classes.

With this discussion of M stars behind us, we might appreciate better the uncertainties in the calculations of models for S and N-type stars, inasmuch as these later are even cooler and effective temperatures, surface gravities, and compositions are much less well known than for the M stars. We recall the early attempts to match the broad-band photometry and high-resolution infrared observations of the N-type carbon star UU Aur (N3) with predictions from the theoretical Q models of carbon stars (Querci and Querci 1976). While fair agreement was obtained between the predictions of the model and the observations, different effective temperatures and chemical compositions were required to match the photometry and the spectroscopy. One might be tempted to say that these models have historical interest only, but we must remember that present day carbon star models have not been improved significantly since this pioneering work!

Surprisingly, in attempting to compare calculations of models for carbon stars with observations, one is immediately frustrated by a lack of observational data. Effective temperatures are known from lunar occultations for only a handful of carbon stars (Ridgway et al. 1980b), and several of these are quite uncertain because of the large dereddening corrections. Effective temperatures have been estimated for an additional 25 stars from an application by Tsuji (1981b) of the infrared photometry method to L band photometry, corrected for molecular absorption. Comparisons with the few values of T_{eff} from angular diameters indicate that effective temperature can be deduced in this way for N-type carbon stars with an accuracy of about ± 200 K. Worse yet are the R stars, for which no angular diameters have been measured and temperatures have been estimated by a variety of schemes (Dominy 1982).

Goebel et al. (1980) find encouraging agreement between spectrophotometric observation from 1.2 - 30 μm of the N-type star Y CVn and both continuum slope and band strengths predicted from the Q models.

Because it is a bright, well observed N-type carbon star with a fairly accurate angular diameter, the star HD 223075 (TX Psc) is a good standard star. Infrared spectrophotometry has been obtained by workers at NASA Ames Research Center (Goebel, et al. 1984; private communication) and scans in the visual region have been made by Fay and Honeycutt (1972). The observations of Wing (1967) help fill a gap between the visual and the infrared data. Unfortunately, the flux curve is incomplete in the crucial region near the flux peak -- and this is one of the best observed N-type carbon stars. Theoretical fluxes from a model by Johnson (1982), with T_{eff} fixed by angular diameter measurements (Ridgway et al. 1980b) and a ratio of C/O from Johnson et al. (1982), are in fair agreement with the observations over much of the spectrum, but there is disagreement near the flux peak. In particular, there are two peaks in this region (somewhat mutilated by molecular bands): the blackbody peak for this effective temperature and the H^- peak at 1.65 μm (due to the H^- opacity minimum). In the model, the H^- peak is significantly stronger than the blackbody peak whereas in the star the reverse is true. To obtain better agreement appears to require either additional molecular opacities in the region of, or just longward of, the H^- peak or a reduction of the H^- opacity. The latter could be accomplished either through a reduction in the electron density because of underionization of the electron donors (though this runs counter to the expected direction of non-LTE effects (Auman and Woodrow 1975)) or to a deficiency of hydrogen (models for slightly hydrogen-deficient atmospheres match the observations quite well, as shown by Alexander et al. (1984)).

What is the situation for S stars? Excellent infrared spectrophotometry has been obtained for a number of S stars at NASA Ames Research Center, and these must form the basic observations against which model predictions are compared. For the warmer S stars, the fluxes predicted by appropriate models (Johnson 1982) and the observed fluxes are in excellent agreement. For the cooler S stars the fit is not quite as good but is still sufficient to calibrate a temperature scale for the stars by means of the technique of model fitting and IR photometry (Augason et al. 1984). The temperatures for 10 non-Mira S stars are found in this way to range from 2700 - 3700 K.

From the (in some cases meager) tests available, we find that models for the <u>normal oxygen-rich red giants of type K and early M</u>

predict emergent fluxes which yield reasonable fits to observed spectrophotometry and color temperatures, leading us to believe these are fairly accurate. Flux curves of the warmer S stars are also fit quite well by the models, while the cooler S stars are fit slightly less well. The predictions of the coolest carbon star models fit only fairly well the observations of either spectrophotometry or high-resolution spectra of N-type carbon stars, and we believe these should be used with caution. Except for self-consistency, no test has been made of R star models.

V. A BROADER VIEW OF STELLAR ATMOSPHERES

A stellar atmosphere -- magnificent as it is -- does not exist entirely alone in the universe, and it is worthwhile to consider the atmosphere of a star in a broader context. What is inside the atmosphere? What is outside? At the center of a star is a tiny region called the core, which contains the hot regions of nuclear reactions. Energy flows out of the core to warm the atmosphere, and the core is the source of nuclear contamination which occasionally reaches the surface, causing people to gather together in conferences such as this. Next comes the inner atmosphere, sometimes referred to as the envelope.

Between the inner and outer atmospheres lies the narrow shell called the photosphere. Its importance arises from the fact that it is the optically accessible region of the star. The outer atmosphere extends to interstellar space and includes such interesting phenomena as chromospheres, TR's, coronas, shocks, dust shells, masers, gas shells, former stellar material, escaping stellar material, and other fascinating phenomena.

The stellar atmosphere provides both a proper upper boundary condition for interior calculations and the means through which computer-generated stars can be matched with real stars. Traditionally, stellar atmospheres and stellar interiors have been two mutually exclusive fields, each of which had its own way of deciphering the stellar universe. In the regions of the PRG stars the two fields are forced together. It is past time for workers in the two fields to unite. Calculations of interiors and evolution must begin to include as upper boundary conditions the results of stellar atmosphere calculations including accurate molecular opacity effects. A study of the effect of accurate outer boundary conditions on the evolution of cool giant stars is urgently needed. At the same time atmospheric model makers must become better aware of the regimes of temperature, gravity (luminosity), and chemical composition which incorporate current results from stellar interiors and evolution.

VI. SPECIAL TOPICS

1. Spherical atmospheres

It has been recognized for at least fifteen years that sphericity effects could be important in determining the struture of cool giant stellar atmospheres. In order to include the effects of spherical extension, three changes must be introduced to the standard plane-parallel approach to computing stellar atmospheres. First, the equation of radiative transport must be modified by the inclusion of a curvature term; second, the acceleration of gravity can no longer be treated as a constant since the atmosphere may now extend over an appreciable fraction of the total stellar radius; finally, a third input parameter, in addition to the effective temperature and the acceleration of gravity (now specified at a conveniently selected optical depth), must be provided. The stellar radius at the optical depth for which the gravity is given is usually used. But this in turn offers the possibility to determine directly from the spectrum the three basic parameters luminosity, mass and radius (Scholz and Wehrse, 1982).

In a pioneering venture, Schmid-Burgk (1975) produced grey model atmospheres for models as cool as T_{eff} = 2500 K. He found that spherical atmospheres, compared to comparable plane-parallel models, were as much as several hundred degrees cooler at the surface, a result that has subsequently been confirmed by many investigators employing more sophisticated opacity treatments. Using relatively complete frequency averaged opacities to compute M giant models, Watanabe and Kodaira (1978) concluded that spherical effects should be included in any model for which the atmospheric extension (defined, for example, as the physical depth between $\tau = 10^{-3}$ and $\tau = 1$) is greater than 5% of the stellar radius. They also demonstrated that, in atmospheres with progressively greater fractional extensions, molecules exhibited differential effects. For example, the partial pressure of CN and CO at a given optical depth decreases with increasing sphericity over a wide range of optical depths, while water vapor shows the opposite effect, and TiO increases at some optical depths and decreases at others. Kipper (1982) and Schmid-Burgk et al. (1981) confirmed and extended these conclusions with extensive sets of models. These results suggest that the strength of molecular bands in the spectra of stars of the same effective temperature, surface gravity, and composition may be differentially dependent upon the stellar radius. In a later study, Watanabe and Kodaira (1979) showed that broad-band colors are strongly affected by the degree of extension of the atmosphere. It should be remembered, however, that these model fluxes were computed with frequency-averaged opacities, which greatly over-emphasize the effect of molecular bands in the emergent flux.

In a study of the effects of different chemical compositions in spherical model atmospheres, Wehrse (1981) found that metal deficiency generally increases the extension of the atmosphere and leads to even greater surface cooling. On the other hand, hydrogen-deficient models had much less physical extension than did the other spherical models. Scholz and Tsuji (1984) extended the study of the effects of sphericity to carbon stars. They found that carbon stars are less extended than comparable oxygen-rich stars because of the greater temperature dependence of the opacity in a carbon-rich atmosphere. They also note that the gas pressure is increased in carbon-rich stars (compared to comparable plane-parallel models), but is decreased in oxygen-rich models. Their emergent fluxes show little effect of sphericity in the continuum, but the molecular bands are generally enhanced in spherical models for both oxygen-rich and carbon-rich models.

Up to the present time, all spherical model atmospheres have been computed with frequency-averaged opacities. While differential comparisons among such models can be quite illuminating, the use of such approximate opacities introduces a considerable uncertainty into the structure of the models. The emergent fluxes computed from such models cannot be accurately compared to observations. Spherical models which utilize the opacity sampling technique are presently being computed, and these models can illuminate the differential effects found by earlier investigators and predict observable properties, such as colors. Although these spherical models have not yet reached the level of physical accuracy to be helpful in interpretation of observations, it is already clear that spherical models will be necessary for some stars, and we applaud the pioneers of this field.

2. Chromospheres

Although the presence of chromospheres in at least some PRG stars has long been assumed from the presence in violet spectra of bright lines of Fe II (Bidelman and Pyper 1963), their systematic study has become possible only with the advent of the International Ultraviolet Explorer (IUE) spacecraft. When such studies began, they were naturally concentrated on hotter and brighter stars, and observations of the interesting cool giant stars were left until later. Yet most of these PRG stars are mass-losing variable stars, and chromospheres are bound to receive increasing attention in the future.

Excluding investigations of M giants and supergiants, which have been rather carefully studied observationally but not theoretically (except for a few individual stars, such as Betelgeuse -- for which see, f. ex., Goldberg (1984a)), chromospheric studies of the PRG stars are still exploratory, as only low-resolution spectra are generally available. Spectra of S stars (only a half dozen are available) appear in most cases to be similar to M stars, with relatively strong Mg II lines although the coolest S stars may have weaker Mg II lines (Johnson et al., this meeting). No high-resolution observations have been published except for HD 35155, which shows high excitation emission lines characteristic of a hot plasma -- perhaps indicative of interaction with a hot, compact companion (Johnson and Ake 1984).

Ultraviolet spectra of barium stars have been carefully scrutinized -- more, however, for information about the presumed white-dwarf companion with its possible evolutionary implications, in view of the established binary nature of these stars (McClure et al 1980) than for information about the stellar chromosphere. While the white dwarf is seen in some systems, it has apparently faded beyond detection in others (see, f. ex. Dominy and Lambert 1983).

Long-wavelength IUE spectra of the early R stars (R0-R3) are essentially identical to spectra of G-K giants with no bright lines and Mg II h and k in absorption, while the spectra of the late R stars (R5-R8) show a filling in of h and k lines or a weak emission feature similar to that of N-type stars (Eaton et al 1984). No companions to R stars have been found with IUE (Dominy and Lambert (1983) report one serious attempt), and it appears that R stars have only the normal rate of binarity (McClure, this meeting).

Because of their low temperatures, N-type carbon stars are extremely difficult to observe with IUE (Querci et al 1982). Several have, however, been observed with low resolution (Johnson and O'Brien 1983; Querci and Querci 1984). These show generally an absorption spectrum of atomic and molecular lines above 2850 A with weak emission lines from Mg II, Fe II, and (perhaps surprisingly) C II (2325 A). A remarkable feature of these chromospheres is their variability (Querci and Querci 1984; Baumert and Johnson 1984). It is, in fact, not even certain whether the chromosphere in many PRG stars is a permanent or temporary feature. In early M giants and supergiants, as in warmer stars, it is apparently permanent (Basri and Linsky 1979; Engvold et al. 1984; Steiman-Cameron et al. 1984). Large variations in the strength of the Mg II h and k lines appear, however, in the N-type carbon stars TW Hor (Querci and Querci 1984) and TX Psc (Baumert and Johnson 1984). In the S mira Chi Cyg, the Mg II h and k lines are present at times and absent at other times (Cassatella et al. 1980). Surely shocks play a dominant role in the outer atmospheres of mira variable stars (see, f. ex., Willson and Hill 1979) and may play a role in other PRG as well, for almost all of them are variable.

Observations reveal variable chromospheric emission lines in the spectra of many PRG stars. As of this meeting, however, few theoretical analyses have been attempted (Avrett and Johnson 1984). We expect this will become a fruitful field of research over the next few years.

In summary, models for photospheres of PRG stars are available, but, except for models of warmer M giants and S stars, these are still exploratory. Yet the PRG stars increasingly fascinate us, and intense research will continue into the future. One after another, the puzzles will be solved, and that venture will be an exciting one!

Special acknowledgement is made of the contributions of David Alexander, without whose help this review could hardly have been completed by meeting time. Thanks to Rainer Wehrse and Francois Querci who read and made useful suggestions on the manuscript. Support of NSF and Indiana University for several of the calculations reported here is deeply appreciated.

REFERENCES

Alexander, D. R., Johnson, H. R., Bower, C.D., Lemke, D.A., Luttermoser, D. G., Petrakis, J. P., Reinhart, M. D., and Welch, K. A. 1984, Bul. A. A. S. **16**, 510.

Augason, G. C., Bregman, J. D., Johnson, H. R., and Witteborn, F. C. 1984, Ap. J. (submitted)

Auman, J. R. 1969, Ap.J. **157**, 799.

Auman, J. R. and Woodrow, J. E. J. 1975, Ap. J. **197**, 163.

Avrett, C. A. and Johnson, H. R. 1984, in Cool Stars, Stellar Systems, and the Sun, ed. S. L. Balivnas and L. Hartmann (Springer-Verlag: New York), p. 330.

Ayres, T. R. 1981, Ap. J. **244**, 1064.

Baschek, B. and Scholz, M. 1982, in Landolt-Bornsteins: Numerical Data and Functional Relationships in Science and Technology, New Series, Vol. 2, Subvol. b: Stars and Star Clusters, ed. K. Schaifers and H. H. Voigt (Springer-Verlag: New York), p. 91.

Basri, G. S. and Linsky, J. L. 1979, Ap.J. **234**, 1023.

Bell, R. A., Eriksson, K., Gustafsson, B., and Nordlund, A. 1976, Astr. Ap. Suppl. **23**, 37.

Bidelman, W. P. and Pyper, D. M. 1963, Publ. Astr. Soc. Pac. **75**, 389.

Blackwell, D. E., and Shallis, M. J. 1977, Mon. Not. R. A. S. **180**, 177.

Blackwell, D. E., Petford, A. D., and Shallis, M. J. 1980, Astr. Ap. **82**, 249.

Bohm-Vitense, E. 1982, Ap.J. **255**, 191.

Bohm-Vitense, E. 1958, Zt. f. Ap. **46**, 108.

Brown, A. 1984, in Cool Stars, Stellar Systems, and the Sun, ed. S. L. Baliunas and L. Hartmann (Springer-Verlag: New York), p. 282.

Carbon, D. F. 1979, Ann. Rev. Astr. Ap. **17**, 513.

Carbon, D. F. 1984, in Methods in Radiative Transfer, ed. W. Kalkofen (Cambridge U. Press: Cambridge), p. 395.

Cassatella, A., Heck, A., Querci, F. Querci, M., and Strickland, D. J. 1980 Proc. Second European I.U.E. Conf, Tubingen, Germany, ESA SP-157, p. 243.

Chan, K. L., Sofia, S., and Wolff, C. L. 1982, Ap.J. **263**, 935.

Chiu, H. Y., Adams, P. J., Linsky, J. L., Basri, G. S., Maran, S. P., and Hobbs, R. W. 1977, Ap. J. **211**, 453.

Deguchi, S. 1980, Ap. J. **236**, 567.

de la Reza, R. 1984, in Atmospheres of M, S, C Stars, ed. H. R. Johnson and F. Querci (NASA-CNRS) (in press).

de la Reza, R. and Querci, F. 1978, Astr. Ap. **67**, 7.

Dominy, J. F. and Lambert, D. L. 1983, Ap.J. **270**, 180.

Dominy, J. F. 1982, Ph.D. thesis, University of Texas.

Draine, B. T. 1981, in Physical Processes in Red Giants, ed. I. Iben and A. Renzini (Reidel: Dordrecht), p. 317.

Eaton, J. A., Johnson, H. R., O'Brien, G. T., and Baumert, J. H. 1984, Ap. J. (in press).

Elitzur, M. 1980, Ap. J. **240**, 553.

Engvold, O., Kjeldseth Moe, O., Jensen, E., Jordan, C., Stencel, R., and Linsky, J. 1984, in Cool Stars, Stellar Systems, and the Sun, ed. S. L. Baliunas and L. Hartmann (Springer Verlag: New York), p. 359.

Eriksson, K., Gustafsson, B., Jorgensen, U. G., and Nordlund, A. 1984, Astr. Ap. **132**, 37.

Eriksson, K., Gustafsson, B., Jorgensen, U. G., and Nordlund, A. 1985, (in preparation). (EGJN)

Fay, T. D. and Honeycutt, R. K. 1972, A.J. **77**, 29.

Glassgold, A. and Huggins, P. 1984, in Atmospheres of M, S, C Stars, ed. H. R. Johnson and F. Querci (NASA-CNRS) (in press).

Goldberg, L. 1984a, in Cool Stars, Stellar Systems, and the Sun, ed. S. L. Baliunas and L. Hartmann (Springer-Verlag: New York), p. 333.

Goldberg, L. 1984b, in Atmospheres of M, S, C Stars, ed. H. R. Johnson and F. Querci (NASA-CNRS) (in press).

Gray, D. F. 1976, The Observation and Analysis of Stellar Photospheres (Wiley: New York).

Gustafsson, B. 1981, in Physical Processes in Red Giants, ed. I. Iben and A. Renzini (Reidel: Dordrecht), p. 25.

Gustafsson, B., Bell, R. A., Eriksson, K., and Nordlund, A. 1975, Astr. Ap. **42**, 407.

Gustafsson, B. and Nissen, P. E. 1972, Astr. Ap. **19**, 261.

Heasley, J. N., Ridgway, S. T., Carbon, D. F., Milkey, R. W., and Hall, D. N. B. 1978, Ap. J. **219**, 970.

Henyey, L., Vardya, M. S., and Bodenheimer, P. 1965, Ap. J. **142**, 841.

Johnson, H. R. 1974, NCAR Techn. Note STR-95 (Boulder).

Johnson, H. R. 1982, Ap. J. **260**, 254.

Johnson, H. R. 1984, in Atmospheres of M, S, C Stars, ed. H. R. Johnson and F. Querci (NASA-CNRS) (in press).

Johnson, H. R. and Ake, T. B. 1984, in Cool Stars, Stellar Systems, and the Sun, ed. S. L. Baliunas and L. Hartmann (Springer-Verlag: New York), p. 362.

Johnson, H. R., Bernat, A. P., and Krupp, B. M. 1980, Ap. J. Suppl. **42**, 501. (JBK)

Johnson, H. R., Goebel, J. H., Goorvitch, D., and Ridgway, S. T. 1983, Ap. J. **270**, L63.

Johnson, H. R. and Krupp, B. M. 1976, Ap. J. **206**, 201

Johnson, H. R. and O'Brien, G. T. 1983, Ap. J. **265**, 952.

Johnson, H. R., O'Brien, G. T., and Climenhaga, J. L. 1982, Ap. J. **254**, 175.

Kalkofen, W. (ed.) 1984, Methods in Radiative Transfer (Cambridge U. Press: Cambridge)

Keenan, P. C. 1982, Publ. Astr. Soc. Pac. **94**, 299

Kipper, T. 1982, Model Calculations of Stellar Spectra, Acad. Sciences, Estonian SSR, p. 3

Kneer, F. 1983, Astr. Ap. **128**, 311.

Kurucz, R. L. 1970, Smithsonian Astrophys. Obs. Special Rpt. 309.

Kurucz, R. L. 1979, Ap.J. Suppl. **40**, 1.

Latour, J., Spiegel, E. A., Toomre, J., and Zahn, J. P. 1976, Ap. J. **207**, 233.

Latour, J., Toomre, J. and Zahn, J. P. 1981, Ap. J. **248**, 1081.

Lester, J. B., Lane, M. C. and Kurucz, R. L. 1982, Ap. J. **260**, 272.

Linsky, J. L. 1980, Ann. Rev. Astr. Ap. **18**, 439.
Linsky, J. L. 1982, in Advances in Ultraviolet Astronomy: Four Years of IUE Research, ed. Y. Kondo, J. M. Mead, and R. D. Chapman, NASA-CP 2238, p. 17.
Manduca, A., Bell, R. A., and Gustafsson, B. 1981, Ap. J. **243**, 883.
Marcus, P. S., Press, W. H., and Teukolsky, S. A. 1983, Ap.J. **267**, 795.
McClure, R. D., Fletcher, J. M., and Nemec, J. M. 1980, Ap. J. **238**, L35.
Mihalas, D. 1978, Stellar Atmospheres (2nd edition), (Freeman and Co.: San Francisco).
Nelson, G. D. 1980, Ap.J. **238**, 659.
Nordlund, A. 1974, Astr. Ap. **32**, 407.
Nordlund, A. 1976, Astr. Ap. **50**, 23.
Nordlund, A. 1980, in Stellar Turbulence, I.A.U. Colloq. 51, ed. D. F. Gray and J. L. Linsky, (Springer-Verlag: New York).
Nordlund, A. 1982, Astr. Ap. **107**, 1.
Pecker, J. C. 1973, in Problems of Calibration of Absolute Magnitudes and Temperature of Stars, I.A.U. Symp. No. 54, ed. B. Hauck and B. E. Westerlund (Reidel: Dordrecht), p. 173.
Peytremann, E. 1974, Astr. Ap. **33**, 203.
Piccirillo, J. 1980, Mon. Not. R. A. S. **190**, 441.
Piccirillo, J., Bernat, A. P., and Johnson, H. R. 1981, Ap.J. **246**, 246.
Praderie, F., Spicer, D. S. and Withbroe, G. L. 1981, Proc. 11th Adv. Course, Swiss Soc. Astr. Ap. (Saas-Fee)
Querci, F. and Querci, M. 1975, Astr. Ap. **39**, 113.
Querci, M. and Querci, F. 1976, Astr. Ap. **49**, 443.
Querci, M., and Querci, F. 1984, Astr. Ap. (submitted).
Querci, F., Querci, M., and Tsuji, T. 1974, Astr. Ap. **31**, 265. (QQT)
Querci, F., Querci, M., Wing, R. F., Cassatella, A. and Heck, A. 1982, Astr. Ap. **111**, 120.
Ridgway, S. T., Joyce, R. R., White, N. M., and Wing, R. F. 1980, Ap.J. **235**, 126. (RJWW)
Ridgway, S. T., Jacoby, G. H., Joyce, R. R., and Wells, D. C. 1980b, A.J. **85**, 1496.
Saxner, M. and Gustafsson, B. 1984, Astr. and Ap. (in press)
Scalo, J. M. and Miller, G. E. 1979, Ap. J. **233**, 596.
Schmid-Burgk, J. 1975, Astr. Ap. **40**, 249.
Schmid-Burgk, J., Scholz, M., and Wehrse, R. 1981, Mon. Not. R. A. S. **194**, 383.
Schmitz, F. and Ulmschneider, P. 1981, Astr. Ap. **93**, 178.
Scholz, M. and Tsuji, T. 1984, Astr. Ap. **130**, 11.
Scholz, M. and Wehrse, R. 1982, Mon. Not. R. A. S. **200**, 41.
Steiman-Cameron, T. Y. and Johnson, H. R. 1984, Ap. J. (submitted)
Steiman-Cameron, T. Y., Johnson, H. R., and Honeycutt, R. K. 1984, in Future of Ultraviolet Astronomy Based on Six Years of IUE Research, NASA, p. 124.
Strecker, D.W., Erickson, E.F., and Witteborn, F.C. 1979, Ap. J. Suppl. **41**, 501.

Strom, S. E. and Kurucz, R. 1966, Jour. Quant. Spectr. and Rad. Transf. **6**, 591.
Thomas, R. N. 1983, Stellar Admospheric Structural Patterns, NASA SP-471 (NASA-CNRS).
Tsuji, T. 1976 Publ. Astr. Soc. Japan **28**, 543.
Tsuji, T. 1978, Astr. Ap. **62**, 29.
Tsuji, T. 1981a Astr. Ap. **99**, 48.
Tsuji, T. 1981b, J. Astrophys. Astr. **2**, 95.
Tsuji, T. 1983, Astr. Ap. **122**, 314.
Tsuji, T., 1984, Astr. Ap. **134**, 24.
Ulrich, R. K. 1976, Ap. J. **207**, 564.
Watanabe, T. and Kodaira, K. 1978, Publ. Astr. Soc. Japan **30**, 21.
Watanabe, T. and Kodaira, K. 1979, Publ. Astr. Soc. Japan **31**, 61.
Wehrse, R. 1981, Mon. Not. R. A. S. **195**, 553.
Willson, L. A. and Hill, S. J. 1979, Ap. J. **228**, 854.
Wing, R. F. 1967, Ph.D. Thesis, U. Calif. Berkeley.
Woodrow, J. E. J. and Auman, J. R. 1982, Ap. J. **257**, 247.
Zuckerman, B. 1980, Ann. Rev. Astr. Ap. **18**, 263.

DISCUSSION

WEHRSE - To what extent do the helium rich models differ from the hydrogen ones of higher gravity ?

JOHNSON - They are of course related, but we have calculated series at only one value of gravity.

KEENAN - Could not some of the scatter in the plots of T_e against spectral type or Ti O strength or colors be caused by metal deficiencies in some of the M type giants ? Such effects would not be detected by types or colors that depend entirely upon total TiO absorption.

JOHNSON - Yes. Composition could well play an important role. I am not aware however of any research which has demonstrated variations in composition among nearby angular-diameter M giants.

SEDLMAYR - Did you include dust opacities for the calculation of your coolest atmospheres, because they might show effective dust production at the outermost layers ?

JOHNSON - Not yet. Therefore I distrust the coolest models.

JØRGENSEN - Inclusion of a "veil" of HCN vibration - rotation absorption in the infrared will make models with a given T_{eff} more blue. Calculation on a T_{eff} = 2500 K C-star model shows the 1.6 μ H^-area flux to stay approximately constant and the Planck maximum area flux to rise. This is qualitatively the same effect as the hydrogen deficient models.

JOHNSON - This is interesting. As mentioned, molecular bands not accounted for in the models may provide an alternative solution to the problem of the shape of the H^- peak, but I was thinking of bands in the region of 1.65 μ. The "veil" of lines at longer wavelengths is another interesting alternative.

CAYREL - I am always a little bit uneasy with the expression "hydrogen deficient star". Because abundances are usually expressed with respect to hydrogen, it sounds as if one was saying that 1 is not equal to 1. If a star is hydrogen deficient by a factor of 10, does it mean that $[He/H]_{\odot}$ = 1.0 and $[M/H]$ = 1.0 ? If so, is there any reason to assume a parallel variation of He/H and M/H ? Would it not be more logical to assume that a "hydrogen deficient star" is a star for which $X < X_{\odot}$, $X + Y + Z = 1$ and Z has a value which has no physical reason to vary parallel to Y ?

JOHNSON - This exploratory models were computed with the number fraction of metals held constant. As the fractional abundance of hydrogen was decreased, the fractional abundance of helium naturally increased to keep the total constant. Does that answer your question ?

JØRGENSEN - The hydrogen is affecting the TiO strength through H_2O influence on the atmospheric structure. I noticed that there was the biggest difference the F - T_{eff} relation of strong and weak TiO bands at low T_{eff}, where the H_2O has the greatest effect. It would be interesting to see the TiO strength for models with H/H_o.

JOHNSON - Yes. Such a calculation is yet in the future.

JØRGENSEN - The decrease in H will give a quite big increase in P_g in the models with molecular opacity from the diatomic molecules only. Are your models of hydrogen deficient carbon stars in better agreement with the H_2 quadruple line strengths than are the corresponding solar hydrogen models ?

JOHNSON - Yes. Decreasing hydrogen abundance to 0.01 of its solar value, decreases the molecular column density by about a factor of 20, which is close to that needed to remove the discrepancy.

WING - I am surprised by the apparent correlation between TiO residuals and chromospheric activity. The relation between TiO strength and color temperature is very tight for giant stars and I have noticed that the stars with (slightly) strong TiO tend to have weak CN, suggesting that it is an O/C effect.

JOHNSON - We are also surprised. Because the correlation is based only on a few stars, it must be confirmed by additional observations. Among the compiting explanations, an abundance effect (C/O) appears most likely at the moment.

MODEL ATMOSPHERES FOR M (SUPER-) GIANTS WITH DIFFERENT ABUNDANCES OF THE HEAVY METALS AND THE CNO GROUP

R. Wehrse
Institut f. Theoretische Astrophysik
Im Neuenheimer Feld 294
D 6900 Heidelberg
F.R.G.

It is now generally agreed that the atmospheres of luminous M stars have geometrical extensions (defined by $d=R(\tau=10^{-4})/R(\tau=1)-1$) which are no longer negligible compared to the stellar radius. It has also been shown that the actual degree of extension depends strongly on the composition (Wehrse, 1981).

In order to study in detail the interaction between abundances and geometry as well as the consequences for the temperature structure and the emitted flux distribution we have calculated model atmospheres with helium/hydrogen abundance ratios in the range 0.1...1.0, abundances of the CNO group $10^{-4}...10^{-2}$ (corresponding to 0.1...1.0x solar) and of the heavier metals $1.2x10^{-5}...1.2x10^{-3}$ (i.e. again 0.1...10x solar). The relative abundances within the groups are considered to be solar.

The main assumptions in the model construction are the same as in the 1981 set: spherical symmetry, hydrostatic and radiative equilibrium, LTE, molecular band absorption by OH,CH,CN,MgH,CaH,SiH,CO,TiO and H_2O, no atomic line blanketing. The models show that e.g. for a star with a luminosity $L/L_{O}=1000$, a radius $R/R_{O}=117$, a mass $M/M_{O}=0.5$ (corresponding to an effective temperature $T_{eff}=3000K$ and a gravity log g=0.0) and He/H=0.1 the geometrical extension may vary between 10 and 30% and the temperature at $\tau=10^{-4}$ differ by more than 600K.

The largest (smallest) extensions and lowest (highest) temperatures at $\tau=10^{-4}$ are found in models with high (low) CNO abundances and low (high) metal content. When both groups are varied by the same amount the extension stays essentially constant.

The variations can qualitatively be understood in the following way: When the abundances of the heavy metals are reduced, the absorption by H^- (the dominant opacity at $\tau>1$) becomes smaller and therefore $R(\tau=1)$ becomes smaller, thus increasing d. Since models with larger d have a larger geometrical dilution of the radiation field, they have lower surface temperatures. This in turn increases the formation of water vapor, which acts as an additional coolant and which lets the temperatures at low depth continue to decrease until all oxygen is used up. Since in the outer parts H_2O is the most important opacity source, the radius $R(\tau=10^{-4})$ is increased during this process and d gets again larger.

M. Jaschek and P. C. Keenan (eds.), Cool Stars with Excesses of Heavy Elements, 293–294.

If the CNO abundance is reduced, $R(\tau=10^{-4})$ gets smaller and thereby dimishes the geometrical dilution of the radiation and the formation of H_2O so that in turn the outer layers become hotter and d is reduced.
For higher metal abundances or lower CNO contents the same type of arguments can be given to explain resulting changes in the opposite direction.
Since the dissociation constants are extremely temperature sensitive these changes in the stratifications often overcompensate the direct effects of abundance changes so that e.g. the TiO bands may be stronger when the Ti abundance is reduced.
A detailed description of the models and their interpretation will be given elsewhere.

Reference

Wehrse, R., 1981, Mon. Not. R. astr. Soc. 195, 553

DISCUSSION

RICHER - Can you predict what will happen in a C rich model when the metals are reduced and extension allowed for ?

WEHRSE - From the few C rich model atmospheres, which take extension into account (c.f. Scholz and Tsuji, 1984 ; Johnson et al, 1984), one gets the impression that geometric effects are much smaller in these stars. Therefore one is left with the "normal" blanketing effect.

LLOYD EVANS - Observations of red giants in globular clusters ordered by the Zirin scale (but Pilachowski's !) of metal abundance shows that, for stars of given infrared colors, TiO weakens as metal abundance is reduced.

WEHRSE - The effects I described depend strongly on the actual mass, luminosity and radius of the star, eg. they vanish if the atmosphere becomes compact. One therefore should check whether or not we are referring to similar atmospheres.

C, N, O, AND THEIR ISOTOPE ABUNDANCES IN COOLEST STARS OF THE RED GIANT BRANCH

Takashi Tsuji
Tokyo Astronomical Observatory
University of Tokyo
Mitaka, Tokyo, 181 Japan

ABSTRACT. A quantitative analysis based on model atmospheres is applied to high resolution IR spectrum of α Herculis(M5Ib-II) observed by P.Connes with his Fourier Transform Spectrometer. Analysis on CO, OH, and CN(first detected in such a cool oxygen-rich giant) gives: (1) [C/H] = -0.7 ± 0.3, [N/H] $=+0.8\pm0.4$, [O/H] $=-0.2\pm0.4$, where [M/H] $=\log(M/H)_{*}-\log(M/H)_{\odot}$, and (C+N+O)/H is solar within 15%. (2) $^{12}C/^{13}C=13\pm5$. (3) $^{16}O/^{17}O=180\pm80$. This shows that the products of CNO tri-cycles are mixed to the surface, but actual process of the mixing is still to be identified. For this purpose, similar analysis is being extended to a larger sample of red giant stars.

1. As is evidenced by the presence of M, S, C, and other peculiar stars, large abundance variations may be expected in the coolest part of the red giant branch. Abundance determination in these evolved stars is important not only for our understanding of stellar evolution at late stages but also for knowing contributions of low and intermediate mass stars to the chemical enrichment of the interstellar medium. In spite of these importances, the progress has been rather slow because of the well known difficulty in the quantitative analysis of the spectra of cool stars. Recently, however, some progresses in related fields contributed to push such an attempt(see e.g., Tsuji,1979) and we show in the following that abundance determination in very cool stars can to some extend be possible.

2. The observed material used in this analysis is the high resolution infrared spectrum of α Her observed by P.Connes; the resolution is 0.017cm^{-1} and S/N=40(Connes & Michel,1974). Before abundance can be determined, some physical parameters of the star should be known:

T_{eff} ; Measured angular diameter and model-atmosphere analysis of observed flux give consistent result of $T_{eff}=3250\pm100$K(Tsuji,1978).

log g; Some arguments suggest that $M_{bol}=-5.0\sim-3.5$ and $M/M_{\odot}=1.7\sim5$ (Reimers,1977). Then, it can be estimated that log g=0.0 ±0.4.

v_{tur} ; As most lines that can be measured with sufficient accuracy are more or less saturated, the value of the turbulent velocity is the most critical factor in our abundance analysis. For this purpose, CO first overtone transitions that cover the wide intensity range with well

M. Jaschek and P. C. Keenan (eds.), Cool Stars with Excesses of Heavy Elements, 295–299.

defined gf-values(e.g.,Chackerian and Tipping,1983) are useful. In Fig.1, detailed abundance analysis on individual CO line for three different values of v_{tur} reveals that v_{tur}=3Km/s gives the consistent result for all the values of observed equivalent widths. This value corresponds to the so-called micro-turbulence. It is found, however, that spectral line widths can be reasonably understood without extra broadening; the so-called macro-turbulence is neglected in the following analysis. Throughout this paper, model atmospheres of T_{eff}=3200K,log g=0.0, and v_{tur}=3Km/s are used; some uncertainties in thermal structure at the uppermost layer still exist(Tsuji,1983). As a reference,solar composition by Lambert(1968) is used, but final result will be compared with that of Lambert(1978).

Now, with these physical parameters, spectral synthesis method is applied to the second overtone band of CO, Δv=-1 sequence of CN Red System, and a relatively less saturated complex of OH 4-2 band, which are abundance indicators of C, N, and O, respectively. Interestingly, CN Red System, analyzed by Cerny et al(1978), is clearly identified in such a cool oxygen-rich giant, in consistent with our previous prediction(Tsuji, 1978). Molecular intensity data have been calibrated on solar and umbral spectra(Tsuji,1984). In Fig.2, predicted CO spectrum based on the solar CNO abundances appears to be too strong(top). Although CN appears to be

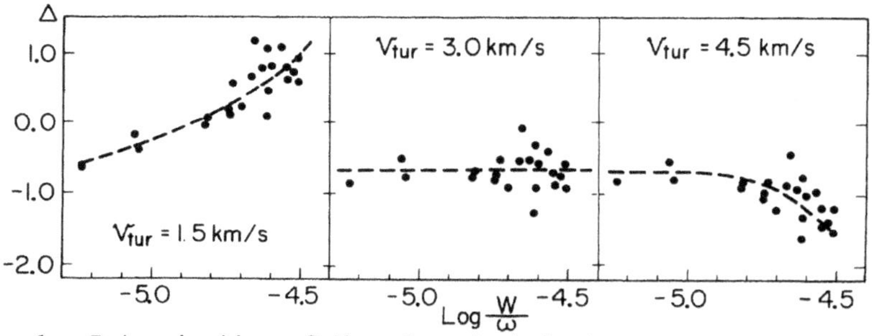

Figure 1. Determination of the micro-turbulent velocity in α Her.

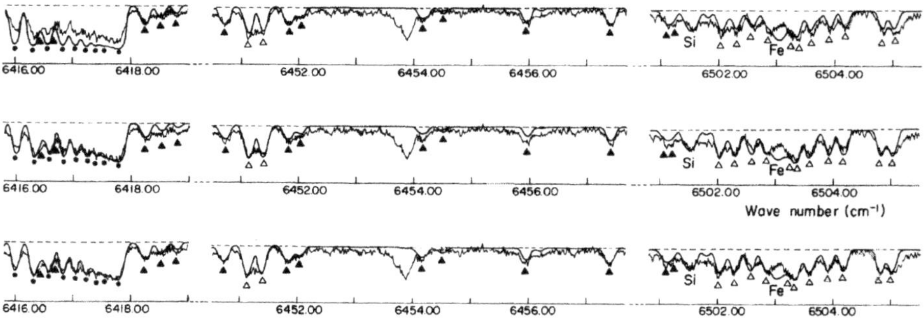

Figure 2. Observed spectrum of α Her(thin solid lines) is compared with predicted ones(heavy solid lines) for CO(dot), OH(open triangle) and CN (filled triangle). top: solar CNO abundances. middle: [C/H]=-0.5,[N/H]=+0.5, and [O/H]=0.0. bottom: [C/H]=-0.6, [N/H]=+0.9, and [O/H]=-0.1.

nearly matched, this may be a fortuitous effect of decreased carbon and increased nitrogen in α Her. Then, we assume [C/H]= -0.5, [N/H]= +0.5 and [O/H]= 0.0, but agreement between observaton and prediction are not so good(middle). After some trials, reasonable fits are found for [C/H]=-0.6, [N/H]=+0.9, and [O/H]=-0.1(bottom). This carbon abundance is consistent with the result from Fig.1: $\Delta = \Delta\log N_{CO} = \Delta\log(C/H) = [C/H] = -0.7$.

In Fig.3, observed equivalent widths of $^{12}C^{16}O$, $^{13}C^{16}O$ and $^{12}C^{17}O$ are plotted against predicted weak line equivalent widths. Relative shifts of these curves-of-growth give $^{12}C/^{13}C=13 \pm 5$ and $^{16}O/^{17}O = 180 \pm 80$. In Fig.4, spectral synthesis analysis is applied to $^{12}C^{16}O$(note less saturated lines such as 2-0,R92-93) as well as to $^{12}C^{17}O$(2-0, R26-33), and best fit is obtained for $^{16}O/^{17}O$=200, in consistent with the result from Fig.3. Similar analysis on $^{12}C^{13}O$ revealed that $^{16}O/^{18}O$ may be not much different from the solar system value of 490. Our $^{16}O/^{17}O$ is smaller by factor of 2-3 as compared with that by Maillard(1974) and Geballe et al(1977), while our $^{12}C/^{13}C$ and carbon abundance show good agreements with those by Hinkle et al(1976) and by Geballe et al(1977),respectively.

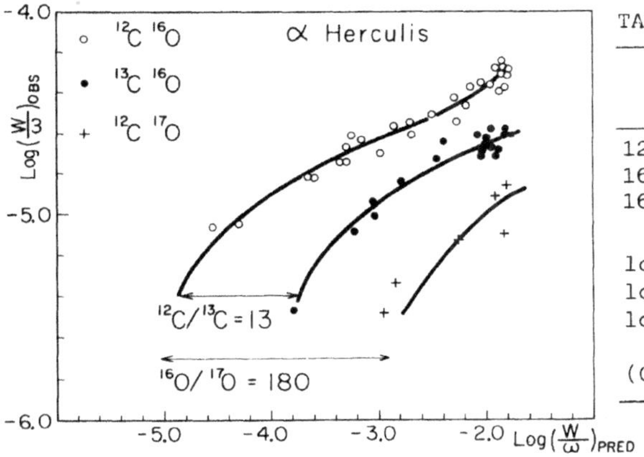

Figure 3. Curve-of-growth analysis of C and O isotope abundances.

TABLE I

	α Her (M5Ib-II)	solar sys. the sun*
$^{12}C/^{13}C$	13± 5	89
$^{16}O/^{17}O$	190±80	2700
$^{16}O/^{18}O$	490:	490
log(C/H)	-4.0±0.3	-3.33
log(N/H)	-3.2±0.4	-4.01
log(O/H)	-3.3±0.4	-3.08
(C+N+O)/H	1.2×10^{-3}	1.4×10^{-3}

*Lambert(1978)

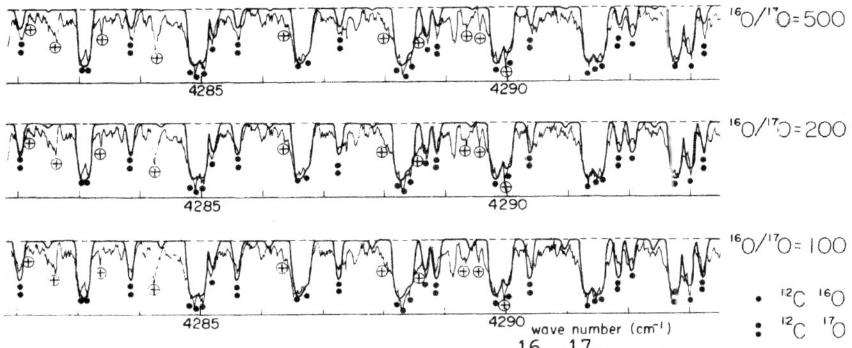

Figure 4. Spectral synthesis analysis on $^{16}O/^{17}O$ ratio.

3. The results summarized in Table I clearly indicate that the products of the CNO tri-cycles have been mixed to the surface in α Her. However, the difficulty to explain the low $^{12}C/^{13}C$ ratio in red giants by the first dredge-up theory is well known(e.g.,Dearborn et al.,1976). In this connection, importance of $^{16}O/^{17}O$ ratio should be emphasized,and oxygen-isotope analysis is being extended to a larger sample of oxygen-rich giants on the basis of KPNO FTS spectra(Tsuji,1984): Large excess of ^{17}O over ^{16}O comparable to that of α Her is found only in a few most luminous giants(e.g., SW Vir, RX Boo) and moderate enhancement in others(e.g., g Her, RZ Ari). In most M-giants that are less luminous than $M_{bol}\sim -5.0$, however, detection of $^{12}C^{17}O$ is marginal and $^{16}O/^{17}O$ ratio may be not much different from the solar system value of 2700. This shows marked contrast to $^{12}C/^{13}C$ ratio which is below 10 in all the M giants surveyed(12 stars of MOIII-M8III). Although rather low $^{16}O/^{17}O$ ratio can be predicted by the first dredge-up theory(e.g.,Shadick et al., 1980; Dearborn,1982), it may be difficult to explain why ^{17}O abundance shows rather sudden increase at about $M_{bol}\sim -5$. Then, it seems to be more reasonable to assume that the products of the new element synthesis in the red giant stage, such as H-shell burning, could have been mixed to the surface, even though such a mixing is generally forbidden by the standard theory of stellar evolution(e.g., Sugimoto & Nomoto,1974). On the hand, if stellar mass is relatively high($\sim 3M_{\odot}$), a possibility of the second dredge-up can be considered(e.g., Becker & Iben,1979).

ACKNOWLEDGEMENTS. I am most grateful to P.Connes for making available his high resolution IR spectrum of α Herculis. My thanks are also due to Drs.R. Tipping and C.Chackerian,Jr for unpublished intensity data on CO isotopes.

REFERENCES

Becker,S.A.,Iben,I.Jr.:1979, Astrophys. J. 232, 831
Cerny,D.,Bacis,R.,Guelachvili,G.,Roux,F.:1978, J. Mol. Spectros. 73, 154
Chackerian,C.Jr.,Tipping,R.:1983, J. Mol. Spectr. 99, 431
Connes,P.,Michel,G.:1974, Astrophys. J. Letters 190, L29
Dearborn,D.S.P.:1982, private communication
Dearborn,D.S.P.,Eggleton,P.P.,Schramm,D.N.:1976, Astrophys. J. 203, 455
Geballe,T.R.,Wollman,E.R.,Lacy,J.H.:1977, Publ.Astron.Soc.Pacific 89,840
Hinkle,K.H.,Lambert,D.L.,Snell,R.L.:1976, Astrophys. J. 210, 684
Lambert,D.L.:1968, Monthly Notices Roy. Astron. Soc. 138, 143
Lambert,D.L.:1978, Monthly Notices Roy. Astron. Soc. 182, 249
Maillard,J.P.:1974, Highlights of Astron. 3, 269
Reimers,D.:1977, Astron. Astrophys. 61, 217
Shadick,S.J.,Falk,H.J.,Mitalas,R.:1980, Monthly Notices Roy. Astron. Soc. 192, 493
Sugimoto,D.,Nomoto,K.:1974, in Late Stages of Stellar Evolution ed. R.J.Tayler, p.105
Tsuji,T.:1978, Astron. Astrophys. 62, 29
Tsuji,T.:1979, New Zealand J. Sci. 22, 415
Tsuji,T.:1983, Astronom. Astrophys. 122, 314
Tsuji.T.:1984, in preparation

DISCUSSION

WEHRSE - M. Scholz and I have recently analyzed FTS data from London, Ontario for alpha Her by means of our extended models. Taking the basic parameters from the literature (in particular T_{eff} = 3000 K) we get the same microturbulent velocity ξ = 3 km s^{-1}, but the carbon abundance log ε_C = 8.6.

TSUJI - Because of uncertainities of the thermal structure in the upper atmosphere I think that the carbon abundance as low as log ε_C = 8.3 is quite possible. However my analysis is consistent in showing some deficiency of carbon, as compared with the solar abundance. I think that T_{eff} = 3000 K may be too low.

BERGEAT - What sources did you use for bolometric luminosities in your first diagram ?

TSUJI - I have used two sources : Wilson - Bappu effect and Eggen's moving groups parallax.

GRIFFIN R.R.M. - A point concerning the C^{12}/C^{13} ratio : I would like to point out that the value for Arcturus is only about 7, so your theory for the enrichment of C^{13} in strange objects should also include its enrichment in non-peculiar objects.

TSUJI - By the extensive works of Dr Lambert's school and of others, evolution of C^{12}/C^{13} ratio through the subgiant to the first red giant branch is well established. An interesting point is that C^{12}/C^{13} seems to reach minimum (near the equilibrium value of the CN cycle) at K-giants including Arcturus and normal M giants. Then, C^{12}/C^{13} ratio again increases through the asymptotic red giant branch as shown by Dr Hinckle (Mira type stars and S-type stars) and by Dr Lambert (C-stars) at this colloquim.

HCN IN STELLAR ATMOSPHERES
- *a quantum mechanical calculation*

U.G. Jørgensen
Astronomiska Observatoriet
Box 515
S-751 20 Uppsala
Sweden

ABSTRACT. CI-CASSCF calculations of transitions between the 360 lowest vibrational energy levels in HCN are discussed, together with results of its application to model stellar atmospheres of cool carbon stars.

1. INTRODUCTION

At approximately T_{eff} = 8000 K molecular absorption has no effect on the modelstructure of stellar atmospheres (Kurucz, 1979). In the interval T_{eff} = 6000 - 3000 K diatomic molecules (CO, TiO, ZrO, CN, C_2, ...) increase in importance for the stellar structure, and somewhat below 3000 K polyatomic molecules (H_2O, HCN, ...) are expected to dominate the atmospheric opacity.

In the late sixties J. Auman (1966) calculated 2.3 million rotational lines from telluric and laboratory measured transitions in water vapour, and applied these to cool M stars where he found the H_2O absorption coefficient to exceed the continuous absorption by a factor of up to 10^4.

One of the most abundant polyatomic molecules in carbon stars is HCN. Unfortunately the laboratory data on HCN are sparse compared to the H_2O data and the few measured oscillator strengths are almost all from bands of little importance for stellar atmospheric structure. Order of magnitude estimates (Eriksson et al., 1984) suggest that the expected great number of weak HCN bands in the 0.8 - 3.0 micron region would be as important for cool carbon stars as was water vapour in M stars. Detailed quantum mechanical calculations of the band strengths have now confirmed these estimates, and model atmospheres have been constructed (Fig. 1).

It is seen that the model structure of cool carbon stars is changed drastically when including the HCN opacity. Details in the flux distribution are changed considerably and also the overall flux distribution is changed.

M. Jaschek and P. C. Keenan (eds.), Cool Stars with Excesses of Heavy Elements, 301–305.

2. THE AB INITIO CALCULATIONS

In harmonic oscillator theory HCN could be regarded as two diatomic molecules, HC and CN. The H-C and the C-N stretchings are two vibrational modes. A third is the CH-CN bending. The transition from the lowest vibrational state, v = 0, in each mode to the next, v = 1, is the fundamental transition. The transitions v = 1 to v = 2, v = 2 to v = 3 etc are the corresponding "hotbands". The overtones $\Delta v = 2$, $\Delta v = 3$, ... and the combination bands, where two or more modes are active "simultaneously", are not allowed in the harmonic theory, and, hence, can only be calculated from a full quantum-mechanical investigation.

In collaboration with J. Almlöf, M. Larsson and P. Siegbahn I have used the CI-CASSCF (Configurational Interacting, Complete Active Space, Self Consistent Field) theory to calculate *ab initio* energy and dipole moment for a grid of stretched and bended configurations of the HCN molecule. 90 configurations were calculated in full valence and 384 in a somewhat smaller active space. From these a polynomial surface was fitted in energy and dipole moment. Solution with respect to the hamiltonian gives eigen-functions and -energies. The eigenfunctions were integrated with respect to the dipole- moment surface to give the dipole-moment derivative $<\Psi_{low}|\mu|\Psi_{up}>$ between the lowest 360 eigen-states. From these the integrated band strengts, S_m^n, of a transition from energy level m to n can be calculated as

$$S_m^n = \frac{8\Pi^3 N_A G_m^n}{3hc} \; \nu_m^n \; |<\Psi_m|\mu|\Psi_n>|^2 \; \frac{e^{-Em/kT}}{Q_{vib}} \; (1-e^{-\nu_m^n hc/kT})$$

where G_m^n is the total number of allowed transitions between level m and n, ν_m^n the energy difference and Q_{vib} the vibrational partition function. The results are in good agreement with observations, and will be published in detail later (Jørgensen et al., 1984).

When constructing a model atmosphere it is necessary to consider not only the transition from the ground level, but also a great number of hotbands for each band. This is especially true in the warmer layers of the atmosphere, where the strength of the first hotbands can even exceed the strength of the ground-level transition. This is because the transition moment for each hotband can be very different from one another in the quantum-mechanical treatment, and because the multiplicity generally increases with increasing hotband level. For the same reason the total strength of a given band (including the corresponding hotband series) is generally not constant with temperature.

3. THE MODEL ATMOSPHERES

In collaboration with B. Gustafsson and K. Eriksson I have calculated model atmospheres for a series of cool carbon stars. In an earlier work (Eriksson et al., 1984) we have calculated model atmospheres including HCN opacity under the assumption of a simple power law relation between the band strength of fundamentals, overtones and combination bands; and by use of the double harmonic oscillator approximation, to calculate the band strength of hotbands relative to ground-level transitions. These results are shown for a T_{eff} = 2500 K model, marked "old" in Fig. 1. Also shown in Fig. 1 is the model marked "no HCN", which has no HCN opacity included, the model marked "3μ" with HCN opacity from the strongest transition only (incl. its hotbands - this is the 3 micron band observed in all carbon stars), and finally the new model marked "CI-CASSCF" which include HCN opacity calculated by the CI-CASSCF method.

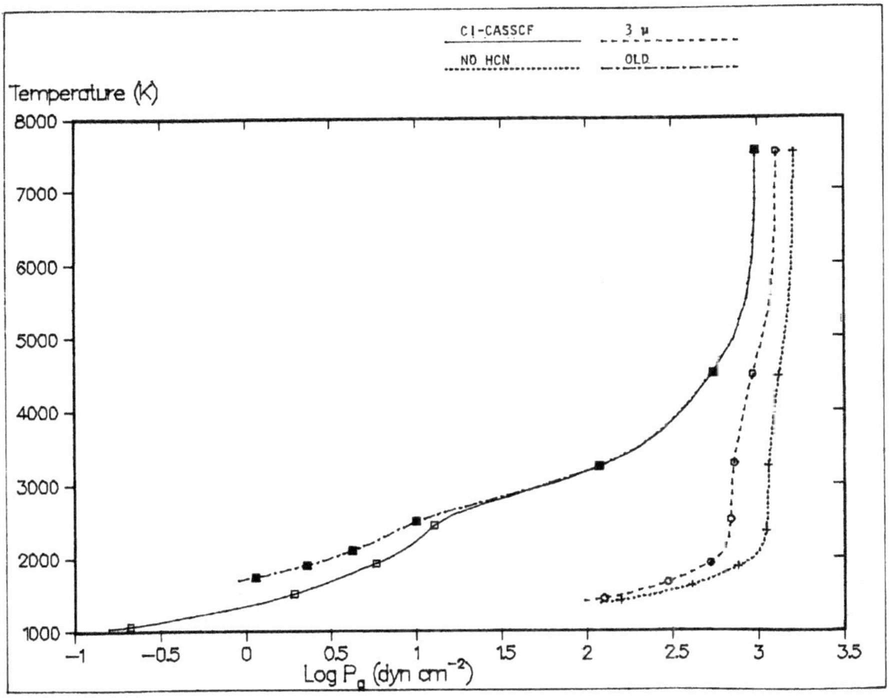

Fig. 1: T = T(Pg) calculated for a model atmosphere with T_{eff} = 2500 K log g = -1 and solar chemical abundance except for carbon and nitrogen enhanced so $\log(N_C)$ = 9.05, $\log(N_N)$ = 8.29. The four different models shown are calculated on the assumption of respectively
1) no HCN opacity included: "no HCN"
2) HCN opacity from the 3 micron absorption only: "3μ"
3) HCN opacity as calculated in our earlier work: "old"
4) HCN opacity as calculated from the CI-CASSCF theory: "CI-CASSCF"
The marks represent values of log τ_{Ros} = -4, -3, ..., 2.

The relative similarity between the "old" and "CI-CASSCF" models largely confirm our conclusions based on the older models. As mentioned in our earlier paper, models with full HCN absorption included, produce a continuous change in the model structure when considering cooler and cooler models, whereas models with no HCN absorption show a rather rapid change when T_{eff} decrease below approximately T_{eff} = 2900 K.

It is furthermore concluded from Fig. 1, that it is not the strongest HCN absorption feature (the ν_3 transition at 3 micron) that affects the structure, but rather the very great number of weak smeared-out vibration-rotation transitions shortward of the 3 micron absorption, which act as a "veil" in the atmosphere. For this reason it is necessary to calculate a very great number of bands to get the right absorption coefficient at stellar atmospheric temperatures. The CI-CASSCF model in Fig. 1 is, hence, based on band strength calculations of up to 280,000 bands, distributed on 65.5 million individual rotational lines. In hydrogen deficient carbon stars the C_3 molecule may play a similar role for the atmospheric structure as does HCN in the models shown here. *Ab initio* calculations of the C_3 absorption are in progress, and also calculation of the C_2H_2 absorption is under consideration, as well as extensive laboratory measurements to compare with the calculated bandstrengths.

4. ACKNOWLEDGEMENT

I am very greatful to Dr. K. Eriksson for very important computational help, and to Dr. M. Saxner for a last minute assistance in ODF-summation.

5. REFERENCES

Auman J.R., Jr., 1969, Astrophys. J., **157**, 799
Eriksson K., Gustafsson B., Jørgensen U. G., Nordlund Å., 1984, Astron. Astrophys., **132**, 37
Jørgensen U. G., Almlöf J., Larsson M., Siegbahn P.E.M., 1984, to be submitted to J. Chem. Phys.
Kurucz R.L., 1979, Astrophys. J. Suppl., **40**, 1

DISCUSSION

BENNETT - Molecular equilibrium calculations at least using Tsuji's dissociation constants, indicate that the C_2H molecule is nearly as abundant as the commonly recognized polyatomic molecules HCN and C_2H_2. Have you ever considered the possible effect of this species on the opacity and atmospheric structure of your carbon star models ?

JORGENSEN - We considered HCN being the most important of the polyatomic molecules at higher temperatures ($\sim$ 2500 K - 3500 K). At lower temperatures, or at higher C/O we expect that C_2H_2 will dominate over HCN. The "ab initio" calculations of the HCN opacity costed more than 1000 cpu hours at the VAX 780 computer, so the number of molecules that have been considered so far are limited. The most important effect of the HCN opacity was to deliver a thin blanket to the atmosphere, to fill out the most transparent areas of the spectrum. Adding another "thin blanket" may not have the same great effect, although it would certainly be nice to know the effect of many more molecules in the atmosphere of late type stars in general. C_2H as well as C_3H and C_3 are abundant molecules in our equilibrium calculations, too. It is not known today whether they could contribute with a high opacity, too.

WING - Your inclusion of CHN opacity is a major step toward the computation of realistic synthetic spectra for the cooler C-stars. There are however other problems. For example, Gustafsson and I recently found (unpublished) that a small change in the microturbulent velocity parameter produce a large effect on the synthetic colors computed for C-stars on my narrow-band filter system.

JORGENSEN - As it was also stressed by Johnson in his review paper of yesterday, the C-stars models are still mainly in the developing state where models made by different groups have been compared with one another only. The next step must be extensive tests against observations, and there are several factors yet to be explored, like sphericity and microturbulence. As concerns the effect of the microturbulence on the molecular blanketing, higher microturbulence broadens the lines which may often make them act more as a blanket.

DUST FORMATION IN STELLAR WINDS

H.-P. Gail and E. Sedlmayr
Institut für Astronomie und Astrophysik
Technische Universität Berlin
Ernst-Reuter-Platz 7
1000 Berlin 10
F.R. Germany

ABSTRACT. Classical nucleation theory is applied to grain formation in C-rich cool stellar winds of heavily obscured objects. The effectivity of the formation process and its implications for the structure of the shells are discussed.

1. INTRODUCTION

The outermost layers of cool stellar atmospheres are known to be sites of copious dust formation. Depending on the abundance ratio of carbon to oxygen, either carbon ($[C]/[O] > 1$) or silicate-like ($[C]/[O] < 1$) grains are expected to be formed, respectively.

We have applied classical nucleation theory for the case $[C]/[O] > 1$ to the atmospheres of cool giants in order to explain the formation of their dust shells and winds.

2. FORMATION AND GROWTH OF GRAINS

Adopting classical nucleation theory, grain formation in an expanding atmosphere generally can be described by the following system of moment equations:

$$\frac{\partial K_i}{\partial t} + \text{div}\left\{ \underline{v} K_i + \underline{v}_D \cdot K_{i+1} \right\} = q_i \qquad (i = 0 \ldots 3) \tag{1}$$

where K_i is the $N^{i/3}$-moment of the cluster distribution function (Fix, 1969; Gail et al., 1984 = Paper I), $\underline{v}$ the mean material velocity and $\underline{v}_D$ the drift velocity of the clusters relative to the carrier gas. The source terms q_i $(i = 0 \ldots 3)$ originating from the divergence of the nucleation currents in cluster space are discussed in Paper I. q_i for $i \geq 1$ always can be split into two contributions describing cluster formation and growth, respectively, whereas q_o is equal to the dust production rate.

Details concerning our assumptions about the thermodynamic proper-

M. Jaschek and P. C. Keenan (eds.), Cool Stars with Excesses of Heavy Elements, 307–312.

ties of the grain material and the chemistry of the gas phase are described in Paper I.

3. APPLICATIONS

3.1 The Dust Production Rate

The dust production rate q_o is a very sensitive function of the supersaturation ratio $S = p_m/p_{sat}$, with p_m being the partial pressure of the monomers and p_{sat} the saturation vapour pressure of the clusters which itself depends strongly on temperature. Fig. 1 shows lines of constant dust production rate per hydrogen atom in the logp-T-plane. A sharp transition at $\ln S \approx 3$ separates the regions of negligible and very efficient dust production. For this reason, effective dust formation under stellar conditions is limited to a temperature regime below $T \approx 1300$ K. This demonstrates clearly that condensation temperatures derived by thermodynamic equilibrium calculations are completely inadequate.

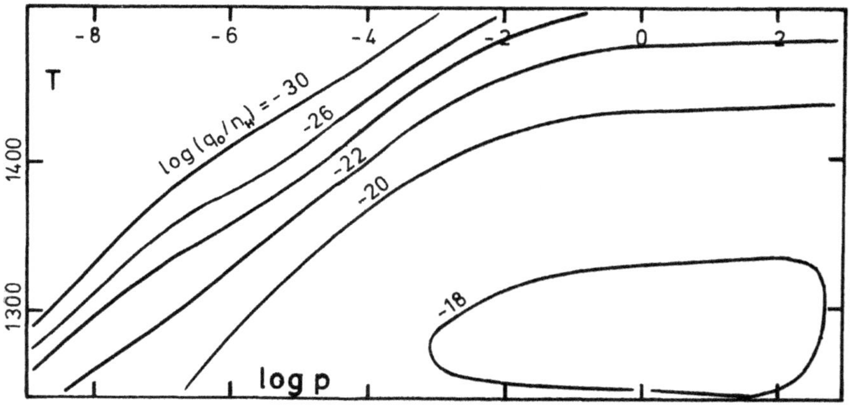

Fig. 1 Dust production rate per hydrogen atom according to classical homogeneous nucleation theory

3.2 Dust Driven Winds

The origin of supersonic winds and mass loss of late type stars is a yet unsolved problem. One possible mechanism is radiation pressure on dust grains. In the following we present the results of a self consistent model calculation of dust condensation in a stellar wind which itself is driven to supersonic velocities by radiation pressure.

Assuming a spherically symmetric stationary wind, the hydrodynamic equations are:

$$\dot{M} = 4\pi r^2 \rho v; \quad v \frac{dv}{dr} = -\frac{1}{\rho}\frac{dp}{dr} - G\frac{M_*}{r^2}(1 - \delta) \qquad (2)$$

where δ accounts for radiation pressure on dust (see Lucy, 1976). For the temperature stratification, we assumed a power law

$$T(r) = T_{*} \left(\frac{R_{*}}{r}\right)^{-\alpha}. \quad (3)$$

From the results of Rowan-Robinson and Harris (1983) a value $\alpha = 0.7$ is derived for the inner part of the dust shell where dust formation and acceleration of the wind occurs.

A simple estimation shows that for mass loss rates $\dot{M} > 10^{-7}\ M_{\odot}/yr$ the drift velocity is negligible small. The coupled system of equations (1), (2) and (3) has been solved numerically using Lagrangean coordinates for eqs. (1).

The following stellar parameters have been adopted for our calculations:

$$M_{*} = M_{\odot},\ L_{*} = 2 \cdot 10^{4}\ L_{\odot},\ T_{*} = 2300\ K,\ \dot{M} = 2 \cdot 10^{-5}\ M_{\odot}/yr,$$

$$[C]/[O] = 1.6 \text{ and cosmic abundances otherwise.}$$

The value 1.6 has been chosen in order to obtain a terminal velocity of 15 km/s in agreement with observations (e.g. Zuckerman, 1977).

The results are shown in Fig. 2:

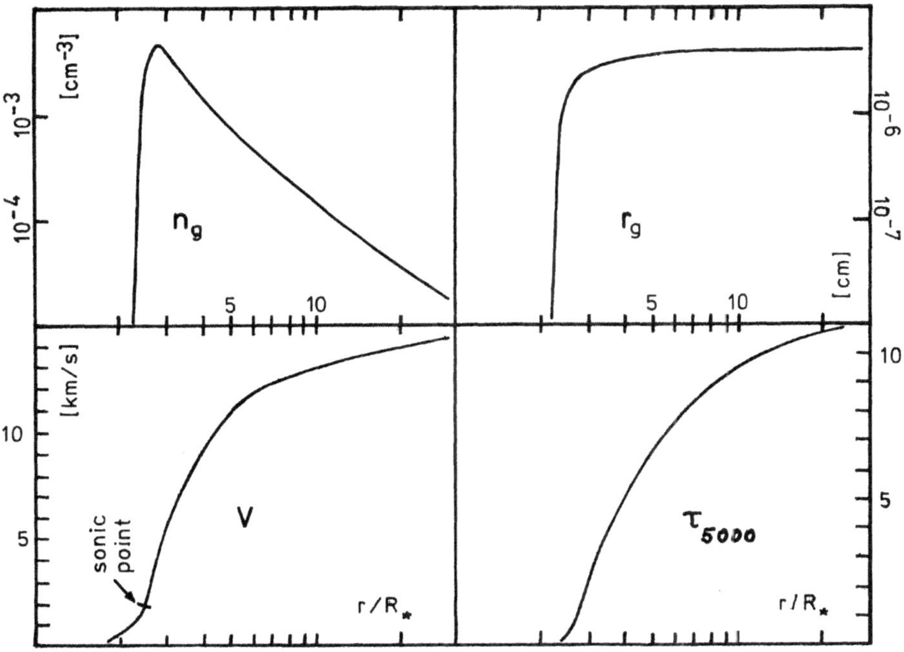

Fig. 2 Grain density n_g, grain radius r_g, hydrodynamic velocity v of the wind and optical depth of the shell vs. radial distance

These results demonstrate that the supersonic velocity fields and the optical depths observed at heavily obscured objects might be explained as consequence of dust formation and its momentum coupling to the gas.

References

Gail, H.-P., Keller, R., Sedlmayr, E.: 1984, *Astron.Astrophys.* **133**, 320.
Fix, J.D.: 1969, *Mon.Not.R.astr.Soc.* **146**, 37.
Lucy, L.B.: 1976, *Astrophys.J.* **205**, 482.
Rowan-Robinson, M., Harris, S.: 1983, *Mon.Not.R.astr.Soc.* **202**, 797.
Zuckerman, B., Palmer, R., Morris, M., Turner, B.E., Gilra, D.P., Bowers, P.F., Gilmore, W.: 1977, *Astrophys.J.* **211**, L97.

DISCUSSION

COWLEY - Could you comment on experimental constraints on the accuracy of the classical nucleation theory ? What are the uncertainties of the rates of condensation ?

SEDLMAYR - This is indeed the crucial point of classical nucleation theory, which uses the concept of surface energy. This might be very uncertain for very small clusters. Here we have adopted empirical values given by Draine. Nevertheless I think that the nucleation rate may be uncertain we think by order of magnitude or even more.

BERGEAT - You get rather small grains (i.e. radius $\simeq$ 0.04 μm) which are only absorptive grains, longward of $\lambda \simeq 0.5$ μm. I wonder how could CN bands be observed near $\lambda \simeq 1$ μm in the case of IRC + 10216 (very high opacity) if larger grains, i.e. effective scatterers, are absent.

SEDLMAYR - I do not know whether CN bands have been observed. If dust grains are present, as the α results show, they provide essentially the absorption coefficients and determine the large optical depth. The dust extinction coefficient is calculated by Mie theory including both absorption and scattering. Its wavelengths dependance shows approximatively a λ^{-1} law, which fits well to the observations.

BIDELMAN - What is the highest temperature at which you can get appreciable graphite formation ? There are optical polarization observations that seems to indicate graphite formation in fairly warm stars.

SEDLMAYR - From the dust production rate plotted in the p-T plane, one infers a significant rate only for temperatures smaller than 1300 K. This is due to the well known fact that in practice nucleation only occurs above $\ln S \simeq 3$.

If you take into account turbulence, a mixing of the grains into hotter layers is possible. We do not know how these hot stars are able to produce dust if they have no wind.

CATCHPOLE - We recently observed mild obscuration in the cool carbon star R For, which could be explained by 0.15 μm graphite particles. Is it easy to make particles of this size ?

SEDLMAYR - The observed particle size is fully in the range our calculation provide. What size you obtain in a particular case depends essentially on the supersaturation ratio and on the growths conditions. Furthermore our size is only an average value defined by the first moment of the distribution function.

TSUJI - As Dr Zuckerman has shown, carbon stars have extensive CO envelopes and loose mass at higher rate than in oxygen rich stars. Recently we found that even an ordinary carbon star such as V Aql with little IR excess shows rather large mass loss of $\frac{dM}{dt} \sim 10^{-6} \frac{M_\odot}{\text{year}}$ by CO observation at 45 m of the Nobevama Radio Observatory by Unno, Kaifu, Ukita, Cho. Then I wonder if your model could be applied to explain mass loss in such a case with a small amount of dust around the stars ?

SEDLMAYR - As you see from the nucleation rate, the dust formation depends strongly on the actual cooling track, which might greatly modify the final grain density in the wind. The models presented are to explain the structures of the heavily abscured objects.

CAYREL - The opinion is frequently met that radiatively driven winds are more likely to exist on hot stars when radiative flux is 10^4 ($\sim T_{eff}$) larger than in a cool giant. However from the simple equations :

$$g_s = \frac{G\,M}{R^2} \qquad \text{(static gravity)}$$

$$g_r = \frac{\pi}{c}\int_0^\infty K\nu\ F\nu\ d\nu = \frac{\pi}{c} \langle K\rangle_F \cdot F \qquad \text{(radiative acceleration)}$$

$$L = 4\,\pi\ R^2\ F \qquad \text{(luminosity definition)}$$

The Eddington's condition $g_r > g_s$ (radiative instability) translate into

$$\frac{\pi}{c} \langle K\rangle_F\ F > \frac{GM}{R^2}$$

or

$$\langle K\rangle_F > \frac{4\ c\ G\,M}{L}$$

So far a star of given mass and luminosity the condition deals only with the mean opacity per gram, which has every opportunity to be larger in a cool star full of molecular opacity than in a hotter star where it is practically equal to the diffusion σ_e by free electrons.

Also the Lucy-Solomon mechanism (radiative acceleration producing velocity gradient decreasing in turn the radiative acceleration) has every opportunity of being more efficient in the presence of many molecular ions.

So a larger value of T_{eff} is not at all a necessary condition for having a radiation driven wind (nor sufficient either).

BENNETT - One of the major problems with classical nucleation theory applied to formation of graphite grains is our lack of knowledge of the grain surface tension. At the time of Woodrow's models several years ago, the value of this parameter was uncertain by more than an order of magnitude. Has the situation improved since then ?

SEDLMAYR - There are empirical values for the "surface energy" which we have taken from Draine's publication. Of course there are still severe conceptual difficulties.

VI. - BINARY SYSTEMS. RADIAL VELOCITY CURVES

BINARY SYSTEMS AMONG THE PECULIAR COOL STARS

Robert D. McClure
Dominion Astrophysical Observatory
Herzberg Institute of Astrophysics
5071 W. Saanich Road
Victoria, B.C. V8X 4M6

ABSTRACT. Radial velocity observations with an accuracy of about 0.5 km s^{-1} have been obtained over the last five years for samples of strong CN, BaII, CH, sgCH, and R stars. These data indicate that those stars that exhibit excesses of s process elements as well as carbon, the BaII and CH stars, are probably all binary systems. Those stars, on the other hand, that show no excess of s process elements, the strong CN and R stars, show the normal binary frequency for K giants. These results are discussed in terms of possible mixing scenerios. The most popular explanation for the BaII and CH stars is that their atmospheres were contaminated by material transferred from a previously more massive companion, although this hypothesis has several formidable problems. The R stars, on the other hand have probably mixed at the helium core flash.

1. STELLAR SAMPLES

During the last five years, a series of precise radial velocity measurements have been obtained at Dominion Astrophysical Observatory (DAO), of various cool stars that have strong carbon features. These observations were made with the 1.2m telescope coudé radial-velocity spectrometer which was shown by Fletcher et al. (1982) to have an accuracy of better than 0.5 km s^{-1}. The samples of stars in this program include BaII stars, CH and sgCH stars, the hotter R type carbon stars (R0-R5), and strong CN or "4150" stars. A random sample of normal K giants has also been monitored over the same period of time by Harris and McClure (1983), and provide a standard to which the binary frequency of the other peculiar giant star samples can be compared.

The stars discussed here are all believed to occupy approximately the same region of the HR diagram. That is, they lie on or near the red giant branch. Other groups of brighter peculiar giants such as the S and cooler R and N carbon stars were not included, since such stars are known to be pulsationally unstable. Velocity variations due to atmospheric motions will not complicate the velocity data for the present stellar samples which were obtained to investigate binary star frequencies. This program, therefore, involved only those peculiar stars

M. Jaschek and P. C. Keenan (eds.), Cool Stars with Excesses of Heavy Elements, 315–329.

that lie below the tip of the first ascent giant branch, and says nothing about those stars that may be on the upper asymptotic giant branch (AGB).

By way of introduction, I will discuss the spectral and population characteristics of the four groups of stars. A more detailed description can be found in the review by McClure (1984).

1.1 Strong CN Stars

These stars originally referred to by Roman (1952) as "4150" stars, appear to be normal K giants, except for an anomalous excess CN absorption, and possibly anomalously strong metal lines. There has been wide debate as to whether they have an excess heavy element abundance (e.g. Branch et al. 1978). It is possible that the excess CN absorption can be explained by CN processing and mixing near the base of the giant branch, and s process element abundance does not appear to be abnormal. The CN strong stars appear, from their kinematics, to be an old disk population (Janes 1975). The stars in the present sample were taken from a list compiled by Schmitt (1971) who used MK dispersion spectra to isolate them. Only those stars in Schmitt's list that have a DDO CN index equal to or greater than 0.10 mag (McClure 1970; McClure and Forrester 1981) were included.

1.2 BaII Stars

These stars have abnormally strong CN, CH, and s process elements (Bidelman and Keenan 1951). Recent abundance analyses of BaII stars by Smith (1984) shows that N is somewhat enhanced, C about normal, and O slightly depleted. Since C is depleted in CNO cycling, the abundances indicate that triple α carbon has been added into the atmospheres of these stars as well as s process elements. The C/O ratio, however, is probably less than unity so that carbon swan bands are not particularly strong, although they are present for some of the more extreme cases. Based on kinematics, the BaII stars appear to be an old disk population (Eggen 1972), and the only such stars found in a cluster (McClure et al. 1974) are in the old disk cluster NGC2420.

There are also marginal cases of the BaII star phenomenon. Bond (private communication) suggests a distinction should be made between two groups of stars that exhibit less extreme characteristics of BaII stars. Stars in the first group, the "mild BaII stars", have BaII and SrII enhanced slightly but CN and CH about normal as found by Sneden et al. (1981). The stars in the second group, the "marginal BaII stars" have SrII and CN and CH strong, but BaII not abnormally enhanced. The latter is a sign of lower neutron exposure s process.

1.3 CH Stars

These stars are population II, high velocity stars which exhibit many of the same characteristics as their population I counterparts, the BaII stars, except that they are metal poor, and in addition, the carbon features are often stronger. They have enhanced CH, CN, C_2, and enhanced s process elements (Keenan 1942). CH stars are found on the giant branches of several globular clusters as well as in the field. In addition to CH stars on the upper giant branch, Bond (1974) has discussed a group of fainter stars with similar characteristics which he refers to as subgiant CH stars. Luck and Bond (1982) have determined chemical abundances for numerous sgCH stars and found a large range in metal abundances. Kinematically they also appear to be a mixture of halo and disk population. They appear to have very low luminosity and form, perhaps, a low luminosity and hot extension to the CH and BaII stars in the HR diagram.

1.4 R Stars

The hotter (R0-R5) carbon stars are similar in luminosity and kinematics to the BaII stars (Eggen 1972). In spectral characteristics, they have much stronger carbon features including strong C_2 bands indicating and C/O ratio greater than unity. They differ from BaII and CH stars, however in having no enhancement of s process elements (Yamashita 1972, Dominy 1984).

Table 1 summarizes the spectral and population characteristics of the stars being discussed here.

TABLE 1

STELLAR GROUP	POPULATION	C_2	S PROCESS	BINARIES
CN STRONG	OLD DISK	NONE	NORMAL	NORMAL
BaII	OLD DISK	USUALLY NONE	STRONG	ALL
MARGINAL BaII	OLD DISK?	NONE	WEAK TO STRONG	?
CH	HALO	WEAK TO STRONG	STRONG	HIGH (ALL?)
sgCH	HALO, OLD DISK	NONE (TOO HOT)	WEAK TO STRONG	HIGH (ALL?)
R	OLD DISK	STRONG	NORMAL	NORMAL

2. VICTORIA RADIAL VELOCITY OBSERVATIONS

The success of this program is due to the invention of the radial velocity spectrometer by Griffin (1967), and the implementation of such an instrument by Murray Fletcher at the coudé focus of the 1.2m telescope at DAO (Fletcher et al. 1982). Because the stars discussed here are giants, and especially in the case of the CH stars which are population II stars and therefore very large giants, only a very small velocity variation can be expected if they are binary systems. The high accuracy of a coudé instrument, combined with the efficiency of the photoelectric detector is needed in order to look for such small velocity variations in a large sample of stars. It is not surprising, therefore that binaries were not looked for among such stars using old photographic techniques.

My interest in this project was generated by the suggestion that globular clusters (and dwarf spheroidal galaxies) which contain CH stars tend to have low central concentration (McClure and Norris 1976). The only reason for this effect seemed to us to be that CH stars might be binaries, and binaries might have formed more easily or may not have been disrupted as easily in a loose globular than in more highly concentrated ones. A diagram of concentration class versus absolute magnitude (≡ mass) for globular clusters is shown in Figure 1. Those systems

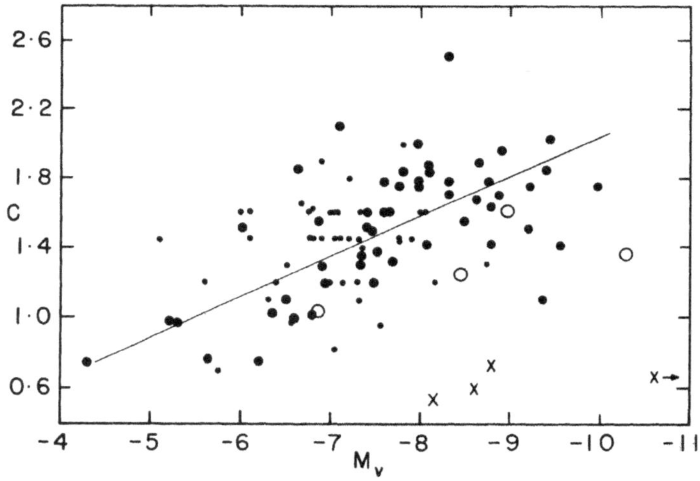

Figure 1. Concentration parameter $c = \log r_t/r_c$ vs. absolute visual magnitude for globular clusters (dots) from data in Harris and Racine (1979). Small dots represent less certain data. Clusters containing CH stars are represented by open circles, dwarf spheroidals containing CH stars by crosses. The line is a least squares solution to the relation.

that contain CH stars are plotted as open circles. Dwarf spheroidal galaxies which contain stars similar to the CH stars are represented by crosses. Note that the line representing a least squares solution to the correlation lies well above (in concentration class) all objects that are known to contain CH stars. After coming to the DAO six years ago, therefore, the author began a program to check out this hypothesis. The halo population stars are less common that the population I BaII stars and are on average considerably fainter. Since the BaII stars are so similar in sprectral characteristics and the instrument at first available was limited in efficiency a start was made with these brighter stars.

2.1 BaII Stars

It was soon obvious that many of the BaII stars exhibited velocity variations. The observations to date are summarized in Figure 2, which shows velocity as a function of Julian date for each star. It is interesting to note that all stars in the sample that have BaII star features stronger than Ba1 on the Warner (1965) scale show velocity variations. McClure (1983) has summarized the orbital characteristics for those stars that have observations over more than one velocity cycle. On average, the orbital data for these stars indicate that separations between binary components are about 2AU, and if the primaries are about 1.5 $\underline{M}_{\odot}$ as thought to be the case for BaII stars, the secondaries are a fraction of a solar mass as could be true if they were white dwarf companions. As pointed out by Dominy and Lambert (1983), the average separation for the complete sample is likely to be somewhat larger since those stars for which orbits have not yet been calculated are ones with longer period.

The case of the marginal or mild BaII stars is not yet determined. As Griffin (1982) suggests, the binary frequency among such stars is large. However, many, including three stars in Figure 2 that are classified as Ba1 show no velocity variations. A severe problem with the marginal BaII stars is that many, such as in the list of MacConnell et al. (1972) are not BaII stars at all, but only stars that were suspected from objective prism plates.

2.2 CH and sgCH Stars

The population II CH stars have now been shown also to have a very high frequency of binaries as illustrated by the velocity data of Figure 3. Although it is perhaps too soon to say that every CH star is a binary, the very high frequency indicated by the observations to date (> 70%) lead one to believe that with further observations, the CH stars will all turn out to be binaries just like the case for the classical BaII stars. It seems reasonable to conclude that the CH stars are just population II equivalents of the BaII stars. Although they tend to show stronger C_2 bands, this may just be due to their lower metal and oxygen abundance.

The very limited coverage that has been obtained so far for sgCH stars (see Figure 3) also shows every indication of velocity variations.

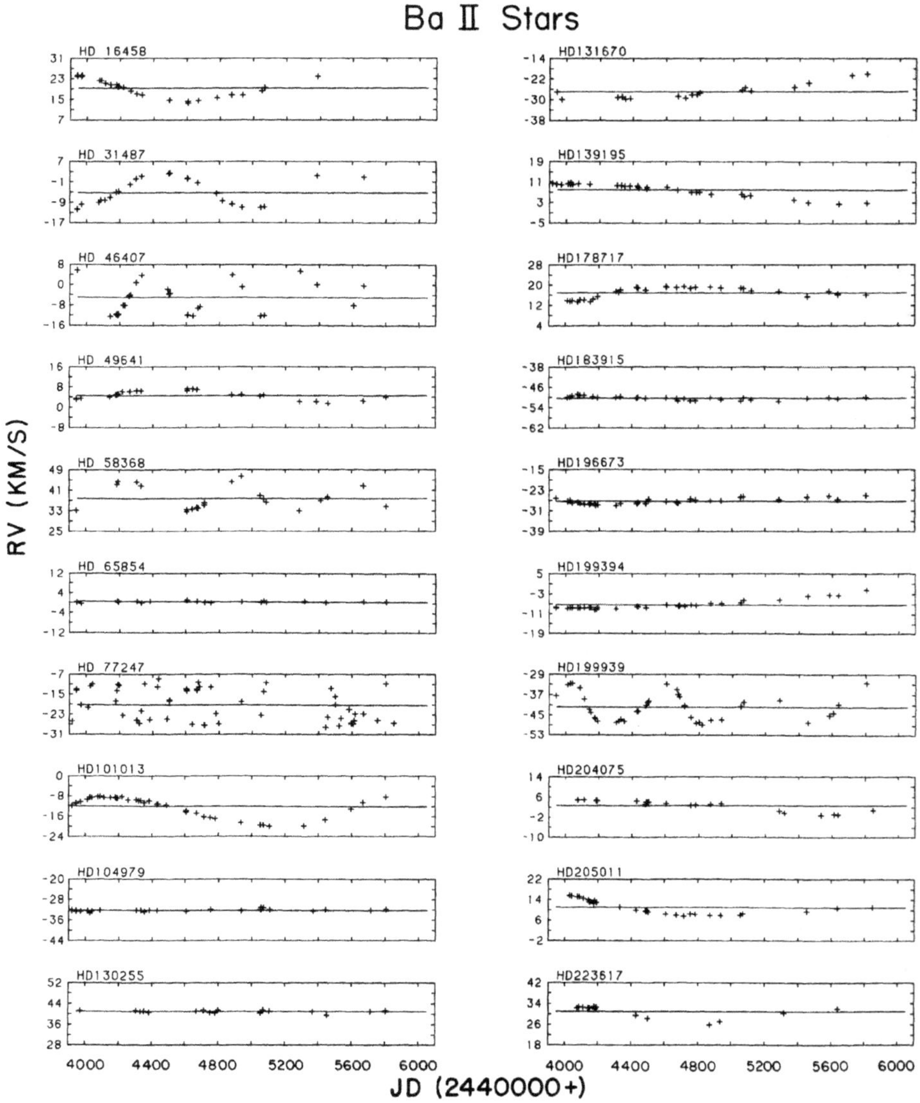

Figure 2. Radial velocities as a function of Julian Date for BaII stars. The horizontal line represents the mean of velocities measured over this interval.

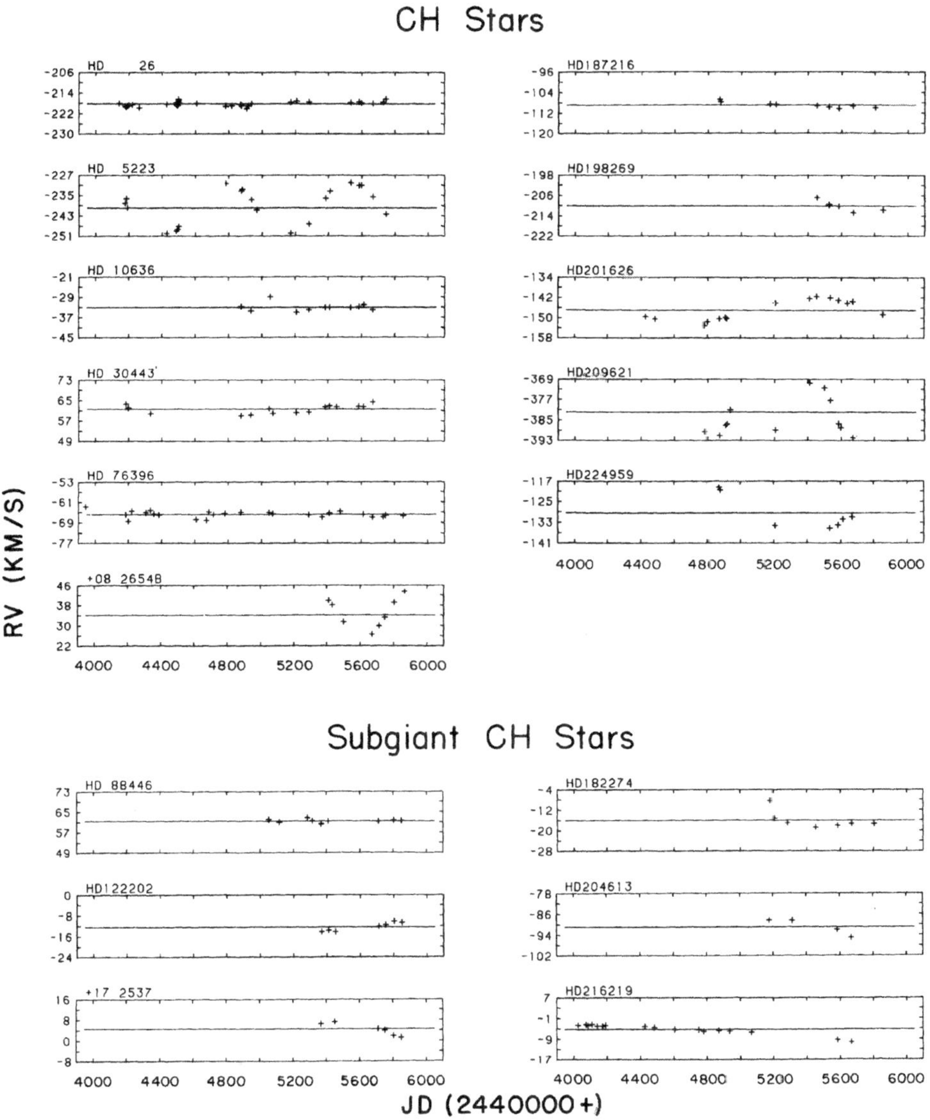

Figure 3. Same as Figure 2 for CH and sgCH stars.

It is not unreasonable to speculate that these stars too will turn out to be binary systems.

2.3 Strong CN Stars

The author, in collaboration with Chris Aikman has been monitoring a sample of 54 strong CN stars over the last four years at DAO. To date we have a total of 357 velocity observations, or an average of 6.6 observations per star spread out over the four years. Because the velocities are very constant for most, the data are illustrated in Figure 4 in the form of a histogram showing numbers of stars as a function of velocity dispersion. A majority of stars (70%) have $\sigma < 0.75$ km s^{-1} indicating that most strong CN stars are not binaries. The few stars with large residuals probably represent a normal binary frequency for K giants (see Harris and McClure 1983).

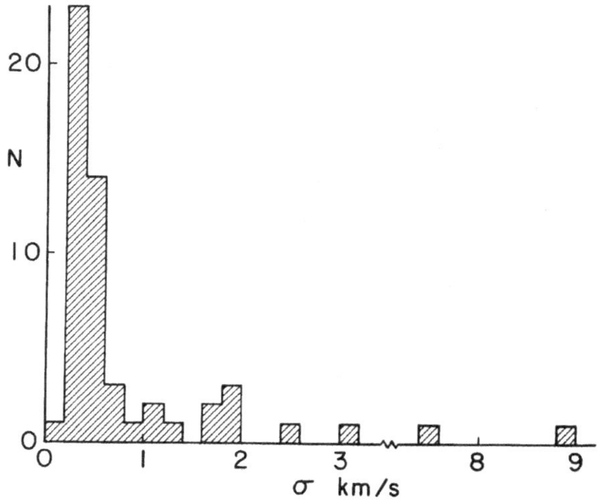

Figure 4. Histogram of the number of CN strong stars as a function of the dispersion in velocities measured.

2.4 R0-R5 Carbon Stars

A surprising new result shown by Figure 5 is that the hot R stars, unlike the BaII and CH stars do not show a high frequency of binaries. There are indications of velocity variations for two or three of these stars, but this would only indicate a normal frequency of binaries for a giant star population. A considerably larger sample of R stars is currently being monitored now at DAO, but it appears unlikely that the R stars will exhibit anything but a normal binary frequency.

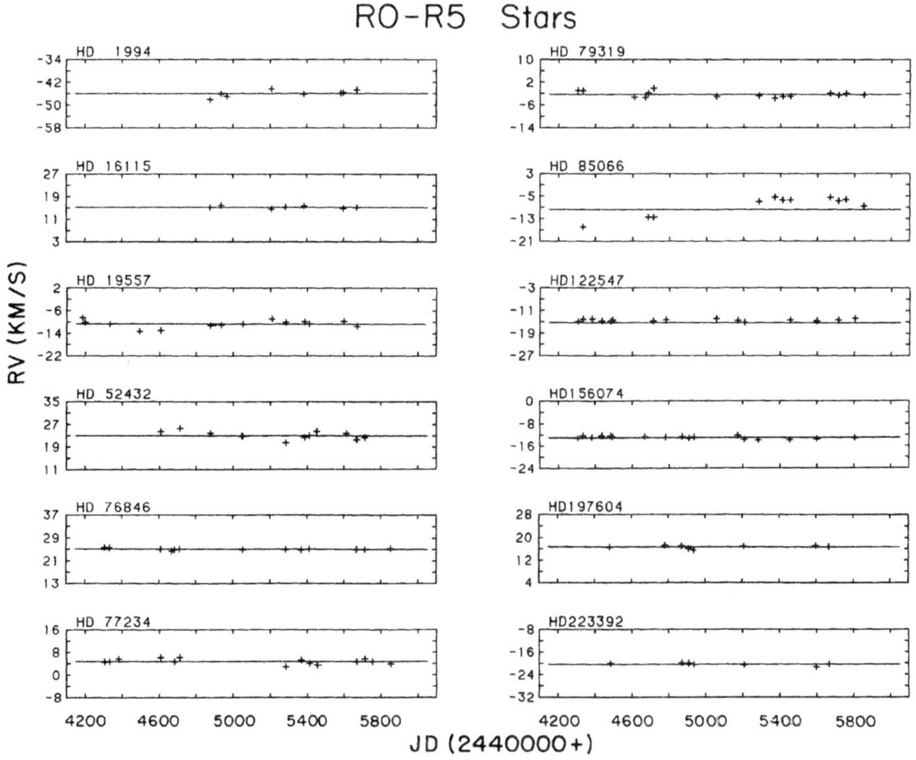

Figure 5. Same as Figure 2 for RO-R5 carbon stars.

2.5 Summary

The data on multiplicity of peculiar cool stars is summarized in the final column of Table 1. It appears that those stars for which s process elements are enhanced, the BaII and CH stars are binary systems, whereas those stars in which s process elements are not enhanced are not binary systems.

3. RELATIONSHIP BETWEEN MULTIPLICITY AND PECULIAR ABUNDANCES

For the stars discussed here, the luminosities indicate that most must be fainter than stars in the double shell phase of the AGB. For those which show enhancements of s process elements it appears that they are binary systems. There are perhaps five possible explanations for this result as follows:

1) There has been mass transfer from a secondary which was once a more massive star that underwent helium shell flashing on the AGB and mixed carbon and s process elements to the surface.
2) Tidal effects from a companion has caused the BaII or CH star to mix.
3) The BaII or CH star has lost mass onto a companion, thereby exposing interior layers that have been enriched in heavy elements.
4) A primordial effect has caused both the BaII or CH star to mix heavy elements to the surface, and has caused the stars to form in binary systems.
5) There is no relationship between multiplicity and the peculiar abundance in these stars.

Since all classical BaII stars appear to be binaries it seems that possibility number 5 above can be ruled out. Possibility number 4 cannot be ruled out, but it will be very difficult to prove. It is conceivable that excess angular momentum could cause both the formation of a binary system and the mixing of a star, but this is just speculation. If possibility number 3 were operating, it would seem reasonable that many BaII stars would have companions of near equal brightness whereas none have been found. Concerning possibility number 2, it is very difficult to believe that a companion that is separated by more than 2 AU can have any significant tidal effect on the deep interior of the BaII or CH star to cause it to mix.

This brings us to the mass transfer case which is the hypothesis that has been discussed in the literature most often. It is very appealing because mixing of s process elements and carbon has been shown (Iben 1972; Truran and Iben 1977) to be feasible in intermediate mass AGB stars as they undergo helium shell flashes. If the contaminated atmosphere of such a star can be transferred onto its companion, and then the original primary evolves to become a white dwarf this may leave a BaII or CH stars as the new primary. We should observe the evidence for the white dwarf companions in the ultraviolet where the hotter star is brightest. It appeared very promising, therefore when Böhm-Vitense (1980) did find such a UV excess in the IUE spectra of the bright BaII star ζ Cap, and several bright marginal BaII stars (Böhm-Vitense et al. 1984).

There are numerous problems, however, that must be addressed concerning the mass transfer hypothesis. Dominy and Lambert (1983) have obtained numerous IUE spectra of BaII and marginal BaII stars, but found evidence for white dwarf companions in only a few. They argue that the cooling times for white dwarfs are short compared to the lifetime on the giant branch, so that the white dwarfs could not have cooled beyond the observing limit of IUE. Böhm-Vitense et al. (1984) however, sum up the statistics concerning white dwarf companions and note that all BaII or marginal BaII stars of apparent magnitude equal to or brighter than $M_V = 4.7$ have observable companions. This seems to be good reason to suspect the cooling rate argument against the mass transfer hypothesis.

A second problem that at first sight appears to rule out the mass transfer hypothesis is that discussed by Tomkin and Lambert (1979) concerning the observed Mg isotope ratios in the BaII star HR774 which appear normal. If the source of neutrons for s processing is the ^{22}Ne(α,n)^{25}Mg reaction which should operate during thermal pulses in intermediate mass AGB stars, a large overabundance of ^{25}Mg and ^{26}Mg relative to ^{24}Mg should result as pointed out by Scalo (1978).

A third problem that is often discussed is the lack of dwarf BaII stars. If mass transfer occurs, it should occur onto dwarf companions, as well as giants. Although a dwarf star is relatively very small, it has a much smaller envelope to dilute the contaminating material. Why do we not see dwarf BaII stars? Perhaps we just don't recognize them. We do see the sgCH stars some of which may be BaII stars just at the turnoff of the main sequence. If we were to see a later main sequence star with enhanced abundance of carbon and s process elements it might not look much different than a normal star. The violet CN band whose strength is a signature of a giant BaII star, is not normally seen in dwarf stars due to the high surface gravity. Similarly the SrII and BaII lines are very surface gravity dependent, and if these elements were enhanced it might be a small effect, not easily interpreted. The star might just look slightly evolved.

Finally, the recent observations of Lithium abundances in BaII stars by Pinsonneault et al. (1984) presents a difficulty for the mass transfer hypothesis (Trimble 1984). Even though Li is easily destroyed in all but the coolest outer layers of an evolving star, the BaII stars have the same Li abundance as normal K giants of the same mass (1-2 $M_{\odot}$). Thus the material transferred from the companion would have to have had a Li/H ratio appropriate to the mass of the recipient rather than the donar which is hypothesized to have been more massive.

Some of the above problems appear to be formidable. On the other hand none of the other possibilities appear to be as attractive. With mass transfer the production of s process elements and mixing is an understood phenomenon. With the other four possibilities we can only speculate that these processes occur.

4. FURTHER INVESTIGATIONS

McClure (1984) discussed the possibility that there may be a continuous gradation of BaII stars which have enhanced s process elements and slightly enhanced carbon through the CH-like stars which have stronger carbon features to the hotter R stars (RO-R3) which have strongly enhanced carbon but no s process element enhancement. The R stars now appear to have a normal frequency of spectroscopic binaries. It now seems likely, therefore that the R stars are the result of a very different process of stellar evolution than the BaII stars. The conclusion of Dominy (1984) that the abundance anomalies are probably due to the result of mixing at the helium core flash appear to be more likely for the R stars.

The results reported here leaves the question of whether other types of stars that have s process enhancement need be binary systems too. Are the CH-like stars just BaII stars that have a C/O ratio somewhat larger? Or, do the least luminous S or SC stars form a cool extension to the BaII star sequence. If so, these groups of stars should also show evidence for binary motion, and it will be important to begin velocity observations of them to look into these possibilities. On the other hand, the brighter and cooler S stars and N stars that also show enhancement of s process elements are light variables. To determine whether these are binaries would be a formidable task because of the unstable nature of their atmospheres, and velocity variations because of this.

Finally, the question of the mild or marginal BaII stars needs to be investigated. Many of the so called marginal BaII stars may not be anomalous at all. Others may just be stars in which there is a small primordial enhancement of heavy elements, while still others may be true but mild examples of the classic BaII stars. A detailed investigation into the anomalous nature of individual marginal BaII stars combined with radial velocity observations will be a necessary step into determining their place within the groups of cool stars with excesses of heavy elements.

5. REFERENCES

Bidelman, W.P., and Keenan, P.C. 1951, Ap.J., **114**, 473.
Böhm-Vitense, E. 1980, Ap.J. (Letters), **239**, L79.
Böhm-Vitense, E., Nemec, J., and Proffitt, C. 1984, Ap.J., **278**, 726.
Bond, H.E. 1974, Ap.J., **194**, 95.
Branch, D., Bonnell, J., and Tomkin, J. 1978, Ap.J., **225**, 902.
Dominy, J.F. 1984, Ap.J. Suppl., **55**, 27.
Dominy, J.F., and Lambert, D.L. 1983, Ap.J., **270**, 180.
Eggen, O.J. 1972, M.N.R.A.S., **159**, 403.
Fletcher, J.M., Harris, H.C., McClure, R.D., and Scarfe, C.D. 1982, Pub. A.S.P., **94**, 1017.
Griffin, R.F. 1967, Ap.J., **148**, 465.
Griffin, R.F. 1982, Observatory, **102**, 82.
Harris, H.C., and McClure, R.D. 1983, Ap.J. (Letters), **265**, L77.
Harris, W.E., and Racine, R. 1979, Ann. Rev. Astr. Ap., **17**, 241.
Iben, I. Jr. 1975, Ap.J., **196**, 525.
Janes, K.A. 1975, Ap.J. Suppl., **29**, 161.
Keenan, P.C. 1942, Ap.J., **96**, 101.
Luck, R.E., and Bond, H.E. 1982, Ap.J., **259**, 792.
MacConnell, D.J., Frye, R.L., and Upgren, A.R. 1972, A.J., **77**, 384.
McClure, R.D. 1970, A.J., **75**, 41.
McClure, R.D. 1983, Ap.J., **268**, 264.
McClure, R.D. 1984, Pub. A.S.P., **96**, 117.
McClure, R.D., and Forrester, W.T. 1982, Pub. Dom. Str. Obs., **15**, 439.
McClure, R.D., Forrester, W.T. and Gibson, J. 1974, Ap.J., **189**, 409.
McClure, R.D., Norris, J. 1977, Ap.J. (Letters), **216**, L101.
Pinsonneault, M.H., Sneden, C., and Smith, V.V. 1984, Pub. A.S.P., **96**, 239.
Roman, N.G. 1952, Ap.J., **116**, 122.
Scalo, J.M. 1978, Ap.J., **221**, 627.
Schmitt, J.L. 1971, Ap.J., **163**, 75.
Smith, V.V. 1984, Astr. Ap., **132**, 326.
Sneden, C., Lambert, D.L., and Pilachowski, C.A. 1982, Ap.J., **247**, 1052.
Tomkin, J., and Lambert, D.L., 1979, Ap.J., **227**, 209.

Trimble, V. 1984, submitted to Nature.
Truran, J.W., and Iben, I. Jr. 1977, Ap.J., 216, 797.
Warner, B. 1965, M.N.R.A.S., 129, 263.
Yamashita, Y. 1972, Ann. Tokyo Astr. Obs., 13, 169.

DISCUSSION

JØRGENSEN - On the discrepancy between s-process enrichment observed in Ba stars and theory, I would like to draw the attention to a series of papers by Nørgard during recent years, where he points out that C^{12} (p,α) N^{13} (e^+, ν) C^{13} can occur under different circumstances in double shell flashing stars. These estimates may give better agreement with observation of Ba stars than more commonly used models of double shell flashing models.

LLOYD EVANS - ω Cen contains stars which, if seen in the field, would variosly be described as Ba or S stars. These appear to be at the extreme upper end of a wide range of metal abundance in the cluster, with an unusual high ratio of S/Fe. These are presumably not binaries.

BIDELMAN - If the important thing is the heavy element abundance, rather than just that of C, as suggested by the fact that the R stars are not all binaries, one should investigate those N stars that show large overabundances of those elements. Further how certain can one be that the hot companions that we have observed are all truly white dwarfs ? Also one must remember that not all of radial velocity variations that you find need be the result of binary motion. Finally one of your objections to the "induced" tidal effect idea was that the two stars are too far apart. But when the present secondary was the primary, the stars were much more really "close binaries" than now. Further, have the orbital elements remained constant ?

McCLURE - I have kept clear of observing bright carbon stars (N stars) because they are light variables. I do not like the idea trying to sort out velocity variations due to a binary from variations due to the stellar atmosphere pulsating.

I think the ultraviolet excesses observed in several Ba stars with IUE has to be due to white dwarfs. It cannot be from main sequence stars because these would be near the brightness of the Ba stars themselves and would be much more obvious in the spectra. I think that if the velocity variations were not due to binary motions you would see light variations, but I have only observed stars of the type that do not show such variations. Landolt has recently checked this for Ba stars and finds they are definitely constant in luminosity.

JASCHEK - What is the normal percentage of the spectroscopic binaries in normal giants ? Do they differ in the distribution of orbital parameters ?

McCLURE - We have been observing a sample of 40 K-type giants at the Dominion Astronomical Observatory for the last five years or so. These include all the K stars classified as luminosity class III in a certain region of the sky, and included in the "Bright Star Catalog". As I remember we found about 20 % of these are binaries. I think Griffin has found a similar percentage from his radial velocity observations over the years. I cannot remember what the distribution in periods etc... is.

CATCHPOLE - I think it would be well worth observing the S stars. It would be important to obtain photometry, so that you could be confident you had a binary, by checking the periods.

McCLURE - Are there any volunteers to do the photometry ?

KEENAN - If we are right in our suggestion that the Ba stars have more carbon than do the S stars, then the cooler stars to be examined for variable velocity would be the stars at the carbon end of the SC sequence.

CAYREL - Is there any correlation between the Ba spectroscopic strength anomaly (strong s-elements) and the value of the semimajor axis found in your orbits ?

Secondly, can you state that the proportion of the CN stars which are spectroscopic binaries is indeed significantly low with respect to the normal stars ?

McCLURE - With regard to the first question, I really need several more years of observation in order to obtain a more complete distribution of periods and amplitudes of velocity curves. If one ommitted HD 77247 which is a Ba 1 star, and which has by far the shortest period,there mightbe a hint that Ba anomaly increases with decreasing semimajor axis. With regard to the second question, again we need to observe these for a couple more years to say anything certain. The frequency of binaries among the CN stars seems at this point to be exceedingly low and it will be very interesting if this holds up with additional observations.

RICHER - I think your first point is no longer a problem. The Ne^{22} mechanism may not operate in the carbon stars one know about. For example the ones in the Clouds do not seems luminous enough to have had the Ne^{22} mechanism operating.

MALANEY - It is your opinion that large scale mixing at the helium core flash is no longer a valid explanation for the production of CH subgiants ? How many of these stars have you observed, and how many were found to be variable ?

McCLURE - No, I cannot say this for certain. Out of ten CH subgiants I found two so far to have certain variability, but most others show signs of variability and more observations are needed.

CATCHPOLE - I would just like to point out that Luck and Bond (Ap J **259**, 792 - 1982) find spectroscopic gravities that place the subgiant CH stars on the main sequence. Could this be related to the problem of finding main sequence Ba stars ?

McCLURE - The subgiant CH stars are all F and early G stars. What Bond worries about is that there are none outside this range, whereas if main sequence Ba stars exist, you should find stars of this sort at later spectral types.

WOOD - The bipolar planetary nebulae are often carbon rich, and binarity has been suggested as the cause of the bipolar nature of the nebulae. These objects may well be carbon stars transfering part of their mass onto companions that will later be recognized as Ba and other peculiar stars.

GRIFFIN R.E.M. - In view of the apparent lack of low luminosity Ba stars, would Prof. Keenan kindly comment on the fact that stellar luminosities are largely estimated from the strengths of Sr II lines which are closely linked to Ba II line strengths ?

KEENAN - I think that we would not be likely to classify a dwarf G8 to K5 Ba star as a subgiant or giant, because the general appearance of a dwarf spectrum (very strong Ca 4227, weakness of λ 4326, etc...) is quite different from that of a more luminous star. In other words, inconsistences of the usual luminosity criteria would tell us that something was wrong. On the other hand I could not guarantee that a subgiant B star might not occasionally be mistaken for an ordinary giant.

LU - From radial velocity period can you determine orbital elements and masses of the components ?

McCLURE - We have not enough observations to determine the mass, but very probably the mass of the companion of Ba stars is small.

LU - Perhaps to sort out weak from strong Ba stars the (48 - 41) vs. (41 - 42) diagram of the DDO system can be used, specially for those stars classed by Keenan and independently observed by photometry.

COWLEY - A good way never to find dwarf Ba stars is to call them something else, like subgiants (or metallic line stars !).

THE FIRST BINARY ORBIT FOR AN S-TYPE STAR — HR 1105

R.F. Griffin
The Observatories
Madingley Road
Cambridge, England

The fifth-magnitude object HR 1105 is the brightest genuine S-type star. The variability of its radial velocity has been known since 1922, the year in which type S was first recognized. Its orbit could have been found, to a first approximation, half a century ago merely by inspection of the published radial velocities. However, photoelectric measurements made in the last few years, mainly with the original radial-velocity spectrometer in Cambridge, have now allowed a good orbit to be determined. The period is 596 days and the orbit is nearly circular, with an amplitude of $8\frac{1}{2}$ km/s.

It is not known whether the incidence of duplicity among S stars is abnormal, as it is among barium stars; but since the causes of the S-type spectrum appear to be safely established there seems no reason to expect S stars to show abnormalities in properties unrelated to their evolutionary situation.

Details of this work will be published in the 1984 October issue of The Observatory Magazine as Paper 58 in the series on 'Spectroscopic Binary Orbits from Photoelectric Radial Velocities'.

DISCUSSION

LLOYD EVANS – Does the orbital radius for HR 1105 give a maximum stellar radius in agreement with values which would follow from currently known luminosities for S-stars ?

GRIFFIN – I think so ; the period is much longer than the periods of many other systems containing late type giants.

M. Jaschek and P. C. Keenan (eds.), Cool Stars with Excesses of Heavy Elements, 331.

A SEARCH FOR COMPANIONS TO IRREGULAR VARIABLES OF TYPE S

Benjamin F. Peery, Jr.
Howard University
Department of Physics and Astronomy
Washington, DC 20059
United States of America

ABSTRACT. IUE archival spectra of S-type stars that may be cool analogues of Ba II stars have been inspected for evidence of hot companions. Although the currently accessible data for 9 stars are of marginal quality, spectra of the brightest candidate, HR 1105, show strong evidence for the presence of a hot companion of low luminosity.

1. RATIONALE FOR THE RESEARCH

The properties of some stars of type S suggest that they may indeed be cooler analogues of the Ba II stars, an early speculation of Burbidge and Burbidge (1957). Like the Ba II stars, their spectra show overabundances of elements favored by s-process production, but with faint indications of the presence of technetium, if at all (e.g., Figure 1). The light of these S-type stars appears to be characterized by irregular fluctuations of the order of a tenth of a magnitude, or less (Eggen 1972); the strength of technetium lines thus conforms to its coarse anti-correlation with stellar variability reported by Peery (1971). Keenan (1954) compiled an early list of stars of type "MS" that may be weakly s-processed-enhanced members of the group of stars under consideration here.

Some of the growing information about Ba II stars prompts one to search for evidence for duplicity among the subgroup of S- and MS-type stars considered herein: reports of the high frequency of occurence of radial-velocity variations among Ba II stars; IUE observations of 16 Ba II stars to date, of which at least four reveal hot, degenerate companions; and the intriguing suggestion that a Ba II star is characteristically a member of a binary system with a degenerate companion, to which the Ba II star owes its heavy-element peculiarities through earlier mass exchange with its evolved companion (e.g., McClure 1984; however, see also the contrary views of Dominy and Lambert (1983)).

M. Jaschek and P. C. Keenan (eds.), Cool Stars with Excesses of Heavy Elements, 333–336.

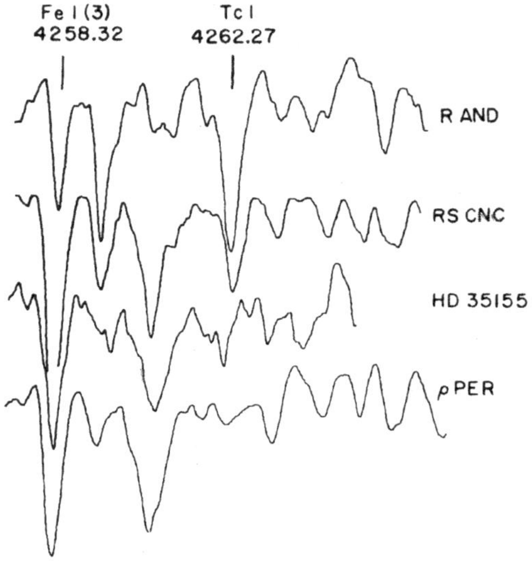

Fig. 1. - Spectra of heavy-element stars in the region of the 4262 A line of Tc. R And: S6,6e, LPV; RS Cnc: M6S, irregular variable; HD 35155: S4,1, non-variable; Rho Per: M4 II-III, no Tc, included for comparison.

Hackos and Peery (1968) have shown that the M4S star 57 Peg has a companion of spectral type approximately A3 (Figure 2). Recently, Johnson and Ake (1983) have reported that IUE spectrograms of HD 35155 (S3,2) show strong emission lines of such ions as C IV and N V. One of their interpretations of the data is that HD 35155 has a hot collapsed companion, analogous to the Ba II binary systems Zeta Cap (Bohm-Vitense 1980) and 56 Peg (Schindler et al. 1982).

TABLE I

Star	Type
HR 1105	S3
HD 35155	S2
HD 49368	S3
HD 58881	S4
HD 121447	S2
HR 6702	S6
HD 198164	S0
HR 8714	S4
57 Peg	M4S

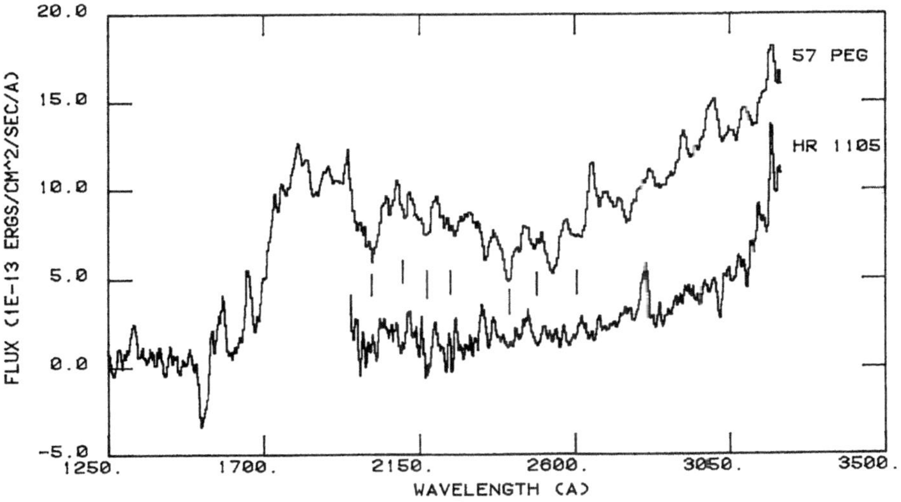

Fig. 2. - Low dispersion IUE spectrograms of 57 Peg and HR 1105. The A-type companion dominates the spectrum of 57 Peg shortward of the Mg II emission at 2800 A. Correspondences between spectral features are indicated in the shorter wavelength region where contributions from the cool components are negligible. (Observation of 57 Peg by R. F. Wing; observation of HR 1105 by T. B. Ake.)

2. RESULTS

I have inspected the spectra of the 9 candidate S- and MS-type stars that are presently in the archives of the IUE satellite (Table 1). The spectrograms are generally quite noisy, and further observations must be made. A spectrogram of the bright candidate star HR 1105 is the best of the lot, and it has been added to Figure 2 for comparison with that of 57 Peg.

Although the spectrum of HR 1105 at first appears to be largely noise at wavelength shortward of the blended Mg II lines at 2800 A, more careful scrutiny reveals correspondences with features in the spectrum of the hot companion of 57 Peg. It appears that HR 1105 has an underluminous companion with a temperature comparable to that of a star of spectral type A.

I am indebted to Drs. Hollis Johnson and Thomas Ake for informative discussions about these stars. It is a pleasure to record my thanks to the staff of the Regional Data Analysis Facility at Goddard Space Flight Center for their skillful assistance. This research has been supported by NASA grant NAGW-545.

3. REFERENCES

Bohm-Vitense, E. 1980, Ap. J. (Letters), 239, L79.
Burbidge, E. M., and Burbidge, G. R. 1957, Ap. J., 126, 357.
Dominy, J. F., and Lambert, D. L. 1983, Ap. J., 270, 180.
Eggen, O. J. 1972, Ap. J., 177, 489.
Hackos, W., Jr., and Peery, B. F., Jr. 1968, A. J., 73, 504.
Johnson, H. R., and Ake, T. B. 1983, Proceedings of the Third Cambridge Conference on Cool Stars, Stellar Systems, and the Sun.
Keenan, P. C. 1954, Ap. J., 120, 484.
McClure, R. D. 1984, Pub. Astr. Soc. Pacific, 96, 117.
Peery, B. F., Jr. 1971, Ap. J. (Letters), 163, L1.
Schindler, M., Stencel, R.E., Linsky, J. E., Basri, G. S., and Helfand, D. J. 1982, Ap. J., 263, 269.

DISCUSSION

WING - Speaking of possible companions of HR 1105, has any one noticed that the M2 II star HR 1155 has the same visual magnitude and radial velocity and is about one degree away in the sky ?

PEERY - I have not looked at it, and I am unfamiliar with it.

BIDELMAN - I have not noticed definitive changes in ZrO bands in 57 Peg, but on some of our older plates the seemed to be at least weakly present.

THE PROPORTION OF BINARIES HAVING A DEGENERATE COMPANION : IMPLICATIONS ON THE FORMATION OF BARIUM STARS

J.L. Halbwachs
C.D.S. - Observatoire de Strasbourg
France

SUMMARY : - The theoretical proportion of barium stars was derived from the frequency of binaries with degenerate secondary components, using a suggestion by Mc Clure et al (1980). It appears that the proportion of Ba II stars should be of about 0.5 % among dwarfs and more than 1 % among giants. The frequency of dwarf stars among Ba II stars later than F6 should be about 5 %.

1. INTRODUCTION

Mc Clure et al (1980) have suggested that barium stars are produced in binary systems in the following way : When the primary component reaches the end of the giant stage, the extension of its atmosphere becomes so large that it overflows its Roche lobe ; mass transfer occurs and the barium cooked in the interior of the star finally appears in the atmosphere of the secondary component. The initial secondary becomes a Ba II star and the initial primary becomes a degenerate object.

This model was actually checked on two points : a) Mc Clure et al (1980) and Mc Clure (1983) have found that many Ba II stars are spectroscopic binaries with low mass ratios and b) Böhm-Vitense et al (1984) have detected evidences for degenerate companions in the spectra of some Ba II stars. In this last paper was also discussed the proportion of Ba II stars that can be expected, and the apparent lack of Ba II stars among dwarfs was point out. However it was not considered to be significant and this point will be re-considered here.

According to Mc Clure's model, the proportion of Ba II stars depends on : a) the proportion of binary systems with separations such that mass transfer occurs at the end of the life of the primary and b) the proportion of binaries with a degenerate component. These two points will be considered separately.

2. THE PROPORTION OF BINARIES WITH A DEGENERATE COMPONENT

2.1. Definitions and fundamental data

A sample of binaries with different ages may contain :

M. Jaschek and P. C. Keenan (eds.), Cool Stars with Excesses of Heavy Elements, 337–343.

2.1.1. - "Young" systems constituted by dwarf or giant components ; the primary (= brighter) components are the more massive ones ;

2.1.2. - Systems for which the initial primary has become a degenerate object : these systems may be Ba II stars ; the primary components are now the initial secondaries ;

2.1.3. - Systems with two degenerate components ; this case is without importance for the problem treated here, because these objects are practically unobservable.

We want to estimate "R", the frequency of systems with one degenerate component (case b) among all binaries with dwarf or giante primaries (cases a and b). We shall also consider "r", the ratio between the systems with one degenerate component (case b) and the "young" systems (case a) ; R may be derived from r by the relation :

$$R = r / (1 + r) \qquad (1)$$

"r" depends on the stellar birthrate, the initial mass function, the duration of the life of the stars and the initial mass ratios of the systems. We shall now develop a model based upon different assumptions for these four functions.

- The stellar birthrate is assumed to be constant since the origin of the galactic disk, which is 10.10^9 years old (according to Neckel, 1975).

- The masses of the initial primaries obey the Salpeter's law (1955) : $\Psi\ (M)\ \alpha M^{-2.35}$. This law was verified for instance by Miller and Scalo (1979) for stars with masses lying between 0.5 and 10 M_o.

- Stars with initial mass M are on the main sequence during the time : $T_V\ (M) = 10\ \ M^{-3.25}\ 10^9$ years ; after that, they are giant during $T_{III}\ (M) = 1.5\ M^{-2.75}\ 10^9$ years (following Neckel (1975) and according to the results of Mengel et al).

With regard to the forth function the situation is more complex and we shall consider it in the next section.

2.2. Dependence of "R" upon the initial mass ratios

In order to derive analitically R for a sample of binaries with the same mass ratio, an approximation is needed for the expression of the duration of the complete life of the stars : $T\ (M)\ \alpha\ \ M^{-3.25}$. Moreover, only primaries with masses so heavy that their lifes are shorter than the age of the galactic disk are considered. According to our assumptions, this limit is 1.045 M_o. Under these conditions, it is easy to derive r (q) for all binaries together as well as for dwarf or giant systems separately.

2.2.1. - Binaries with giant and dwarf primaries. In such case, one obtains easily :

$$r\ (q) = q^{1.35} - q^{4.60} \qquad (2)$$

2.2.2. - Binaries with dwarf primaries. I assume that the total life of the stars is 1.2 longer than the time spent on the main sequence. This implies :

$$r(q) = \mathrm{Max}\,(0\ ,\ q^{1.35} - 1.2\ q^{4.60}) \qquad (3)$$

2.2.3. - Binaries with giant primaries. I assume now that the duration of the giant stage is 15 % of all the life of the stars (this is the case for stars with M=1.25 $M_{\odot}$, which are the progenitors of the majority of the giants according to Neckel) ; then :

$$r(q) = q^{1.35}\ \mathrm{min}\,(0.15\ ,\ 1 - q^{3.25})\ /\ 0.15 \qquad (4)$$

The frequencies R derived from equations 1 to 4 are plotted in figure 1.

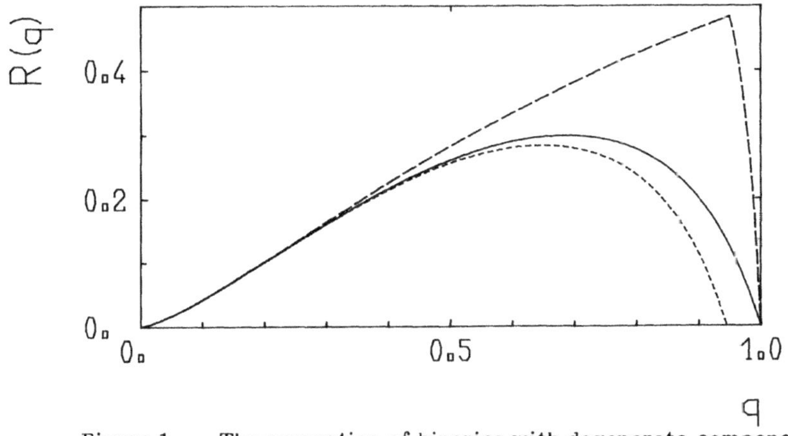

Figure 1. - The proportion of binaries with degenerate components related to the mass ratios ; - unbroken line : case of binaries with dwarf or giant current primaries, considered together ; - dashed line : case of binaries where the current primaries are giants ; - dotted line : case of binaries where the current primaries are dwarfs.

It appears that the proportion of systems with degenerate components is maximum for binaries with mass ratios in the range (0.3, 0.9).

We may now return to our main purpose, which is the proportion of stars with degenerate companions among binaries with all mass ratios.

2.3. Dependence of R upon the spectral types of the current primaries

R may be computed now from the fundamental data given above, completed by the distribution of mass ratios found by Halbwachs (1983). This calculation, made by sampling experiment, provides the results of table 1. Instead of listing the results as a function of mass, I have preferred to list them by spectral types. These were derived from masses by means of the relation given by Schmidt-Kaler (1982).

TABLE 1

Spectral types of the current primaries	Proportions of systems with one degenerate component
A5 - F0 V	17 %
F0 - G2 V	17 %
G2 - K0 V	15 %
K0 - M0 V	10 %
Late - type giants	25 %

3. THE PROPORTION OF BINARIES WITHIN A CONVENIENT RANGE OF SEMI - MAJOR AXES

In order to produce a barium star, a binary must have components separated that much that mass transfer cannot occur when the primary is ascending the giant branch for the first time ; according to Sweigart and Gross (1978), the radius of a star at the top of the giant branch is always about 0.67 AU. On the othar hand, the primary component of a binary fill in its Roche lobe when its radius, R_1, the separation between the components, a_{min}, and the mass ratio $q = M_2/M_1$, are such that

$$a_{min} = R_1 / (0.38 - 0.2 \log q) \qquad (1)$$

(see Heintz, 1978). As we have see above, stars with degenerate companions are produced by binaries with initial mass ratios lying usually between 0.9 and 0.3. For both values a_{min} differs little, so that I have used an average value of a a_{min} = 1.5 AU.

The separation between the components must however be lower than an upper limit, otherwise the overflow of the Roche lobe, at the end of the life of the primary, is not efficient enough to transfer heavy elements like barium to the secondary component. The value of this limit is difficult to derive from physical assumptions, but may be estimated by considering the Ba II SBs found by Mc Clure. The maximum period of these systems is about 1700 days, and it seems then that the maximum separation to consider is a $_{Max}$ = 4 AU. Since the detection of SBs with long periods is rather difficult, this value may however be an underestimate.

Having chosen the limits in semi-major axes, the proportion of binaries lying between them may now be estimated. Be assuming the distribution of semi-major axes found by Heintz (1969), the proportion is about 7 %. On the other hand, Halbwachs (1983) has found that the proportion of wide binary or multiple systems among primaries and single stars is 60 % and that half of the remaining "single" stars are in fact spectroscopic binaries. The actual proportion of binaries or multiple systems is therefore 80 % and the proportion of stars that will become Ba II stars should be 6 %.

4. THE THEORETICAL PROPORTION OF BARIUM STARS

Since the ratio of stars that will become Ba II stars is 0.06 while the proportion of dwarf binaries that actually contain a degenerate companion is about 10 to 17 % the proportion of Ba II stars among dwarf should be about 0.5 to 1 % ; this is rather more than the 0.1 % derived by Böhm et al from the study of the degenerate companions of Ba II stars.

For giants, the proportion of binaries with degenerate components is somewhat larger and the predicted proportion of barium stars is finally 1.4 %. This result agrees well with the observed frequency, which is about 1 % according to Mac Connell et al (1972).

Thus it seems that Mc Clure's hypothesis explains what has been observed, but it must be stressed that the hypothesis also predicts a certain number of dwarf Ba II stars which have not been observed. Therefore Mc Clure's hypothesis can only be fully accepted if the so far unobserved dwarf Ba II are discovered. The proportion of dwarfs among Ba II stars may be estimated in the following way : According to the Bright Star catalogue (Hoffleit and Jaschek, 1982) and its supplement (Hoffleit, 1983), the number of giants brighter than 6.5 mag. and of spectral type later than F6 is about 7 times larger than the dwarf within the same range of spectral types and magnitudes. The proportion of dwarfs among Ba II stars later than F6 should thus be about 5 %. This percentage is uncomfortably large, as compared with the 0 % detected.

ACKNOWLEDGEMENT

It is a pleasure to thank prof. C. Jaschek for suggesting the subject and for valuable discussion. Many thanks are also due to M.M. Holtzer and Dr M. Jaschek.

REFERENCES

Böhm-Vitense E. - Nemec J. - Proffitt C. : 1984, Astrophys. J. **278**, 726
Halbwachs J.L. : 1983, Astron. Astrophys. **128**, 399
Heintz W.D. : 1978, in "Double stars", Reidel - Dordrecht-Holland
Heintz W.D. : 1969, J. Roy. Astron. Can. **63**, 275
Hoffleit D. : 1983, Supplement to the Bright Star Catalogue, Yale University Observatory
Hoffleit D. - Jaschek C. : 1982, the Bright Star Catalogue, 4th ed. , Yale University Observatory
Mac Connell D.J. - Frye R.L. - Upgren A.R. : 1972, Astron. J. **77**, 384

Mc Clure R.D. : 1983, Astrophys. J. **268**, 264
Mc Clure R.D. - Fletcher J.M. - Nemec J.M. : 1980, Astrophys. J. **238**, L 35
Mengel J.G. - Sweigart A.V. - Demarque P. - Gross P.G. : 1979, Astrophys. J. Suppl. Ser. **40**, 733
Miller G.E. - Scalo J.M. : 1979, Astrophys. J. Suppl. Ser. **41**, 513
Neckel H. : 1975, Astron. Astrophys. **42**, 379
Salpeter E.E. : 1955, Astrophys. J. **121**, 161
Schmidt-Kaler T. : 1982, in Landolt Börnstein VI 2b Springer Verlag
Sweigart A.V. - Gross P.G. : 1978, Astrophys. J. Suppl. Ser. **36**, 405

DISCUSSION

CATCHPOLE - Do the semi-major axes you assume agree with the separations that would be predicted from Mc Clure's periods ?

HALBWACHS - Yes, the two limits of the range of the semi-major axis that I used agree quite well with the lower and upper periods found by Mc Clure.

Mc CLURE - I would think that a dwarf Ba star would be difficult to detect spectroscopically. The features one usually look at in a Ba star are very surface dependant. CN is not usually seen in a dwarf in blue spectra, and if one saw a spectrum with slightly enhanced Sr and Ba you might just mistake it for a subgiant.

GRIFFIN R.F. - I had like to make the point that the statistics of binary stars with separations of several astronomical units, and therefore periods of several years, are not at all well known. These ranges correspond to the gap which has existed historically between spectroscopic and visual binaries. Some well known Ba and mild Ba stars probably fall in the gap - for example 16 Ser and ζ Cyg have periods between 15 and 20 years.

TUTUKOV - Why the duplicity of giants exceeds the duplicity of main sequence stars, although not all main sequence stars in binaries can evolve into giants ?

HALBWACHS - It is a fact that a very close binary can not produce a system constituted by a giant and a white dwarf, the proportions in section II were derived by neglecting all proximity effects - except the fact that the component may become a Ba star when the primary dies. These results are thus restricted to the case of binaries with wide separations ; their application to the problem of Ba stars is already an approximation.

LU - I have two comments :
- A four meter speckle interferometer study will be carried out in a near futur to observe strong and weak Ba stars in order to assess binary frequency of Ba stars ;

A small number of these stars will be further observed for orbital elements. The numbers of binaries you proposed with a giant primary component that have a degenerate secondary can then be tested.

– My second comment : The dwarf Ba stars predicted by this study and others may be there but, simply spectroscopists have not searched for them. However it is agreed that they are not numerous.

FORMATION OF HELIUM STARS WITH EXTENDED ENVELOPES IN BINARIES

A.V. Tutukov
Astronomical Council of the USSR - Academy of Sciences - USSR

I. Iben
Department of Astronomy - University of Illinois - U.S.A.

ABSTRACT : - We present a short preliminary of some results related to the formation of some helium rich stars.

Evolution of binaries of initial masses of components 3 ÷ 12 $M_\odot$ has been computed taking into account the mass exchange between components. Evolution has been studied beginning from the main sequence up to the formation of the cooling helium, carbon-oxygen or oxygen-neon dwarf. This report is a preliminary discussion of some results related to the formation of at least some helium rich stars.
The evolutionary track of the 7 $M_\odot$ primary is displayed on figure 1 as an example

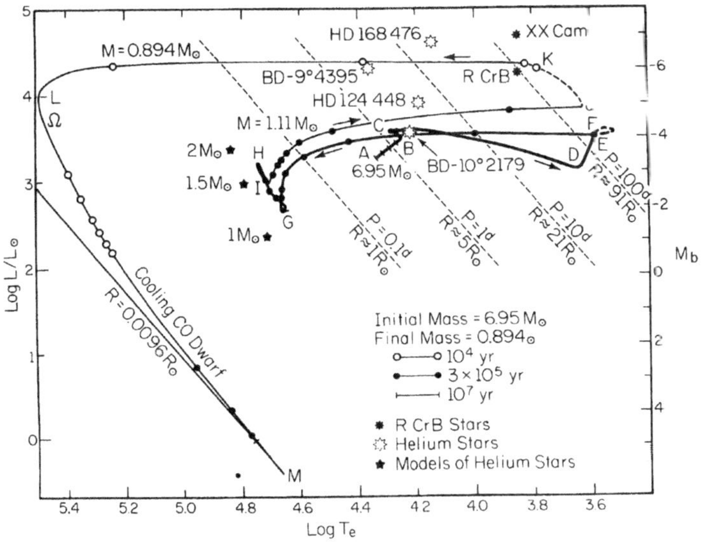

Fig. 1. The L-Te diagram of the ∿ 7 $M_\odot$ close binary component. The position of several helium and R Cr B stars is marked.

M. Jaschek and P. C. Keenan (eds.), Cool Stars with Excesses of Heavy Elements, 345–347.

In the course of evolution of binary components with initial masses 4 - 12 $M_\odot$, helium stars with masses 0.60 - 2.50 $M_\odot$ arise. The exhaustion of helium in their cores induces the envelope expansion, so helium rich stars with effective temperatures $3.10^3 - 5.10^4$ K can be formed naturally. Extended helium envelopes have a low (if any) abundance of hydrogen and a specific distribution of CNO elements (see as an example Figure 2).

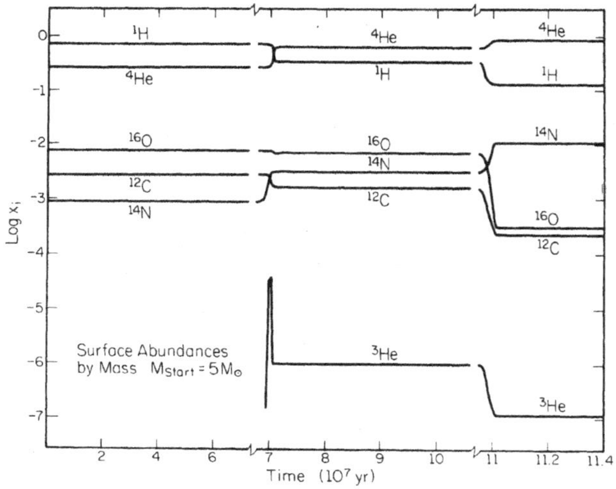

Fig. 2. The evolution of the surface chemical composition of ∿ 5 $M_\odot$ close binary component.

The family of helium models on the HR diagram overlaps well the position of helium-rich stars according to Hunger (1975). The frequency of the formation of helium stars with extended envelopes descended from binaries with initial masses of components ∿ 7 - 12 $M_\odot$ and initial semi-major axis ∿ 20 - 2 000 $R_\odot$ is about ∿ 10^{-2} yr^{-1}. But helium shell sources around the CO core are thermally stable. This probably excludes the enrichment of the helium envelope by carbon, which is necessary to explain R Cr B stars, or by s-elements, which is usual for some of the red giants and supergiants like Ba stars.

There are other ways to form the helium - rich extended envelope. If the semi-major axis of the final close binary consisting of CO and He dwarfs is lower than about 3 $R_\odot$, the radiation of gravitational waves leads to the merging of components. The partial mixing and the shell helium ignition can explain the formation of extended envelopes and, possibly, the overabundance of carbon which is necessary to explain the R Cr B stars.

The accretion of hydrogen - rich matter can also form the helium - rich shell about a CO dwarf. The periodical burning of this helium can also explain the formation of extended helium - rich envelopes of R Cr B stars. The absence of a very close satellite would help to bring about the long ($\sim 10^4$ yrs) nuclear timescale of the helium burning, avoiding the destructive formation of a common envelope.

A part of helium stars with extended envelopes could be a result of evolution of the nuclei of some single planetary nebula during the final helium flash. Since the frequency of the R Cr B star formation is only about 10^{-2} yr^{-1} only about one hundredth part of all planetary nebula nuclei could follow this path. The most promising masses seem to belong to the lightest CO dwarfs with the most massive helium shells to produce the giant size after the shell flash.

The presence of a close satellite gives the possibility of the mass exchange between components. If the mare evolved component is an AGB star with s-elements in its envelope, the mass exchange will enrich the envelope of the "unevolved" component with s-elements also. The accreted mass is determined by the ratio of thermal time scales of the components during the accretion. Thus the giant component can sometimes accrete enough mass to display anomalies in its chemistry. This process possibly may be applicable to barium stars.

The complete result of the work will be published in the Astrophysical Journal.

REFERENCE

Hunger K. : 1975, in "Problems of stellar atmospheres and envelopes"- Eds Baschek B., Kegel W.H. and Traving G. 157 - Springer Verlag.

DISCUSSION

MALANEY - I think it would be interesting to extend this work to the production of heavier elements with perhaps lower mass stars. I do not think that a large PN is expected since this would lead to low luminosity. I would also point out that a few of the hydrogen deficient stars are known to be pulsating. What was the original envelope compositions of your binary system ?

TUTUKOV - Normal main sequence composition.

MIRA'S ROLE IN RR TEL REVISITED

A. Heck[1] & J. Manfroid[2,+]

1 - Observatoire Astronomique, Strasbourg, France
2 - Institut d'Astrophysique de l'Université de Liège, Belgium
+ - Research Associate, National Fund for Scientific Research

ABSTRACT. uvby data, IUE FES measurements and AAVSO visual estimates of the very slow nova RR Tel show that the object has gone through periodic variations in visual light since it started fading in 1949 after the outburst it incurred in 1944. The period is however not constant and varies between 350 and 410 days, confirming Mayall (1957)'s statement and in contradiction with the conclusions of later investigators. Synchronized variations of a Mira in the infrared and of a blue component indicate probably that the Mira mass outflow varies in phase with its pulsations. Large variations occur in visual light on a time scale of a few weeks.

1. INTRODUCTION

RR Tel is a very slow nova for which one outburst has been recorded and whose lightcurve before that outburst is one of the best known. A historical account of the first observations can be found in Heck & Manfroid (1982). Thackeray (1977) gave an excellent synthesis of the successive evolutionary spectroscopic stages in the visible after the outburst. Feast et al. (1983) reviewed infrared data and their implications, while Penston et al. (1983) extensively described ultraviolet spectra collected with the International Ultraviolet Explorer (IUE). Recent interpretations of the nature of RR Tel point to a complex combination involving a Mira variable, a dust component, a hot ionizing source, gaseous emission continua and, possibly, a star of spectral type near A0 or optically thick circumstellar emission from a chromosphere (Feast et al., 1983; Penston et al., 1983).

2. POST-OUTBURST VARIABILITY

Before the outburst (wich occured in November 1944, but remained unnoticed until 1948), the behavior of the star can be appreciated for the years 1889-1947 on the figure published by Mayall (1949) and is described in Heck & Manfroid (1982). Gaposhkin (1945) derived a period of 386.73 days from 40 acceptable pre-outburst observations and this value is to be

M. Jaschek and P. C. Keenan (eds.), Cool Stars with Excesses of Heavy Elements, 349–353.

compared with that obtained by Feast et al. (1983) from JHKL photometric data collected during 1975-1981: 387 days.

The first extensive study of the variability in the visible range for the post-outburst phase has been carried out by Heck & Manfroid (1982, 1984). Fluctuations of the y magnitudes have been found from uvby observations collected with the ESO 50cm and 1m telescopes. From the IUE FES count records (for a description of IUE and the FES, see Boggess et al., 1978 a&b, and Holm & Crabb, 1981), variations of about 0.9m in a band close to Johnson's B have been detected with a period of 395 ± 5 days. Finally, the visual estimates gathered by the American Association of Variable Stars Observers (AAVSO) for the period 1948-1981 show a general decrease in brightness, more pronounced from JD 2,433,000 to 2,435,000, with a couple of "accidents" between JD 2,433,000 and 2,433,200 (increase of $\sim$1m) and between JD 2,438,000 and 2,438,100 (increase slightly larger than 0.5m). For the rest, a periodicity in the variations is clearly visible, even in the steepest descent after the outburst. However the period is definitely variable. According to the intervals considered, the values obtained are distributed between 357 and 410 days.

There were difficulties in determining a period from the observations following the second accident: during about a thousand days, the frequency of the variations seems to have almost doubled. It seems also that the period increased at the time of the outburst during about four cycles. Then it went back to its initial value while following some kind of damped oscillations. The final discrepancy, if we assume an average value of 387 days for the period, amounts to a phase shift of about 70 days of the present variations over the pre-outburst ones. Moreover the various light-curves vary in phase within about ten days.

Such results confirm an almost unnoticed mention by Mayall (1957) of "*a slow variation from 8.9 to 9.3* {apparent visual magnitudes} *with a tendency to a period of 350 to 400 days*" following the outburst, and contradict comments in the literature pointing towards no evidence of variability in the visible range or at least towards no evidence since the outburst of the period found in the infrared (Feast et al., 1983, P. 952; Glass & Webster, 1973, p. 81; Walker, 1977). On the other hand, the steepness of the y variation detected in 1981 and several anomalous values in the AAVSO non-averaged data indicate that large variations can occur in visual light on a time scale of one or two weeks.

3. SOURCE OF THE VARIATIONS

A discussion of RR Tel variability cannot be tackled without keeping in mind that the spectrum of the object displays hundreds of emission lines, some of them very strong. Therefore quite a significant portion of the energy in the range considered in this paper is released through these lines ($\sim$70%, Glass & Webster, 1973, p. 81 - but we suspect this to be highly variable). This has a couple of important implications: first, the likelihood that the main source of the variations in the visible range lie with the lines rather than with the continuum is far from being negligible, and, second, the definition of the passbands (or eye sensitivities in the case of amateur observations) is critical when comparing photometric data from different sources. This last statement is particularly true for Stroemgren's y filter which is fairly free from strong emission lines which, on the contrary, are integrated in eye or plate estimates and thus contribute to a brighter magnitude.

As recalled by Wood & Zarro (1981), several authors pointed out that "*most Mira variables do not have regular periods, but show random cycle-to-cycle variations in period of a few percents as well as abrupt changes of about 2% at intervals of 10-50 years*". Thus the irregularities pointed out in Par. 2 might have been induced by the Mira component, if not expressing its direct contribution to the total visual light.

Commenting the periodicity found in the infrared, Feast et al. (1983) considered the J index compatible with "*a Mira component present in the system with the same period before and after the outburst* ... {and}... *with the hypothesis that this Mira has remained constant at approximately the same level of brightness throughout*". According to the same authors, "*Mayall* {(1949)}*'s lightcurve is consistent with the presence at all times of a Mira component varying from* ~*13m to* ~*17m* {photographic magnitudes} *or fainter, with the addition of another component of radiation whose intensity varies at a different scale*".

However, if the presence of a Mira component in the system seems well established (Webster, 1974; Thackeray, 1977; Allen et al., 1978) through red and infrared spectrographic observations, the continuum in the visible range is not that of an M star. Actually, there is a consensus in the literature that the blue component of the system dominates the wavelength range under consideration in this paper (Feast et al., 1983, p. 952 & 955; Penston et al., 1983, Sect. 4; Thackeray, 1977, p. 28). Therefore, if long-term infrared variations can be attributed to the Mira component directly, it is quite unlikely the case for those reported here, even if they might be related to the Mira through an astrophysical process to be precised when more observational data will be available. The long-term variations we reported here are those of the blue component. Other indications can be found in the fact that their amplitudes keep roughly the same value while the object luminosity decreases. They seem even be slightly smaller in the second part of the curve, contary to the effect expected if the Mira were coming up at that time in that range.

If we assume that the almost perfect synchromism between the variations of the Mira, on one hand, and those of the blue and visual lightcurves, on the other hand, can be extrapolated in the past, we come to the conclusion that the Mira has undergone successive changes in its pulsational behavior (relatively large frequency modifications and possible mode swapping). The link between the Mira and the blue component would be a variable outflow synchronized with the Mira pulsations.

The domination of the blue component and the blue color index ($B-V_{Geneva}$ $\simeq 0.9$) are also consistent with Gaposhkin's judgement (reported by Payne-

Gaposhkin, 1957, p. 157) implying that RR Tel was also a blue object before the outburst. Thus it seems that the blue component has been in the system permanently and has masked completely the variations of the Mira supposed to remain between B ≈ 13 and 17 every time RR Tel has been brighter than about B = 11. This blue component would also be responsible for the suspected fast variations wich might be those alluded by Thackeray (1977, p. 29) and involving a close binary.

4. CONCLUSIONS AND FINAL COMMENTS

Since it started fading in 1949, RR Tel shows periodic variations in the visible range due to its blue component. The period varied approximately between 350 and 410 days, confirming Mayall (1957)'s statement. Besides a general decrease of brightness, the lightcurve is also marked by two sudden increases of the visual magnitude lasting about a hundred days each. The synchronism between the variations of the blue and red components might also indicate that the mass outflow of the Mira varies in phase with the pulsational period.

There are also fast variations in y possibly due to the binary nature of RR Tel. Regarding the long-term variations, a comparison with the infrared data cannot rule out the possibility of a phase shift of less than 10 days between both sets of data for the common time coverage. Additional spectroscopic and photometric observations are necessary to establish more clearly the respective contributions of the emission lines and of the continuum to the short and long-term variations of RR Tel, as well as to adopt a satisfactory model precising the link between the Mira component and the variations in the visible range.

REFERENCES

Allen, D.A. et al. 1978, Monthly Not. Roy. Astron. Soc. 182, 57P

Boggess, A. et al. 1978a, Nature 275, 372

Boggess, A. et al. 1978b, Nature 275, 377

Feast, M.W. et al. 1983, Monthly Not. Roy. Astron. Soc. 202, 951

Gaposhkin, S. 1945, Harvard Ann. 115, 22

Glass, I.S., Webster, B.L. 1973, Monthly Not. Roy. Astron. Soc. 165, 77

Heck, A., Manfroid, J. 1982, Messenger 30, 6

Heck, A., Manfroid, J. 1984, in preparation

Holm, A., Crabb, W. 1981, NASA IUE Newsl. 15, 74

Mayall, M.W. 1949, Harvard Bull. 919, 15

Mayall, M.W. 1957, J. Roy. Astron. Soc. Canada 51, 4

Payne-Gaposhkin, C. 1957, Galactic Novae, ed. North-Holland, Amsterdam

Penston, M.V. et al. 1983, Monthly Not. Roy. Astron. Soc. 202, 833

Thackeray, A.D. 1977, Memoirs Roy. Astron. Soc. 83, 1

Walker, A.R. 1977, Monthly Not. Roy. Astron. Soc. 179, 587

Webster, B.L. 1974, in Stellar Stability and Evolution, eds. P. Ledoux, A. Noels & A.W. Rodgers, D. Reidel, Dordrecht, p. 123

Wood, P.R., Zarro, D.M. 1981, Astrophys. J. 247, 247

DISCUSSION

QUERCI - We have to organize a monitoring of a selected set of such stars by photometry.

HECK - I fully agree. There is a definitive need for surveying both bright and faint long period variables, and both photometrically and spectrographically. During classically short observing runs, what we can do in the specific case of RR Tel is to monitor the fast variation and to precise their slope and origin (lines or continuum).

BIDELMAN - Are there any radial velocity variations with the period over the Mira variables or otherwise ?

HECK - As far as I remember, there has been no correlations pointed out up to now.

VII. - EVOLUTIONARY TRACKS OF THE PECULIAR STARS. NUCLEAR REACTIONS. DREDGE-UP MECHANICS

STELLAR EVOLUTION, NUCLEOSYNTHESIS AND DREDGE-UP IN COOL GIANTS

P. R. Wood
Mount Stromlo and Siding Spring Observatories
Australian National University

ABSTRACT

The results of calculations of helium shell flashes in stars on the asympototic giant branch (AGB) are discussed. These results when compared with observations of cool giants with peculiar abundances in the Magellanic Clouds unambiguously show that there is an evolutionary sequence up the AGB in the sense M→S→C (N and late R) which is caused by dredge-up of carbon and s-process elements at helium shell flashes. The quantitative agreement between the theory of shell flashes and observations is still not good. The Ba and CH stars, which may all have white dwarf companions, are probably the result of mass transfer from N stars in binary systems. The results of calculations of helium core flashes in stars at the tip of the first giant branch are also discussed. The early R stars may result from mixing at the helium core flash.

1. INTRODUCTION

The one feature common to all the peculiar red giants (Ba, CH, R, N, S and R CrB stars) is an excess in the envelope of ^{12}C produced by the triple-alpha process. At the present time, two evolutionary events are known which could possibly cause mixing to the surface of a single star of the products of the 3-α and associated reactions - the helium core flash and the helium shell flash. One of these processes must be the initial source of the abundance patterns seen in each of the peculiar red giant types. However, it is possible that in binary systems, the star which is now seen to have peculiar abundances may not have generated those abundances within itself. It may have acquired them by mass transfer from a companion which had previously generated its own peculiar abundance patterns as a result of a helium core or shell flash. In this article, theoretical models of helium core and shell flashes will be discussed, and the properties of the various types of peculiar red giant will be examined to see if the method of

M. Jaschek and P. C. Keenan (eds.), Cool Stars with Excesses of Heavy Elements, 357–371.

production of each of the stellar types can be identified. A recent comprehensive review on the subject of peculiar red giants is that of Scalo (1981).

2. HELIUM SHELL FLASHES

Helium shell flashes were first discovered in stellar models by Schwarzschild and Harm (1965) and Weigert (1966). Following exhaustion of helium in the core of a star with a mass $\lesssim 9\ M_\odot$ (Paczynski 1971; Becker and Iben 1980), nuclear burning occurs in two shells surrounding an electron-degenerate carbon-oxygen core. Stars in this phase of evolution are often called asymptotic giant branch (AGB) stars because, in globular clusters, such stars merge asymptotically in the H-R diagram with stars on the first ascent of the giant branch. A comprehensive review of the evolution of asymptotic giant branch stars has recently been given by Iben and Renzini (1983).

The helium burning shell in an AGB star does not burn continuously but in bursts (shell flashes) separated typically by 10^4–10^5 years. When the energy released by a shell flash diffuses out of the core of the star to the surface, it can cause the envelope convection zone to move inward in mass and dredge up the products of helium burning, as first shown by Iben (1975a).

The behaviour of the luminosity and the convective regions in an AGB star undergoing a shell flash are shown as a function of time in Figure 1. In this particular model, the main helium burning region occurs at $M_r \sim 0.675\ M_\odot$. Energy is transported upward from this point by a convection zone (the convective shell) which extends almost out to the hydrogen-rich region where hydrogen shell burning had previously been occurring (hydrogen shell burning is extinguished by the helium shell flash). It is a general feature of helium shell flash calculations that the helium shell does not make contact with any hydrogen-rich material. The surface luminosity drops after the hydrogen burning shell has been extinguished but it then rises to values larger than those that existed during the quiescent shell burning phase that preceded the helium shell flash. The base of the envelope convection zone moves in sympathy with the surface luminosity. When the surface luminosity drops, the base of the envelope convection recedes, but when the surface luminosity rises again, it moves inward beyond its quiescent position and dredges up carbon-rich material previously deposited in the region $M_r/M_\odot \sim 0.686 - 0.687$ by the convective shell.

A slight refinement of the above process has been found by Iben and Renzini (1982a,b) wherein a semi-convective zone develops in the carbon-rich material of the extinct convective shell when the surface luminosity rises above its quiescent value. The main effect of this semi-convection is that it carries small amounts of hydrogen down into the carbon-rich regions below the point to which envelope convection penetrates. This has an important influence on subsequent s-process nucleosynthesis. Semiconvection does not cause

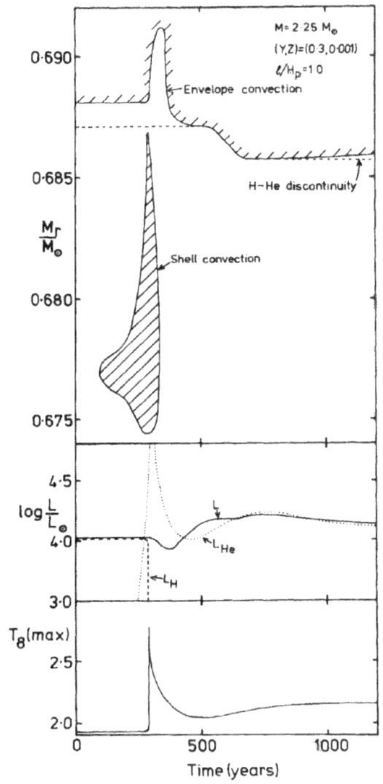

Figure 1. Upper panel: the positions of the convective envelope, the convective shell and the H-He discontinuity as functions of time during a helium shell flash. Middle panel: time dependence of helium burning luminosity (L_{He}), hydrogen burning luminosity (L_H) and surface luminosity (L). Lower panel: the maximum temperature in the core of the star as a function of time.

any more material to be dredged-up than would occur with normal dredge-up.

The material of the extinct convective shell which is dredged up by envelope convection has a composition $X_C \approx 0.25$, $X_{He} \approx 0.75$. For core masses (mass interior to the hydrogen burning shell) $\gtrsim 0.9\ M_\odot$, neutrons produced by the reaction $^{22}Ne(\alpha, n)^{25}Mg$ can produce significant s-process nucleosynthesis with s-process abundances in the solar system ratios (Iben 1975a; Truran and Iben 1977). The important feature of s-process element production in helium shell flashing stars is that material in the convective shell (and therefore the dredged-up material) has been subjected to a variety of exposures with an exponential fall-off; material still in the shell is made up of newly incorporated material that has had only one exposure, as well as material incorporated in previous flashes which has been exposed to 2 or more flashes. It is this particular exposure history that is required in order to build up the observed solar system s-process abundance pattern.

In stars with core masses less than $\lesssim 0.9\ M_{\odot}$, the temperatures reached during the shell flash are not sufficient for the $^{22}Ne(\alpha,n)^{25}Mg$ reaction to liberate many neutrons. However, in stars less massive than ~0.7 $M_{\odot}$ and with low Z, the Iben-Renzini semi-convective process described above provides a means whereby the $^{13}C(\alpha,n)^{16}O$ reaction can be activated in the convective shell (Iben 1984). Since a small amount of hydrogen is mixed down into the carbon-rich extinct convective shell, the reactions $^{12}C(p,\gamma)^{13}N(e^{+}\nu)^{13}C(p,\gamma)^{14}N$ occur when this region heats up again as the helium burning reactions die down after the flash. The ^{13}C so produced is then incorporated into the convective shell of the next flash, at which time the $^{13}C(\alpha,n)^{16}O$ reaction is initiated. The neutron flux produced by this sequence of events is somewhat higher than that allowable in the classical s-process (Iben 1984).

In summary, at the present time, a method exists for producing and dredging up carbon and s-process elements in luminous, massive ($M_{bol} \lesssim -6$, $M \gtrsim 3\ M_{\odot}$) and faint, metal weak, low mass ($M_{bol} \gtrsim -5$, $M \lesssim 0.7$, $Z \approx 0.1\ Z_{\odot}$) AGB stars. No efficient means for producing s-process enhancements has been found for stars between these two regimes. The bulk of the stars in the solar vicinity fall in this no-mans-land!

3. THE HELIUM CORE FLASH

When a low mass star has burnt all the hydrogen in its core, it evolves from the main-sequence up the first giant branch. Provided the mass of the star is $\lesssim 2.5\ M_{\odot}$ (Iben 1967; Sweigert and Gross 1978; Lattanzio 1984), the core does not become hot enough to ignite helium during contraction, and an inert core of almost pure helium develops in which the electron gas is degenerate. As the core grows due to the burning of hydrogen to helium in the surrounding shell, the temperature in the core grows until it becomes high enough for the 3-α reaction to occur. Mestel (1952) showed that the ignition of helium in these conditions would lead to a run-away situation. When energy is released into the core by the 3-α reaction, the temperature increases but the pressure does not (pressure is independent of temperature in the degenerate electron gas) so no hydrostatic readjustment of the star occurs. The highly temperature sensitive 3-α reaction runs away in this situation until electron degeneracy is finally removed by the energy input, and subsequent hydrostatic readjustment of the star decreases the central stellar density and thereby the rate of energy generation.

Many computations through the helium core flash have been performed, beginning with Schwarzschild and Harm (1962). Although the early studies of the helium core flash indicated that helium ignition occurred at the centre of the star, subsequent studies which included neutrino energy losses from the core (e.g., Thomas

1967; Sweigert and Gross 1978) showed that helium ignition occurs in an off-centre shell. There are two important consequences of this fact (1) the lifting of degeneracy in the core does not occur in a single core flash but in a series of flashes, with each successive flash being closer to the centre of the star, and (2) the large energy flux flowing to the surface of the star following each flash can cause envelope convection to come in and dredge-up material processed by the 3-α reaction to the surface of the star (Thomas 1967; Mengel and Gross 1976; Paczynski and Tremaine 1977; Despain 1981). The behaviour of a star undergoing an off-centre helium core flash is qualitatively almost identical to the behaviour of an AGB star undergoing helium shell flashes. In stellar evolution calculations using "standard" input physics, the dredge-up of carbon-enhanced material from the core does not occur during the core flash (e.g., Despain 1981). It is only by artificially altering the input physics to produce helium ignition at the edge of a large core ($\sim$0.4 M_o) that dredge-up can be induced in the models (Paczynski and Tremaine 1977). Two plausible ways in which the core masses in standard models might be increased are (1) by an increase in the neutrino energy loss rates from the core (these are uncertain at present - Ramadurai 1976), or (2) by having a rapidly rotating stellar core (Mengel and Gross 1976). As with the helium shell flash, the convective shell which develops above the main helium burning region does not make contact with the hydrogen-rich layers above (Despain 1982).

All the above calculations have been done assuming that the star was in hydrostatic equilibrium and that convection adjusted instantaneously to local conditions according to the mixing-length theory. Calculations which have tried to relax these assumptions (Edwards 1969; Wickett 1977; Cole and Deupree 1981) have often led to disruption of the star (a supernova explosion is suggested by Deupree and Cole 1983), a result clearly at odds with the numerous types of stars observed in the post core flash stages of evolution. However, the most recent attempts at studying time-dependent convection and hydrodynamics of the helium core flash (Deupree and Cole 1983) indicate that the core flash need not lead to disruption of the star. These time consuming calculations are necessarily approximations themselves (being two dimensional and coarse zoned) but are important indicators of the types of behaviour that might be expected at the helium core flash. The models of Deupree and Cole (1983) with the best input physics (outer boundary far from the stellar centre, non-central helium ignition) indicate that the results of hydrostatic evolution calculations through the helium core flash are approximately correct. Small amounts of elements heavier than ^{12}C and ^{16}O are produced in the hydrodynamic calculations but not in the hydrostatic calculations because of the lower maximum temperatures found in the latter.

All the non-disruptive core flash models described above burn $\lesssim$5% of the helium in the core into carbon. Hence, more material will need to be dredged up at the helium core flash than at helium shell flashes in order to produce a given envelope carbon abundance because the material dredged up at helium shell flashes has X_C ~ 0.25. In some of the core flash models, the temperature is sufficient to produce neutrons for s-process nucleosynthesis via the $^{22}Ne(\alpha,n)^{25}Mg$ reaction after conversion of ^{14}N left over from CNO cycling to ^{22}Ne by α particle reactions e.g., Dupree and Cole 1983 find temperatures of 6 x 10^8 K in their third model with a non-central core flash. However, other models (e.g., Demarque and Mengel 1971; Despain 1983) do not attain high enough temperatures for the $^{22}Ne(\alpha,n)^{25}Mg$ reaction to work effectively. If hydrogen could be mixed into the core by the convective shell then neutrons could be produced via the $^{12}C(p,\gamma)^{13}N(\beta^+\nu)^{13}C(\alpha,n)^{16}O$ reactions in all the models. No such mixing has been found in the hydrostatic models but Deupree and Cole (1983) find that bubbles of hot material from the core may possibly mix with the hydrogen-rich material above.

In summary, the helium core flash models so far produced have shown that under certain conditions it may be possible to produce carbon and s-process elements in the core and to dredge-up these elements to the surface of the star. However, no completely satisfactory models have actually found such processes to occur.

4. OBSERVATIONS OF M, S, N AND LATE R STARS

The rarity of cool giants of peculiar composition in clusters in our Galaxy has greatly hampered their study because of lack of good distance estimates and, therefore, good absolute magnitudes. However, in recent years at least some types of peculiar giants have been found in the Magellanic Clouds and this has helped to clarify the evolutionary status of these stars.

A most important set of observations in the present context have been the observations of sequences of M, S and C giants along the AGBs of intermediate age globular clusters by Bessell, Wood and Lloyd Evans (1983) and Lloyd Evans (1984). At low luminosities on the AGB ($M_{bol} \gtrsim -4.0$), the giants are all pure M types. Going to higher luminosities, there is an interval of $\lesssim 0.4$ magnitudes over which the S stars can be found, and over which the S star characteristics (in particular, the ZrO band at 6474A) increase in strength with luminosity (Lloyd Evans 1984). The clear interpretation of these results is that dredge-up of carbon and s-process elements (one of which is Zr) has begun. The increased strength of the ZrO band probably reflects an increase in the Zr abundance (Piccirillo 1980) although the increasing C/O ratio may also increase the ZrO band strength; in either case dredge-up of processed material is involved in the enhancement of ZrO band. At luminosities higher than those of the S stars in the Magellanic Cloud clusters, all the stars are carbon stars. The infra-red colours and luminosities of these carbon stars (e.g., Frogel and Cohen 1982; Mould and Aaronson 1982; Bessell, Wood and Lloyd Evans

1983), when compared with infra-red colours and luminosities of Galactic carbon stars (e.g., Mendoza and Johnson 1965; Eggen 1972; Scalo 1976; Dominy 1984) indicate that the carbon stars at the tips of the AGBs in the Magellanic Clouds are of spectral type N (or late R). The Magellanic Clouds thus provide unambiguous evidence for the production of S and N stars by dredge-up of carbon by helium shell flashes on the AGB.

An independent piece of evidence for the occurrence of dredge-up in M, S, and N stars is the observation of the s-process element Tc in these stars (Merrill 1952; Little-Marenin and Little 1979). Since the longest-lived isotope of Tc has a half-life of $\sim 10^6$ years at the temperatures found in the envelopes of red giants, the Tc in these stars must have been recently synthesized and dredged-up. A helium shell flash is the only known mechanism by which this could have occurred.

5. DETAILED COMPARISON OF OBSERVATIONS AND THE THEORY OF DREDGE UP

a) Luminosities

Although there is qualitative agreement between observations of M, S and C stars and the theory of helium shell flashes and dredge-up, the quantitative agreement between theory and observation is still not good. The main problem is that the M to C star transition is found to occur at luminosities significantly below the luminosities at which dredge-up of carbon has been found to occur in standard stellar models (e.g., Iben and Renzini 1983, and references therein).

Firstly, it is necessary to be careful which observational and theoretical luminosities are compared. The luminosity of a low mass ($\lesssim 4M_\odot$) AGB star varies by ~0.4 mag over a shell flash cycle (eg. Gingold 1974; Wood and Zarro 1981; Iben 1982) but the star spends most (~75%) of its time with a luminosity near the quiescent luminosity maximum (which occurs immediately before the shell flash). Consider a cluster where there is a transition from M and S stars to C stars as luminosity increases up the AGB. Then the most luminous M star should have a luminosity near the quiescent luminosity maximum immediately preceding the shell flash which will turn the star into a carbon star. On the other hand the least luminous carbon star may well be a star in the luminosity dip which lasts ~25% of the flash cycle following the shell flash (for clusters with more than one C star, the probability that one star is well below the quiescent maximum is obviously high). Thus the best estimator of the luminosity at which a star becomes a carbon star on the AGB of a cluster is the maximum luminosity of the M (or S) stars in the cluster.

The very oldest clusters in the Magellanic Clouds (age$>10^{10}$ years) do not have carbon stars at the tips of their giant branches (e.g., NGC 121- although there is a carbon star in this cluster, it is almost certainly related to the CH stars, not the N type carbon stars). The oldest clusters exhibiting carbon stars (e.g., Kron 3, age ~7x10^9 years) show an M to C transition at M_{bol} ≈ -3.9 and a maximum C star luminosity of M_{bol} ≈ -4.4. Somewhat younger clusters of age ~3x10^9 years (e.g., NGC419, NGC1783, NGC1846) have a more luminous M to C transition at M_{bol} ≈ -4.3 to -4.6 and a maximum C star luminosity of M_{bol} ≈ -5.1. Both the higher luminosity for production of carbon stars and the higher maximum luminosity for carbon stars in these younger clusters are indicative of higher envelope masses compared to clusters such as Kron 3. The higher maximum AGB luminosities are achieved because there is a larger envelope mass which needs to be dissipated by mass loss and by consumption via nuclear burning before the star evolves away from the AGB (it then has negligible envelope mass left). The higher luminosity for carbon star production results from the need for a number of flashes to dredge up enough carbon into the envelope to give C/O>1 there.

Direct evidence for multiple dredge up events before carbon star production in clusters of age ~3 x 10^9 years is provided by the S stars, which are spread over an interval of ~0.4 mag. below the luminosity of the carbon stars. Since theoretical calculations indicate that increasing the mass of an AGB star will enhance the efficiency of dredge-up (Wood 1981), it seems likely that dredge-up begins at least as low down the AGB in the clusters of age ~3x10^9 years as in the oldest clusters with carbon stars i.e., at M_{bol} ≈ -3.9. In this case, stars on the AGB of LMC clusters of age ~3x10^9 years must remain oxygen-rich while dredging up carbon and s-process elements over an interval of ~0.5 mag; this interval corresponds to ~5 shell flashes. The core mass at this stage is ~0.6 M_o while the main-sequence mass of the stars was ~1.3 M_o. After allowing for mass loss, an envelope mass of ~0.5 M_o seems appropriate for the S stars in these clusters. With a metal abundance Z~0.1 Z_o (e.g., Bessell, Wood and Lloyd Evans 1983), and noting that the material dredged up has X_c ≈ 0.25, it can eastily be calculated that ~1.4x10^{-4} M_o of material must be dredged up at each of the ~5 shell flashes that precede carbon star formation. The hydrogen-burning shell typically advances outward through 0.006 M_o between flashes so that λ (the ratio of dredged-up mass to the mass through which the H-burning shell advances per flash cycle) is ~0.02. This is much less than the values found in existing calculations (e.g., Iben 1976, 1984) and indicates that dredge-up in real, solar mass stars is not simply the result of envelope convection extending down to the extinct intershell convection zone where X_c ≈ 0.25. A variety of mechanisms which may induce the dredge-up process in models where it does not exist at present have been listed by Scalo (1981). However, the basic requirement for getting dredge-up at luminosities lower than has been found so far is a stronger shell flash. Since the strength of the shell flash is determined mainly by the interflash period (Sugimoto and Fugimoto 1978), the physics determining the quiescent evolutionary behaviour of the stellar core needs careful examination.

b) CNO and s-process element abundances in S and N stars

Theoretical calculations give diverse predictions regarding expected envelope abundances of CNO and s-process elements. Clearly, the C/O ratio will increase at each dredge-up event. Observational estimates of C/O abundances in N stars are few (e.g., Kilston 1975; Gow 1977) but they indicate C/O < 4 in N stars. There does not seem to be any theoretical limit on the C/O abundance obtainable from shell flashes (Iben 1984 finds C/O ~ 12 after dredge-up in a low Z, 0.7 M_o star).

In the more massive ($M \gtrsim 3\ M_o$) AGB stars, envelope convection can reach down into the region where the CN cycle is operating during quiescent evolution, with the result that dredged up ^{12}C is converted to ^{14}N and the $^{12}C/^{13}C$ ratio will decrease to the equilibrium value of ~3 (Scalo, Despain and Ulrich 1975; Iben 1975a, Renzini and Voli 1981). The few N stars whose nitrogen abundances have been analysed show normal N abundances (Kilston 1975). However, the lack of carbon stars more luminous than $M_{bol} \approx -6$ in the Magellanic Clouds, coupled with the fact that there are S stars in the Magellanic Clouds with $M_{bol} < -6$ and J (^{13}C rich) stars with $M_{bol} \approx -6$, indicates that the CN cycle begins to operate in the convective envelopes of AGB stars with $M_{bol} \approx -6$; in the brighter stars it appears capable of converting most of the dredged-up ^{12}C into ^{14}N between shell flashes (Wood, Bessell and Fox 1983).

S stars in the Magellanic Clouds range in luminosity from the brightest possible AGB stars ($M_{bol} \approx -7.1$) down to the luminosities of the faintest N type carbon stars ($M_{bol} \approx -4$) (Bessell, Wood and Lloyd Evans 1983; Lloyd Evans 1984; Wood, Bessell and Fox 1983). Since these stars have enhanced ZrO bands, it appears they have enhanced Zr abundances (Picirillo 1980) and they are therefore dredging up freshly synthesized s-process elements. The problem of finding a neutron source in theoretical stellar models with $-6 < M_{bol} < -5$ (see section 3) does not seem to exist in real stars.

6. THE Ba, CH AND EARLY R STARS

An important clue to the origin of the abundance peculiarities in each of these groups of stars lies in the fact that their luminosities (Scalo 1981) are significantly less than the values of $M_{bol} \approx -4$ which are required for dredge-up by helium shell flashes. The CH stars have been found on the subgiant branch (Luck and Bond 1982) so that they are too faint to have been produced by the helium core flash either (unless the whole star is mixed at the core flash and sent back down to the main-sequence). A further important characteristic of the Ba and CH stars is that they all appear to be in binary systems (McClure 1983, 1984). IUE observations of a group of Ba stars by Bohm-Vitense, Nemec and Proffitt (1984) showed a high frequency of white dwarf companions among the Ba stars. This result led them to suggest that all Ba stars probably have white dwarf companions. Under these circumstances, it is hard to escape the conclusion that the abundance peculiarities of the Ba stars are the result of mass transfer from the original primary after it has turned into a carbon star. The white dwarf is the remnant of the primary left over after the completion of mass transfer. The binary nature of the CH stars also suggests that they are the result of mass transfer in a binary system.

The Ba stars and CH stars in general have enhanced carbon and s-process element abundances (e.g., Luck and Bond 1982; Warner 1965; Tomkin and Lambert 1979, 1983) as would be expected if these stars have envelopes enriched by mass transfer from an N star. Typically, both Ba and CH stars show C enhancements of ~3 and s-process element enhancements of ~10. In the N stars, C/O ratios are generally <4 (Gow 1977), from which C enhancements <10 can be inferred assuming that surface ^{16}O is not affected by stellar evolution; s-process element enhancements of up to ~10 -100 have been estimated for the N stars (Utsumi 1970; Kilston 1975). Both the C and s-process element enhancements in the Ba and CH stars indicate that a large amount of mass needs to be transferred from the N star to the future Ba star during binary mass exchange. Assuming M~1.5 $M_\odot$, for the Ba star (e.g., Scalo 1981), the convective envelope will occupy the outer ~1 $M_\odot$ during evolution on the giant branch (Faulkner and Cannon 1973). To get C enhancements of ~3 in this envelope will require the addition of $\gtrsim$0.2 $M_\odot$ of N star material. Since the binary N star must lose ~0.8 $M_\odot$ of material before becoming a white dwarf, it can be seen that a considerable fraction of the lost mass must go to the binary companion.

Dominy and Lambert (1983) have argued that the presence of white dwarf companions with age > 10^9 years means that the Ba stars can not have resulted from mass transfer. These authors point out that since the combined red giant and helium core burning lifetimes of low mass red giants are <10^9 years, the mass transfer must have occurred while the Ba star was on the main-sequence. They then point out that main-sequence precursors of the Ba stars are very rare (possible candidates are the Ba rich star ε Ind and the carbon dwarf G77-61). Bohm-Vitense, Nemec and Proffitt (1984) have re-discussed this argument and conclude that main-sequence precursors of Ba stars should indeed be rare, so that there is no need to reject the binary transfer hypothesis on the above grounds.

As well as the two possible main-sequence pre-cursors mentioned above, there are a number of other possible candidates worth noting. Firstly, a second faint, nearby (high proper motion) carbon star has recently been identified by Ratnatunga (1983). And secondly, some or all of the 4 faint carbon stars which have been assigned to the galactic halo (Sanduleak 1980; Phillips and Terlevich 1982; Ratnatunga 1983; Margon et al. 1984) may in fact be nearby dwarf carbon stars. It seems that a set of possible main-sequence precursors for the Ba (and R) stars is being discovered to match the known main-sequence and subgiant precursors of the CH stars (Luck and Bond 1982).

Finally, the early R stars (R0-R4) need discussing. These stars are similar to the Ba stars except that they show enhancements of C only (by a factor of ~5), not of s-process elements (Dominy 1984).

As pointed out by Dominy (1984), this result is difficult to explain in terms of the mass transfer process since the s-process elements should be enhanced. If the R stars and Ba stars are both the result of mass transfer in binary systems, there would need to be a range in s-process element enhancement in N stars. Perhaps the sequence Ba - mild Ba - R star is a sequence resulting from mass transfer from N stars with progressively smaller s-process element enhancements.

There is some evidence that s-process element enhancements are not universally large in N stars. Utsumi (1970) found normal s-process abundances in 4 of 23 N stars, and Kilston (1975) found small enhancements of ~2 in some stars. The R CrB stars may also indicate that s-process enhancements are not universal in N stars, as they have enhanced carbon abundances but normal s-process abundances (Warner 1967; Cottrell and Lambert 1982). The luminosities of these stars, and other hydrogen-deficient carbon stars, can be deduced to lie in the range $-5 \lesssim M_{bol} \lesssim -4$ from photometry of Magellanic Cloud members by Feast (1972) and Richer, Olander and Westerlund (1979). The luminosities of the R CrB stars, together with their lack of surface hydrogen, suggest that they are the stripped cores of AGB stars. The material seen on their surfaces should be representative of the material dredged up into the envelopes of carbon stars. Since the R CrB star surface material is not enhanced in s-process elements, it may be that not all stars that dredge up at helium shell flashes are enhanced in s-process elements. It seems desirable that abundance analyses be made for larger samples of Ba, R and N stars so that a clearer picture of element enhancement can be found. At the present time though, the lack of s-process element enhancement in the R stars does seem to argue against their being the result of mass transfer in binary systems.

The alternative method of producing carbon enhancements in the early R stars is the helium core flash. Dominy (1984) suggests that the abundance patterns in the R stars result from the He core flash only in stars with rapidly rotating cores (so that not all low mass, post core flash stars are carbon stars). To conclude, the R stars appear to be the only group of peculiar cool giants which require the helium core flash for their production; and if the helium core flash needs to be invoked in order to produce the R stars, then it is not easy to discount the possibility that the helium core flash is involved in the production of the Ba and CH stars as well.

REFERENCES

Becker, S., and Iben, I. 1980, Ap. J., 237, 111.
Bessell, M.S., Wood, P.R., and Lloyd Evans, T. 1983, M.N.R.A.S., 202, 59.
Bohm-Vitense, E., Nemec, J., and Proffitt, C. 1984, Ap. J., 278, 726.
Cole, P.W., and Deupree, R.G. 1981, Ap. J., 247, 607.

Cottrell, P.L., and Lambert, D.L. 1982, Ap.J., 261, 595.
Demarque, P., and Mengel, J.G. 1971, Ap. J., 164, 317.
Despain, K.H. 1981, Ap. J., 251, 639.
Despain, K.H., 1982, Ap.J., 253, 811.
Deupree, R.G., and Cole, P.W. 1983, Ap.J., 269, 676.
Dominy, J.F. 1984, Ap. J. Suppl., 55, 27.
Edwards, A.C. 1969, M.N.R.A.S., 146, 445.
Eggen, O.J. 1972, Ap. J., 174, 45.
Faulkner, D.J., and Cannon, R.D. 1973, Ap. J., 180, 435.
Feast, M.W. 1972, M.N.R.A.S., 158, 11p.
Frogel, J.A. and Cohen, J.G. 1982, Ap.J., 253, 580.
Gingold, R.A. 1974, Ap.J., 193, 177.
Gow, C.E., 1977, P.A.S.P., 89, 510.
Iben, I. 1967, Ann. Rev. Astr. Ap., 5, 571.
Iben, I. 1975a, Ap.J., 196, 525.
Iben, I. 1975b, Ap.J., 196, 549,
Iben, I. 1976, Ap.J., 208, 165.
Iben, I. 1982, Ap.J., 260, 821.
Iben, I. 1984, preprint.
Iben, I., and Renzini, A. 1982a, Ap.J.Lett., 259, L79.
Iben, I., and Renzini, A. 1982b, Ap.J.Lett., 263, L23.
Iben, I., and Renzini, A. 1983, Ann.Rev.Astr.Ap., 21, 271.
Kilston, S. 1975, P.A.S.P., 87, 189,
Lattanzio, J. 1984, Thesis, Monash University.
Little-Marenin, I.R., and Little, S.J. 1979, A.J., 84, 1374.
Lloyd Evans, T. 1984, M.N.R.A.S., 208, 447.
Luck, R.E., and Bond, H.E. 1982, Ap.J., 259, 792.
McClure, R.D. 1983, Ap.J., 268, 264.
McClure, R.D. 1984, Ap.J., 280, 231.
Margon, B., Aaronson, M., Liebert, J., and Monet, D. 1984, A.J., 89, 274.
Mendoza, E.E., and Johnson, H.L. 1965, Ap.J., 141, 161.
Mengel, J., and Gross, P.G. 1976, Ap. Sp. Sci., 41, 407.
Merrill, P.W. 1952, Science, 115, 484.
Mestel, L. 1952, M.N.R.A.S., 112, 598.
Mould, J. and Aaronson, M. 1982, Ap.J., 263, 629.
Paczynski, B. 1971, Acta Astr., 21, 271.
Paczynski, B., and Tremaine, S.D. 1977, Ap. J., 216, 57.
Phillips, M.M., and Terlevich, R. 1983, P.A.S.P., 95, 43.
Picirillo, J. 1980, M.N.R.A.S., 190, 441.
Ramadurai, S. 1976, M.N.R.A.S., 176, 9.
Ratnatunga, K.U. 1983, Thesis, Australian National University.
Renzini, A. and Voli, M. 1981, Astr. Ap., 94, 175.
Richer, H.B., Olander, N., and Westerlund, B.E. 1979, Ap.J., 230, 724.
Sanduleak, N. 1980, P.A.S.P., 92, 246.
Scalo, J.M. 1976, Ap.J., 206, 474.
Scalo, J.M. 1981, in "Physical Processes in Red Giants", eds. I. Iben and A. Renzini (Dordrecht: Reidel), p.77.
Scalo, J.M., Despain, K.M. and Ulrich, R.K. 1975, Ap.J., 196, 805.
Schwarzschild, M., and Harm, R. 1962, Ap.J., 136, 158.

Schwarzschild, M., and Harm, R. 1965, Ap.J., 142, 855.
Sugimoto, D., and Fujimoto, M.Y. 1978, Publ. Astr. Soc. Japan, 30, 467.
Sweigert, A.V., and Gross, P.G. 1978, Ap.J.Suppl., 36, 405.
Thomas, H.C. 1967, Zeit, f. Ap., 67, 212.
Tomkin, J., and Lambert, D.L., 1983, Ap.J., 273, 722.
Truran, J.W., and Iben, I. 1977, Ap.J., 216, 797.
Utsumi, K. 1970, Publ. Astr. Soc. Japan, 22, 93.
Warner, B. 1965, M.N.R.A.S., 129, 263.
Warner, B. 1967, M.N.R.A.S., 137, 119.
Weigert, A. 1966, Zeit. f. Ap., 64, 395.
Wickett, 1977, in "Problems in Stellar Convection", eds. E.A. Spiegel and J.P. Zahn (Berlin: Springer Verlag), p.284.
Wood, P.R. 1981, in "Physical Processes in Red Giants", eds. I. Iben and A. Renzini (Dordrecht: Reidel), p.135.
Wood, P.R. and Zarro, D.M. 1981, Ap.J., 247, 247.
Wood, P.R., Bessell, M.S., and Fox, M.W. 1983, Ap.J., 272, 99.

DISCUSSION

BEER - It is not clear that an exponential neutron exposure is the appropriate assumption when considering s-process abundances in an individual star in the solar vicinity. Only a relatively small number of shell flashes may have occurred, whereas many flashes are required before the exponential exposure approximation becomes accurate.

LLOYD EVANS - It is correct that the large number of S stars, over a wide range of M_{bol}, but still no pure S or SC stars in younger LMC cluster (eg. NGC 1718, 1783), is in accordance with current theory ?

WOOD - In order for a star to be pure S or SC, a very small range in O/C is required. The rarity of these stars is probably due to the small range in O/C required.

BERGEAT - Could you give an upper limit in age and a mass range for producing galactic carbon stars by helium shell flashes ?

WOOD - Not reliably.

CATCHPOLE - If I understand you correctly it would be possible to form SC stars on two occasions : once at the point where S stars became C stars for the first time (provided the increment of C/O per shell flash will not jump across this gap). The second occasion would be toward the top of the Clcud asymptotic giant branch where they have previously C stars. Is this correct ?

WOOD - It is in principle correct that a star could dredge up carbon to give C/O > 1 (ie C stars), and for hot bottom burning following dredge up to convert the star back to an O-rich star. Such stars would be N rich and have low C^{12}/C^{13}.

SCHATZ - I understand that the stellar evolution calculations assume a spherical symmetry. It is difficult for me to believe that the helium shell flash starts all over a sphere at the same time. If it starts locally, it might trigger some large scale convection with consequences which are not clear to me. Could you comment on that ?

WOOD - The helium shell flash is far from being a dynamical event (ie sound can cross the core in a time much less than the time scale of the flash). So the flash may be roughly spherical. Some non-spherical processes would help to get the extra mixing need for the s-process.

MALANEY - I would have thought that if the helium shell did ignite locally, then the thermal run-away would quickly overcome the entire shell.

WOOD - I see no reason not to expect the ignition to be a spherically symmetric.

RICHER - If Utsumi's results presented at this conference are correct (that N type stars do not have enhanced s-process elements), is it true that the theory is in difficulty and cannot explain these stars ?

WOOD - The theory is in most trouble trying to explain the enhancements in s-process elements, not the absence of s-process enhancements. Some non standard mixing of hydrogen into carbon rich material at dredge up seems to be required (the Iben-Renzini semiconvection process is one of these), but it occurs only at low z and M.

BIDELMAN - Could you expand a bit on your statement concerning the ordinary R stars ? These are really fairly common objects.

WOOD - It seems we may need to invoke the helium core flash to explain the R stars. In this situation, the core flash may also be involved in production of Ba, CH stars.

LLOYD EVANS - Observations of LMC clusters show a sharp upper limit of $M_{bol} = -5.6$ for the C stars, with no sign of change back to M or S stars. However a possibility is that the C star looses its envelope by this luminosity, as Frogel and Lloyd Evans have both suggested.

WOOD - Stars on the AGB with $M_{bol} < -5.6$ are definitely seen in the Magellanic Clouds (Wood, Bessell and Fox, 1983).

MALANEY - Regarding your statement that the R CrB stars have normal s-process abundances, I do not think enough objects have been studied to justify this. Indeed one of these stars shows large s-process enhancements. However I agree that there is a problem explaining enhancement in one R CrB and not in others. Can I raise a point concerning the Ne^{22} intershell abundance ? If the O^{18} (α,γ) Ne^{22} reaction does not go to completion in 3 α burning, then I have found in some of my calculations O^{18} (p, α) N^{15} reactions competes very significantly with C^{12} (p, γ) N^{13} for the available protons. This reduces drastically the flux from the C^{13} neutron source. Can you say whether the O^{18} (α, γ) Ne^{22} reaction will go to completion within the intershell region ?

WOOD - Yes, I think the high temperatures associated with the helium flash will destroy all remaining O^{18}.

COWLEY - Is there a difficulty with the observations of Kovacs of the carbon abundance in the Ba stars ? He finds a deficiency of C in these s-process enhanced stars. I thought C/O was greater than unity in the Ba stars, but these recent results appear to contradict this.

WOOD - If these particular Ba stars come from mass transfer, from upper AGB stars, in which hot bottom burning has been occurring, then one would not expect enhanced carbon.

JASCHEK - Can you predict the proportion of the different peculiar stars with regard to normal giants ?

WOOD - We can predict the relative numbers of M, S and cool C stars. On the AGB such stars evolve through a constant number of magnitudes per unit time.

ANALYSIS OF Zr AND Tc ABUNDANCES FROM S-STARS USING THE s-PROCESS WITH AN EXPONENTIAL DISTRIBUTION OF NEUTRON EXPOSURES

Hermann Beer and Gerold Walter
Kernforschungszentrum Karlsruhe, IK III
P.O.B. 3640
D-7500 Karlsruhe 1
Federal Republic of Germany

ABSTRACT. The abundances of Zr, Mo, Tc and Ru from S-stars were studied in the frame of the s-process with an exponential neutron fluence distribution. Estimates for the average time integrated neutron flux and the s-process neutron density and temperature have been derived from the abundances of the Zr-isotopes and from Tc, respectively.

1. INTRODUCTION

Atomic nuclei generated by slow neutron capture nucleosynthesis (s-process) provide important information about the late stages of evolution of a star. This is evident as the s-process yields a large variety of isotopes which are distinct fingerprints of their synthesis. Although the analysis of s-process abundances from stellar atmospheres is normally limited to elemental abundances important features of the synthesis can still be extracted.

2. SOLAR ABUNDANCES AND s-PROCESS

The s-process nucleosynthesis applied to solar abundances is a well investigated subject [1]. With the increased accuracies of both solar abundances and neutron capture cross sections more and more details of the fine structure of the s-process have been detected [2-6]. Fig.1 shows a recent fit of solar abundances with the new code SPEED using an s-process characterized by neutron exposures of an exponential form. Three exponential distributions were required: (1) For the bulk of abundances from A=96-200 a distribution with an average time integrated neutron flux τ_0=0.3/mb [7], (2) for the iron slope nuclei additionally a component τ_1~0.06/mb [8], (3) for the main Pb isotopes, especially Pb208, an extra equilibrium s-process nucleosynthesis [9]. The various branchings at Se79, Kr85, Sm151, Dy163, Yb170, W185, and Ir193 are a means to extract astophysically relevant parameters like s-process temperature and neutron and electron density. As in the frame of stellar s-process models [10] the exponential model is a natural consequence of the repeated occurrence of neutron pulses in the He-shell of a red giant star the investigation of solar abundances can be considered as a guideline for the study of s-process abundances in stars.

M. Jaschek and P. C. Keenan (eds.), Cool Stars with Excesses of Heavy Elements, 373–377.

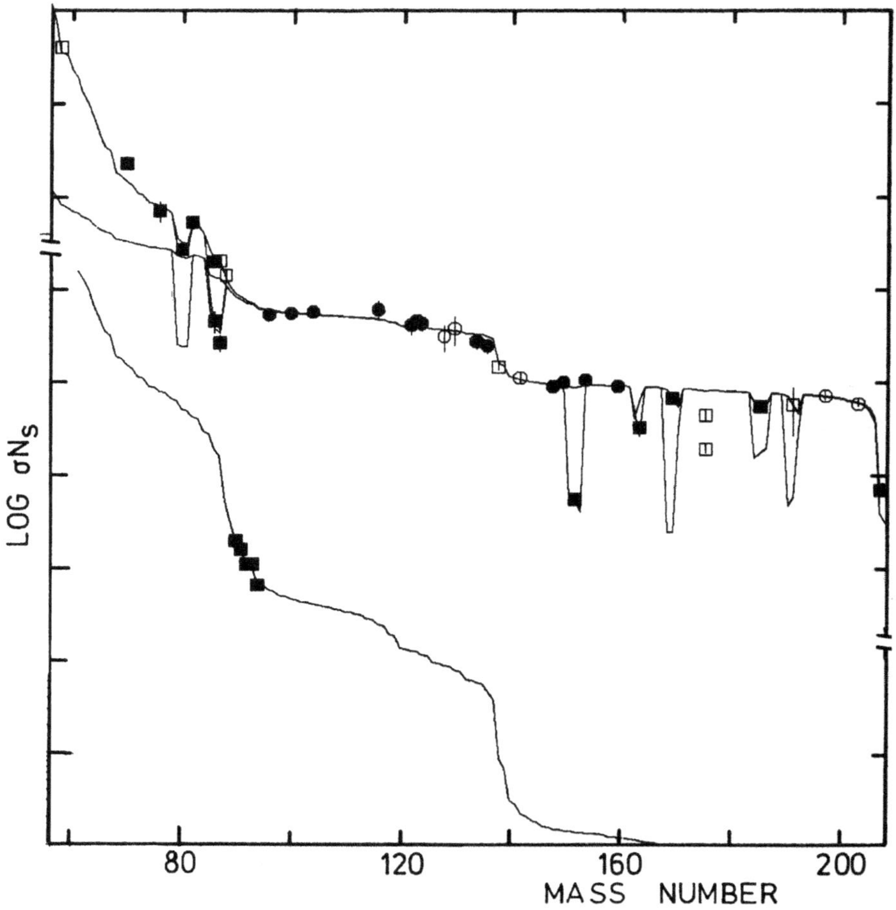

Fig.1 The product of capture cross section σ and s-process abundance Ns as a function of mass number. The symbols correspond to empirical data. Values in full black are used to adjust astrophysical parameters. (above) The curve is a fit to solar abundances. They are reproduced by by three components, a weak and a strong component superimposed to the main component which describes the bulk of the nuclei from A~96 to 204. The weak component gives sizeable contributions only for the iron slope nuclei, and the strong component acts only significantly on the main Pb isotopes. For the respective branchings of each component one common temperature and neutron density are sufficient (For the main component: kT=18-28 keV and the neutron density 0.8 to 1.8 • 10(8) /cm^3).
(below) The s-process required to reproduce the isotopic Zr abundances from the S-stars, R Cyg, V Cnc, and R Gem resembles the weak component of the iron slope nuclei.

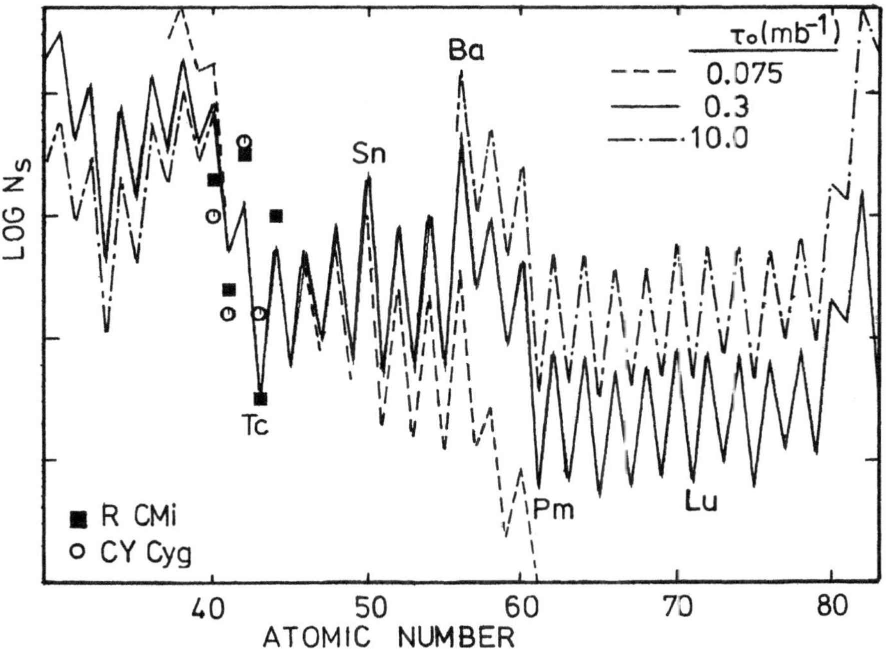

Fig.2 Ns calculated for three values of τo and normalized at Mo. The symbols are stellar data. The normalization between the calculation and the data was taken from ref.[12]. To reproduce Tc from R CMi as shown a branching is required where half of Tc99 is allowed to decay to Ru.

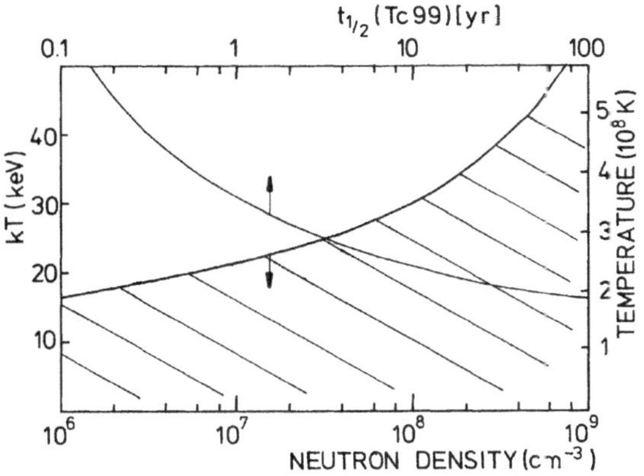

Fig.3 The hatched area is the range of values for the Tc reproduction in our stars. The limiting curve is determined by the abundance of R CMi. The temperature dependence of the Tc99 half-life is shown, too.

3. s-PROCESS IN LATE TYPE RED GIANT STARS

The detection of Tc and the enhancement of s-process elements confirms that the s-process actually occurs in stars. In this work results on Zr90,91,92,93,94,96 [11] and on Tc [12] are studied which are unequivocally the product of a recent stellar synthesis due to the occurrence of Zr93 and presumeably Tc99. Under typical s-process conditions Zr93 is practically stable whereas the Tc99 half-life is a sensitive function of temperature via allowed beta decay [13]. The unbranched s-process from Zr90 to 94 is sensitive to τ_o due to the small capture cross sections near the magic neutron shell 50. The s-process synthesis of Zr96 must arrive via unstable Zr95 (64d). This allows for an estimate of the neutron density. The temperature sensitivity of Tc99 leads to an s-process braching which is both a function of temperature and neutron density. The results of our analysis of Zr-isotopes and Tc are illustrated in Fig.1-3. Zr90,91,92,93,94 yield an average time integrated neutron flux τ_o=0.075/mb. From the upper limit of the Zr96 abundance the neutron density was found to be < 10(8)/cm^3 [14]. From Tc in R CMi and CY Cyg we calculated that during synthesis up to 50% of Tc had time to decay before neutron capture (Fig.2). This gives constraints for the s-process temperature and neutron density (Fig.3).

4. CONCLUSIONS

Our results on the Zr isotopes indicate that (1) the S-stars, R Cyg, V Cnc, R Gem provide no solar like s-process distribution, this is in agreement with predictions for low mass stars [10], (2) a sensitive indicator of solar-like s-process abundance production in a star is Zr96 as its synthesis requires the high neutron densities characterizing Iben's model [10].
The example of Tc demonstrates (Fig.2) that also elemental abundances are useful indicators of astrophysically important parameters. A comparison of abundances around Zr,Sr,Mo with abundances of Ba and the rare earths allows for an analysis of τ_o. Other key elements for s-process temperature and neutron density determination besides Tc are for instance Cu/Zn, Br/Se, Rb/Sr, Pm, and Lu/Hf.

REFERENCES

[1] F.Käppeler, H.Beer, K.Wissak, D.D.Clayton, R.L.Macklin, R.A.Ward, Ap.J. 257,821(1982)
[2] H.Beer, R.A.Ward, Nature 291,308(1981)
[3] R.A.Ward, H.Beer, Astron.Astrophys. 103,189(1981)
[4] H.Beer, F.Käppeler, K.Wisshak, R.A.Ward, Ap.J.Suppl. 46,295(1981)
[5] H.Beer, Ap.J. 262,739(1982)
[6] H.Beer, F.Käppeler, K.Yokoi, K.Takahashi, Ap.J. 278,388(1984)
[7] H.Beer, G.Walter, R.L.Macklin, P.J.Patchett, Phys.Rev.C (in press)
[8] G.Walter, Report KfK 3706(1984), Kernforschungszentrum Karlsruhe
[9] H.Beer, R.L.Macklin (in preparation)
[10] I.Iben jr, A.Renzini, Ann.Rev.Astron.Astrophys. 21,271(1983)
[11] A.C.Zook, Ap.J. 221,L113(1978)

[12] V.V.Smith, G.Wallerstein, Ap.J. 273,742(1983)
[13] K.R.Cosner, J.W.Truran, Astrophys.Space Sci. 78,85(1981)
[14] H.Beer, G.Walter, Astron.Astrophys. 133,317(1984)

DISCUSSION

MALANEY - Do you think that the different exposure parameters you have used indicate that various mechanisms are needed to produce the solar system abundances rather than just one ?

BEER - It is very difficult to say with the present observational data.

MALANEY - Some authors suggest the Ne^{22} source in intermediate mass stars can explain a large range of solar s-process abundances.

BEER - I think this is very difficult to prove.

COWLEY - I have never been convinced that the exponential distribution fit any stellar abundances in a way that was demonstrably better than a single, or combination of a small number of exposures. There is perhaps some room for debate about this, but I have some numerical results with me that you can look at if you'd like.

THE S-PROCESS WITHIN THE R CORONAE BOREALIS STAR U AQUARII

R. A. Malaney
University Observatory
Buchanan Gardens
St Andrews
Fife KY16 9LZ
Scotland

ABSTRACT. This paper attempts to show why a detailed study of U Aquarii, with its peculiar abundance characteristics of an extreme hydrogen deficiency coupled with a strontium and yttrium overabundance of about 100 relative to the sun, could lead to an improved understanding of some aspects of current asymptotic giant branch theory. Early indications using nuclear reaction networks suggest some outstanding problems in U Aquarii with respect to this theory.

1. INTRODUCTION

The evolutionary status of the R Coronae Borealis (RCrB) stars and the origin of their extreme hydrogen deficiency (Warner 1967, Feast 1975) remains unclear. It has been suggested they may be remnant asymptotic giant branch (AGB) stars (Schönberner 1979) and that their hydrogen deficiency is due to the capture of the hydrogen envelope into the interior as the star undergoes a final helium shell flash as it approaches the white dwarf regime (Renzini 1981). Such an event would cause the star to 'loop' back to the supergiant region where the RCrB stars are observed. However, Bond et al, (1979) in their analysis of U Aquarii (an RCrB star with Sr and Y 100 times the solar value) deduced from their observed Sr/Fe and Sr/Ba ratios ([Sr/Ba] = 2 in standard logarithmic notation) that this object had undergone a single neutron exposure, rather than the more usual exponential distribution of exposures (Clayton 1961, Ulrich 1973), and that the event which led to this single exposure most likely occurred at the helium core flash (Cole and Deupree 1981) prior to the horizontal branch.

The aim of this work is to show that by investigating the neutron exposures (via $^{13}C(\alpha,n)^{16}O$) and nucleosynthesis arising from hydrogen mixing at the helium shell flash, then the observed abundances may be used to set some constraints on any post-AGB model of U Aquarii. The implications as to whether these constraints may pose difficulties for such AGB models is briefly discussed.

M. Jaschek and P. C. Keenan (eds.), Cool Stars with Excesses of Heavy Elements, 379–383.

2. METHODS

The model used is the 1.45M⊙ AGB model of Schönberner (1979). The initial abundances (varied slightly from original model) assumed non-zero are shown in table 1. The intershell region was split into 50 equal mass zones and the method of mixing and abundance calculation in these zones is outlined in Cowan and Rose (1977). The reaction network consisted of proton, neutron and alpha particle reactions on all important light nuclei and neutron absorptions on those heavy elements (atomic mass > 22) on the synthesis path. The numerical solution of the network is shown in Arnett and Truran (1969). The effect of the energy generated by these reactions is neglected.

Isotope	Envelope mass fraction	Intershell mass fraction
1H	0.76	0.0
3He	1.0(-4)	0.0
4He	0.2	0.483
12c	3.5(-3)	0.480
14N	1.5(-3)	0.0
160	8.0(-3)	0.02
180	2.0(-5)	0.0
19F	1.0(-6)	0.0
22Ne	8.0(-4)	0.01
56Fe	9.0(-4)	9.0(-4)

Table 1

3. RESULTS AND DISCUSSION

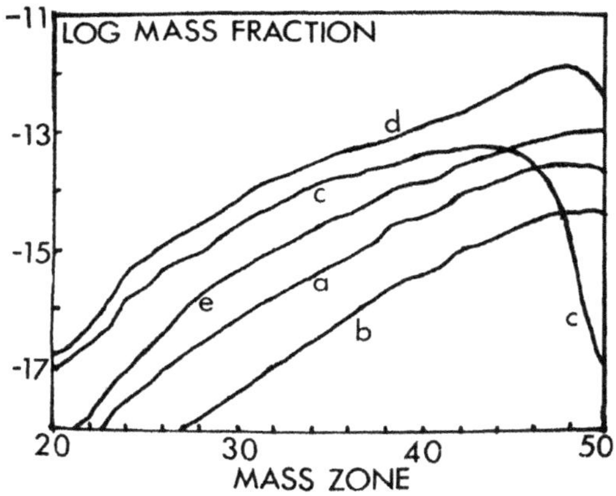

Fig.1 The neutron mass fraction at each zone of the intershell is shown for differing initial abundances, base temperatures and mixing fractions. The hydrogen and helium shells are at zones 1 (not shown) and 50 respectively. The initial abundance changes of the intershell are shown in Table 2.

Line	Temp. (zone 50)	Initial Abundances	Hydrogen mixing (M⊙)
a	2.5×10^8 K	Table 1	2.5×10^{-6}
b	2.5×10^8 K	1, ^{18}O=0.01, ^{22}Ne=0	2.5×10^{-6}
c	2.75×10^8 K	Table 1	2.5×10^{-6}
d	2.75×10^8 K	Table 1, ^{56}Fe= ^{22}Ne=0	2.5×10^{-6}
e	2.5×10^8 K	Table 1	5.0×10^{-5}

Table 2

The full lines in fig.1 show the neutron mass fraction at each zone in the intershell region after the envelope matter has diffused through each zone. The model for each line is given in table 2. The helium shell is at zone 50. The difference of line (b) from (a) is due to the ^{18}O competing with ^{12}C for the available protons. Line (c) shows a rapid drop due to the high destruction rate of ^{13}C at these temperatures and line (d) shows the effect of neglecting neutron absorptions on ^{22}Ne and ^{56}Fe in the network. (e) is for larger mixing.

Although these abundances are high, the lower mixing models will not produce enough neutrons. Since the total number of neutrons available is given by $T \approx N(^{13}C)/N(^{56}Fe)$, where N = no.density, and the fraction captured by ^{56}Fe is given by -

$$f = \sigma N(^{56}Fe)/\sigma N(\text{others}), \qquad \sigma = \text{thermal cross section for neutron absorption},$$

then assuming Sr is the peak in the heavy element distribution, thus requiring about n_c=30 neutron captures on Fe, we want Txf=30. This would then imply, after inserting the appropriate quantities, $N(^{13}C)=1.5\times10^{-3}$, which for these calculations requires $\sim5\times10^{-5}$ M⊙ of mixing (depending slightly on ^{22}Ne abundance). However, fig.2 shows that no dilution factor, d, (U Aquarii must be diluted with unprocessed matter since there is no significant Fe depletion) gives the observed $Sr/Sr_\odot$=100 and [Sr/Ba]=2 of U Aquarii, simultaneously, for n_c=30. But it can be seen that both the correct ratios are found for n_c=13 and d≈2. This would mean that the abundance peak occurs at roughly Zn-Ge rather than Sr. Also since the intershell mass=0.02M⊙ then the dilution factor implies at the time of envelope mixing the envelope had a mass $M_e \approx 0.04$M⊙, of which 5×10^{-5}M⊙ reaches the hotter regions of the intershell.

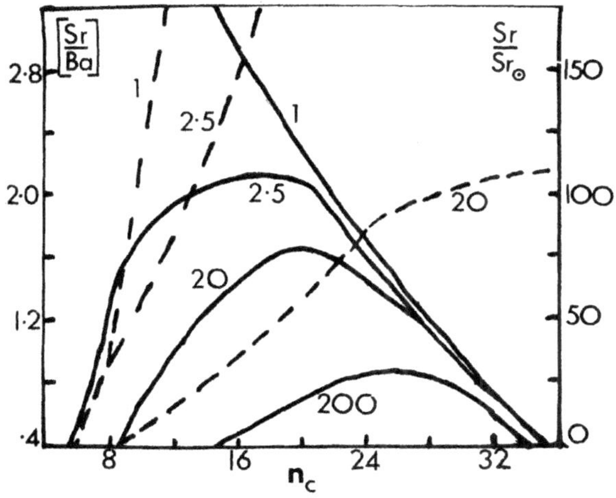

Fig.2 The full and dashed lines show [Sr/Ba] and $Sr/Sr_\odot$ varying with n_c, the number of neutrons captured on ^{56}Fe, for different values of d given by d = $(m_1+m_2)/m_1$, where m_1 and m_2 equal the masses of the exposed (to neutrons) and unexposed regions respectively.

The best agreement found is approximately n_c=13 and d=2.5.

It should be noted that intershell splitting as discussed in Cowan and Rose (1977) may be important in lowering the above calculated value of envelope mass.

In current models of AGB stars this value of M_e=0.04M⊙ would put the star in a well evolved position near the tip of the AGB branch at the time of ingestion. However, the theoretical problem of mixing such a large mass to create the hydrogen deficiency still remains (Schönberner 1979). The 'loop' hypothesis could not be valid for such a mass since $M_e \leq 10^{-5}$ M⊙ is required for this hypothesis (Fujimoto 1977, Schönberner 1979). So U Aquarii may still be left unexplained by current AGB theory, but the situation is not clear. More accurate abundance ratios and a mass determination of U Aquarii are required. The inclusion of various effects into the reaction network such as energy generation, branching, time dependent neutron flux and isomeric nuclei would also provide a more accurate model.

4. CONCLUSIONS

The RCrB star U Aquarii is an interesting object to study in relation to some aspects of AGB theory. It has been shown how constraints for U Aquarii may lead to difficulties in explaining this star by current AGB theory. However, further observational and theoretical work on this object is required before a more positive conclusion can be drawn.

ACKNOWLEDGEMENTS. I would like to thank D.W.N. Stibbs, P.W. Hill and staff at the University Observatory, St Andrews. Also thanks to D. Schönberner, K. Despain, J.M. Scalo, A. Renzini and the United Kingdom S.E.R.C. for financial support through a postgraduate studentship.

REFERENCES

Arnett W.D., Truran J.W., 1969, *Ap. J.* 157, 339

Bond H.E., Luck R.E., Newman M., 1979, *Ap. J.* 233, 205

Clayton D.D., Fowler W.A., Hull T.C., Zimmerman B.A., 1961, *Ann. Phys.* 12, 121

Cole P.W., Deupree R.G., 1981, *Ap. J.* 247, 607

Cowan J.J., Rose W.K., 1977, *Ap. J.* 212, 149

Feast M.W., 1975, *IAU Collog.* 46, 246

Fujimoto M.Y., 1977, *Publ. Astron. Soc. Japan* 29, 331

Renzini A. 1981, *Effects of mass loss on Stellar Evolution*, Ed. C. Chiosi, R. Stalio (Reidel: Dordrecht) p.319

Schönberner D. 1979, *Astr. Astrophys.* 79, 108

Ulrich R.K., 1973, *Explosive Nucleosynthesis*, Ed. Schramm D.N., Arnett W.D., p.139

Warner B., 1967, *M.N.R.A.S.* 137, 119

DISCUSSION

RICHER - This star (U Aqu) has obvious s-process enhancement. We have been told here that H-deficient carbon stars have normal s-process elements. Perhaps we should not be so willing to accept such a blanket statement.

MALANEY - I agree. What is obviously desired is further detailed observational analysis of these objects. We should not rule out observations of s-process enhancements (in other hydrogen deficient stars) been discovered in the future.

BIDELMAN - The other studied R CrB stars of our own galaxy, even those having quite strong C_2 bands, are certainly nothing like U Aqr in their Sr abundance. This must be an unusual situation.

MALANEY - Yes, indeed, the problem of why U Aqr has large s-process enhancements when other R CrB stars do not, is perplexing. Perhaps the reason is an indication of different degrees of hydrogen mixing in these objects. Of course some may say that this could be an indication that U Aqr is a binary system undergoing mass transfer. Further infrared photometric and radial velocity studies are required for such a suggestion to have any firm basis. However I think a binary explanation for these stars creates more problems than solutions.

VIII. – CONCLUSIONS. DISCUSSIONS OF POSSIBLE COOPERATIVE PROGRAMS

CONCLUDING SUMMARY AND REMARKS

R.F. Griffin
The Observatories
Madingley Road
Cambridge, England

In these difficult days, when such money, jobs and instruments as are still available to astronomy are directed almost exclusively to extragalactic work, we stellar astronomers may well see ourselves as a small and beleaguered population. Even the IAU itself reinforces that perception of ourselves: as Dr. Jaschek pointed out in his opening address, not one of the 100-odd IAU Symposia has been concerned with the types of stars that form the subject of this Colloquium, and the IAU has even gone so far as to withhold its blessing from this meeting. In the circumstances, I should particularly like to congratulate the Scientific Organizing Committee — and even more particularly Drs. Keenan and Jaschek, who have borne the main burden of the Committee's work — in standing firm and bringing the Colloquium to such a successful fruition.

Although it has been a little disappointing that some of our distinguished colleagues whose names appeared on the provisional list of participants have not, in the event, materialized in person, yet I should like to pay tribute to this present company by remarking that it must surely represent the most formidable assembly of talent and enthusiasm for our subject that has ever been gathered under one roof. I hope that the significance of this matter will not be entirely lost upon the IAU.

With so many experts present, it might seem invidious of me to mention particular names. As I advance in years I recognize that much of the good research is being done by people of my generation and younger, who indeed constitute the large majority of those here. But this very fact enables me perhaps without undue offence to single out a few of our seniors who have truly been pioneers of major topics of this conference, and whose names have been known and esteemed by the rest of us throughout our astronomical careers — legends in their own lifetimes. It has, therefore, been a special privilege for all of us to meet Professor Fujita, the founding father of the great Japanese school of cool-star astronomy; Dr. Locanthi, who has worked systematically on S-type stars ever since (under her maiden name of Dorothy Davis) she offered the first classification of them fully fifty years

M. Jaschek and P. C. Keenan (eds.), Cool Stars with Excesses of Heavy Elements, 387–392.

ago; and Drs. Keenan and Bidelman, those grand masters of spectral classification, who can infer from mere eyepiece inspection of a tiny spectrogram those nuances of stellar character and composition that the rest of us can scarcely determine by quantitative discussion of detailed spectrophotometry.

In my search for inspiration in the unenviable task of trying to summarize this Colloquium, I have looked back at the papers we received when the meeting was in prospect. It seems useful to recall — and I quote — that:

> "The aim of the Colloquium is to compare progress and results of the several programs that are under way to define the connections between these groups: barium stars, S stars, SC stars, and carbon stars. Many individual studies of these different kinds of stars have been carried out in the past, but only now are we beginning to see how the groups may be related. Among the approaches to be discussed are:
>
> 1. Radial velocity measurements extended over periods of years. The discovery that barium stars are generally members of binary systems implies that spectroscopic determinations of duplicity provide a necessary test of relationship.
> 2. Accurate abundance analyses. Only recently have model atmospheres of stars with banded spectra begun to achieve the necessary precision.
> 3. Distribution in stellar systems. Such powerful tests as the 'grism' permit the test of association to be applied in distant systems.
> 4. Progress in evolutionary theory.
>
> It is the coordination of these approaches that is to be the theme of this Colloquium."

It is apparent that in certain parts of our field we have "over-fulfilled our norm". In the very first talk Dr. Jaschek referrred not only to the four groups of stars already mentioned but to CH, CH-like, subgiant CH, J-type, lithium-rich, weak-G-band, and hydrogen-deficient carbon stars; and from there the meeting never looked back. Moreover, the fundamental connections between the S, SC and carbon stars are already understood to a considerable extent, and interesting speculations have been made concerning the relationships between other groups of cool peculiar stars. There is, on the other hand, one respect in which we have perhaps fallen short of what was hoped of us, and that is in the coordination of the various observational and theoretical approaches. However, upon reflection I think that it is scarcely to be expected that such coordination could already have taken place: it is something that may well occur in the future as a direct result of this Colloquium as the exchange of information, and the new awareness of and respect for one another's work, translates itself into new collaborations and better informed observing programmes.

There is little time — and still less profit — for me to re-summarize the papers that we have already heard. Rather, it seems more appropriate, as well as perhaps more in accordance with the admittedly rather liberal brief given me by the Chairman of the Scientific Organizing Committee, if I remark on a few particular points to which further thought might be given, and lines of investigation that seem especially fruitful just now. These are necessarily personal judgements, and if they do not command approval you will have an opportunity to voice your equally subjective rejection of them in the discussion later!

Before turning to specific topics I want to refer to two rather general matters — the bugbear of observational selection, and the potentially related business of the uses of narrow-band photometry.

In the last resort it is of course impossible to prevent some types of observational selection; but at least we should try to avoid worsening it unnecessarily, and we should certainly be alert to watch for its effects on our analyses. A glance at the poster paper showing some Galactic distributions of objects from the preliminary Catalogue of Stellar Groups indicates the unwisdom of basing a comparison of (say) latitude distributions of different types of stars on such data. It is obvious from the plots that certain parts of the Galactic plane have been searched for carbon stars far more intensively than other parts of the sky; and CN stars appear to be concentrated in southern declinations to an astounding degree. These remarks are, of course, by no means to be construed as a criticism of the Catalogue of Stellar Groups, which simply reports what is in the literature — indeed, the provision of the Galactic-distribution diagrams is extremely helpful for the very reason that it alerts the user of the Catalogue to the inhomogeneity of the compiled data.

In other instances, we have seen how the magnitude distributions of classical and marginal barium stars suggest that the marginal ones are not easily recognized except among the bright stars; and Dr. Keenan has pointed out that the assessment of spectral types of M stars from the absolute strength of the TiO bands may easily obscure possible variations in chemical abundances. Then again, I was rather shocked by Dr. McClure's remark, after he had asked for advice on whether he should monitor a group of S stars for radial-velocity changes, that "if they're not likely to be binaries it's a waste of time to measure them."! In my view it is just as useful and interesting that he has found the frequency of R-type binaries to be normal as it is that he has found the frequency of barium-star binaries to be large; and the extension of his beautiful work to other important groups of stars would be more than welcome whatever its outcome. As a final example, I should like to ally myself with Drs. Lloyd Evans and Cowley who, in different contexts, expressed their distrust of kinematical studies in cases where the homogeneity of the population being studied was not above suspicion.

As far as narrow-band photometry is concerned, my attitude, which I suppose could well be described as one of extreme reservation, is probably coloured by experiences early in my astronomical work. After several attempts to determine astrophysically interesting quantities from stellar spectra by measuring absorption features that were rigorously defined and isolated by spectrometers with ever-increasing resolution, I gave up because the measurements were merely difficult while the extraction of the astrophysically significant results seemed impossible — at least, it eluded me! Upon receiving a preliminary version of the paper, now notorious, that launched the "super-metal-rich" scare many years ago, I related this state of affairs (perhaps rather too frankly) to the authors concerned; they chose at the time to ignore my advice but I believe that they have nevertheless suffered an experience identical to mine! However, where Cambridge and Lick may have failed, the admirable tenacity — and dare I say optimism — at such places as Toronto, Geneva and Copenhagen seems largely to have succeeded. Even so, it is noticeable that narrow-band photometry only gives plausible results for objects within the multi-dimensional volume for which its calibrations are valid, that is, for objects of the expected types; and it does not always flag objects that fall outside the region of validity. Like Dr. Catchpole, I do not trust the use of DDO (or any other) indices to give true luminosities for stars as abnormal as the barium stars. On the other hand, the truly gross differences seen between C and M stars at the wavelengths indicated by Richer et al. in their poster paper appear to me to offer a wonderful opportunity to exploit narrow-band methods in a case where there is little likelihood that unforeseen spectral peculiarities will seriously invalidate the results.

Moving now to less general matters, I should like to say how much firmer our understanding of the M—S—C-star relationships seems to me to have become recently as a result of the discovery of such stars in external galaxies, star clusters, and particularly in the Magellanic Clouds. In this context, perhaps I might be permitted particularly to commend the work of Dr. Lloyd Evans. The immediate comparability in the Magellanic Clouds of the relative magnitudes and spatial distributions of significant numbers of the various types of stars is clearly leading to an understanding of the relationships between the groups, an understanding which would have taken an enormous amount of time and effort to establish from Galactic field stars alone if indeed it would have been possible at all. This is not the only frontier of our subject where, as an observer, I was pleased until this morning to think that the observations had for the time being outrun the resources of the theories to explain them: there are several relationships identified by Dr. Lloyd Evans and others which are clearly of significance but whose significance seemed not yet to be appreciated. An example is the remarkable difference in the relative numbers of M-type and carbon stars between — to take extreme situations — the centre of the Small Magellanic Cloud and the centre of our own Galaxy. Another is Dr. Utsumi's finding of the normality of s-process abundances in J-type carbon stars. However, Dr. Wood's talk this morning restored my faith in the ability of theoreticians to explain at least qualitatively nearly anything that seems to be observed, even including on occasion things that may subsequently turn out not to be so! Seriously, though, Dr. Wood's talk was a revelation to me, and showed how an appreciation of the causes underlying some of the observational relationships that other speakers have discussed opens the way to a really deep understanding of the fundamental processes of stellar and galactic evolution.

A glaring case of the continuing want of a theory to explain the observations is obviously that of the duplicity of barium stars. It has seemed in the last 24 hours, as usual, that no explanation other than contamination by a previously evolving companion star is at all acceptable, and I think I detect some willingness at last to see whether the still profound difficulties with that hypothesis can be overcome. We have been driven to that position only by the lack of a remotely plausible alternative: the four other possibilities which were very properly listed by Dr. McClure would not be seriously entertained by anyone except perhaps the late Professor Zwicky in one of his most morphological moods! However, there is no reason at all for the observers to be idle while the theoreticians catch up with them, even on the subject of barium stars. Among the observational problems urgently awaiting solution are whether the abundance idiosyncrasies of the barium and Am stars are or are not so similar as to be attributable to the same origin; the nature of the secondaries in all the barium-star systems; and the question as to whether main-sequence barium stars exist or not. These problems pose substantial challenges to the ingenuity and patience of observers, and can hardly be expected to be quickly resolved. A relatively minor question, by comparison, is whether there are supergiants which truly show enhancements of barium and strontium alone, or whether such supposed enhancements are artefacts of the model atmospheres employed in the analysis. I recall, with slight embarrassment, making an ex cathedra judgement on that issue soon after it first arose, and I am naturally pleased at the prospect offered by Dr. Lambert of a reprieve for my opinion!

Finally, I should remark briefly on the desirability of cooperative programmes on some of these topics. It has already been pointed out that any measurement of the angular diameter of a photometrically variable star at a lunar occultation should ideally be accompanied by contemporaneous photometry — presumably to be obtained by a collaborator remote from the Earth! Again, radial-velocity measurements of a star which is (or may be) pulsating need to be accompanied by photometry before the significance of any radial-velocity changes can be assessed. An inverse example of the same need may be mentioned in relation to the Quercis' proposal for comprehensive monitoring of some of the groups of cool peculiar stars. One of the groups to which they refer is the hydrogen-deficient carbon stars, which according to my understanding includes objects of the R Coronae type. It may not be generally known that R Coronae itself shows appreciable radial-velocity changes, semi-regular in nature, with a characteristic time-scale of a few weeks — I have measured it on a total of more than 200 nights in the last several years. It also shows photometric variations on the same sort of time-scale — quite apart from its occasional dramatic minima. Such correlation as is possible between the radial-velocity and photometric data suggests that a phase relationship, different from that in Cepheids, may exist between the respective variations; but reliance on casual overlaps between two sets of data, each fragmentary, is hardly the best way of establishing the relationship. If the Quercis' suggestion can be put into effect, therefore, as is without doubt desirable, its scope could usefully be expanded to include the cooperation of the radial-velocity fraternity in respect of certain objects.

These are only a sample of the cooperative programmes which might be undertaken, but I shall now give place to the general discussion for the suggestion of others.

OPENING REMARKS TO THE GENERAL DISCUSSION

William P. Bidelman
Warner and Swasey Observatory
1975 Taylor Rd
E Cleveland OH 44112 - U.S.A.

If we limit ourselves to the topic of this colloquium -namely, cool stars with excesses of the heavy elements- we have essentially only three different classes of objects : (1) the barium stars, (2) the S-type stars, and (3) some, but apparently not all, of the cooler carbon stars. It is quite apparent from our discussions here that we observers have not been sufficiently detailed in our descriptions of the spectra of these objects to enable meaningful statistical investigations to be made. For example, just calling an object a barium star -or even giving a quantitative strength for the barium lines- is certainly not enough information for one to really know what sort of object we are dealing with.

It seems to me that the types of data that we need for all the objects of these groups in order to bring some sort of order out of our present chaos are the following :

1. The object's position in the HR diagram.
2. An index (or perhaps several) of the heavy element abundance.
3. An index of the Fe-group and hydrogen abundance.
4. The C_2 -band strengths or better the C/O ratio.
5. The CN-band strengths or better the N/C ratio.
6. The C^{12} / C^{13} ratio.

Some of these can be estimated with fair reliability from moderate dispersion plates ; others require detailed analysis. At any rate these are some of the parameters that should be specified if we are to have a complete description of these objects. There may be more !

Note : - The discussion that followed was not recorded

M. Jaschek and P. C. Keenan (eds.), Cool Stars with Excesses of Heavy Elements, 393.

ADDRESS TO PHILIP KEENAN ON BEHALF OF THE PARTICIPANTS.

M. McCARTHY S.J.
Specola Vaticana
Citta del Vaticano
Vatican City State - Italy

I have been asked to thank our dear Philip Keenan and I accept to do so most willingly. He won't like it so we shall try to be bright and brief and than begone.

We cannot dedicate these proceedings to Philip because he was the organizer, invited us, and served as chairman of Scientific Organizing Committee.

We cannot give Philip an honorary degree since we are all guests here at the University and because school here is out. It is vacation time in Strasbourg.

We cannot canonize Philip because thank the Lord Philip is very much alive, is very well and is most actively engaged in studying his beloved red giant spectra.

I choose not to embarass Philip with a long account of his vital statistics. We all know of his early studies in astronomy in Arizona, of his Ph. D. and post doctoral work at Chicago. His work with the U.S. Navy during world war II was followed by 30 years of active service at Ohio State, Ohio Wesleyan and Perkins Observatory where now he is enjoying his first decade of active retirement. Just during the past 20 years between 1960 and 1980 he published 15 papers and 4 books all of them heavily weighted in favor of the cool stars.

But we can do something no other group or persons on the entire planet can do. Indeed we can do something which perhaps no other group of persons in this part of our galaxy can do. We can each of us continue to explore and explain these wonderful and evolving magnificent cool red stars. This we can do to honor Philip : keep on working on red stars.

Yes. We can keep on doing what we do best and can help each other to extend, as we did here in peaceful Strasbourg during these four days, our knowledge of the structure and age and origin and future development of the cool red stars.

We can do all these things and honor Philip thereby with supertelescopes with superreceivers and super-image processing too. As we examine the distribution in space and the evolution in time of red giant populations in our nearest spiral neighbors, just as he and we together have extended the study of red giants from

M. Jaschek and P. C. Keenan (eds.), Cool Stars with Excesses of Heavy Elements, 395–396.

the solar neighborhood out to the center of our galaxy and on to its nearby spiral neighbors.

If we do all these things we will know that we continue Philip works since we will be standing on his super giant's shoulders, for his one of the founding parents of cool star astronomy along with Secchi, Huggins, Miss Maury, Pickering, Merrill, McKellar, Nassau, Sanford and Thackeray.

We rejoice with Philip today that we can now with Grisms and CCD's and IUE's and IRAS and soon with Space Telescope extend the fruitful applications of careful classical work by Keenan and Morgan out out to the far infrared and to the deep ultraviolet.

We praise you today Philip.

We honor you today as the teacher of us all. Take on new strength now Pilip and we rise now to thank you.

Macte nova virtute Philip
Sic iter ad Astra.

This way, your way, our way
Leads toward the stars, the big cool still mysterious Red Giant Stars.

AUTHOR'S INDEX

GPSR Compliance

The European Union's (EU) General Product Safety Regulation (GPSR) is a set of rules that requires consumer products to be safe and our obligations to ensure this.

If you have any concerns about our products, you can contact us on ProductSafety@springernature.com

In case Publisher is established outside the EU, the EU authorized representative is:

Springer Nature Customer Service Center GmbH
Europaplatz 3
69115 Heidelberg, Germany

Batch number: 10761599

Printed by Printforce, the Netherlands